Digital War

D1809945

Digital War offers a comprehensive overview of the impact of digital technologies upon the military, the media, the global public and the concept of 'warfare' itself.

This introductory textbook explores the range of uses of digital technology in contemporary warfare and conflict. The book begins with the 1991 Gulf War, which showcased post-Vietnam technological developments and established a new model of close military and media management. It explores how this model was reapplied in Kosovo (1999), Afghanistan (2001) and Iraq (2003), and how, with the Web 2.0 revolution, this informational control broke down. New digital technologies allowed anyone to be an informational producer leading to the emergence of a new mode of 'participative war', as seen in Gaza, Iraq and Syria. The book examines major political events of recent times, such as 9/11 and the War on Terror and its aftermath. It also considers how technological developments such as unmanned drones and cyberwar have impacted upon global conflict and explores emerging technologies such as soldier-systems, exo-skeletons, robotics and artificial intelligence and their possible future impact.

This book will be of much interest to students of war and media, security studies, political communication, new media, diplomacy and IR in general.

William Merrin is Associate Professor in Media and Communication at Swansea University, UK. He is the author of *Media Studies 2.0* (Routledge, 2014) and *Baudrillard and the Media* (2005).

Digital War
A Critical Introduction

William Merrin

LONDON AND NEW YORK

First published 2019
by Routledge
2 Park Square, Milton Park, Abingdon, Oxon OX14 4RN

and by Routledge
711 Third Avenue, New York, NY 10017

Routledge is an imprint of the Taylor & Francis Group, an informa business

© 2019 William Merrin

The right of William Merrin to be identified as author of this work has been
asserted by him in accordance with sections 77 and 78 of the Copyright, Designs
and Patents Act 1988.

All rights reserved. No part of this book may be reprinted or reproduced or
utilised in any form or by any electronic, mechanical, or other means, now
known or hereafter invented, including photocopying and recording, or in any
information storage or retrieval system, without permission in writing from the
publishers.

Trademark notice: Product or corporate names may be trademarks or registered
trademarks, and are used only for identification and explanation without intent to
infringe.

British Library Cataloguing in Publication Data
A catalogue record for this book is available from the British Library

Library of Congress Cataloging in Publication Data
Names: Merrin, William, author.
Title: Digital war : a critical introduction / William Merrin.
Description: Abingdon, Oxon ; N.Y., NY : Routledge, [2018] | Includes
bibliographical references and index.
Identifiers: LCCN 2018008060 | ISBN 9781138899865 (hardback) |
ISBN 9781138899872 (pbk.) | ISBN 9781315707624 (e-book)
Subjects: LCSH: Cyberspace operations (Military science)--History. |
Cyberterrorism--History. | Military art and science--Technological innovations--
History--Case studies. | War--Technological innovations--History--Case studies.
Classification: LCC U163 .M398 2018 | DDC 355.4--dc23
LC record available at https://lccn.loc.gov/2018008060

ISBN: 978-1-138-89986-5 (hbk)
ISBN: 978-1-138-89987-2 (pbk)
ISBN: 978-1-315-70762-4 (ebk)

Typeset in Times New Roman
by Swales & Willis Ltd, Exeter, Devon, UK

Printed and bound in Great Britain by
TJ International Ltd, Padstow, Cornwall

Contents

Boxes

Abbreviations

Political, military and technological terminology is full of abbreviations. I've usually initially given the full name of phenomena with the abbreviation in brackets and then used the abbreviation in later mentions. To help anyone who forgets what the abbreviation is, this is a list of the most commonly-used ones in the book.

9/11	The terrorist attacks on the USA of September 11, 2001
A2/AD	Anti Access/Area Denial
ABC	The American Broadcasting Company
AFB	Air Force base
AFSOC	Air Force Special Operations Command
AI	Artificial intelligence
ANA	Afghan National Army
APT	Advanced persistent threat
AQAP	al-Qaeda in the Arabian Peninsula
AQEA	al-Qaeda in East Africa
AQI	al-Qaeda in Iraq
AQIM	al-Qaeda in the Islamic Maghreb
AR	Augmented reality
ARPA	Advanced Research Projects Agency (later DARPA)
AUMF	Authorization for the Use of Military Force
BBC	The British Broadcasting Corporation
BCI	Brain-computer interface
C+C	Command and control
C2W	Command and control warfare
C3W	Command, control and communications warfare
C3IW	Command, control, communications and intelligence warfare
C4IW	Command, control, communications, computers and intelligence warfare
CBS	Columbia Broadcasting System
CCDCE	Cooperative Cyber Defence Centre of Excellence
CCW	Convention on Certain Conventional Weapons
CENTCOMM	Central Command
CI	Critical infrastructure
CIA	Central Intelligence Agency
CITRU	Counter Terrorism Internet Referral Unit
CNA	Computer network attack
CND	Computer network defence
CNE	Computer network exploitation

CNN	Cable News Network
CNO	Computer network operations
COIN	Counter-insurgency
CONTEST	Counter Terrorism Strategy
COTS	Commercial off-the-shelf
CPA	Coalition Provisional Authority
CVE	Countering Violent Extremism
DARPA	Defense Advanced Research Projects Agency
DCGS	Distributed Common Ground System
DDoS	Distributed denial of service
DHS	Department for Homeland Security
DOD	Department of Defense
DOS	Denial of service
DMZ	Demilitarized Zone
EDT	Eastern Daylight Time
EEG	Electroencephalography
EIT	Enhanced interrogation techniques
EST	Eastern Standard Time
EW	Electronic warfare
FATA	Federally Administered Tribal Areas
FBI	Federal Bureau of Investigation
FISA	Foreign Intelligence Surveillance Act
FLIR	Forward-looking infrared radar
FPS	First-person shooter
FSA	Free Syrian Army
GA	General Atomics
GCHQ	Government Communications Headquarters
GCS	Ground control station
GPS	Global positioning system
GRU	Main Intelligence Directorate
HMD	Head-mounted display
HUD	Heads-up display
HULC	Human Universal Load Carrier
HUMINT	Human intelligence
HVT	High-value target
IAEA	International Atomic Energy Agency
IDF	Israeli Defense Forces
IED	Improvised explosive device
IO	Information operations
IP	Intellectual property
IS	Islamic State
ISF	Iraqi Security Forces
ISI	Islamic State of Iraq
ISI (PAKISTAN)	Inter-Services Intelligence
ISIL	Islamic State of Iraq and the Levant
ISIS	Islamic State of Iraq and Syria
ISR	Intelligence, surveillance and reconnaissance
ISAF	International Security Assistance Force

IT	Information technology
ITN	Independent Television News
IW	Information warfare
JCS	Joint Chiefs of Staff
JSOC	Joint Special Operations Command
JWICS	Joint Worldwide Intelligence Communications System
KLA	Kosovo Liberation Army
LAWS	Lethal Autonomous Weapons Systems
LW	Land Warrior
MAD	Mutually Assured Destruction
MOD	Ministry of Defence
MP	Military Police
NATO	North Atlantic Treaty Organization
NBC	National Broadcasting Company
NBC	Nuclear, biological, chemical
NCSC	National Cyber Security Centre
NCW	Network-centric warfare
NGO	Non-governmental organization
NIPRNet	Non-classified Internet Protocol Router Network
NORAD	North American Aerospace Defense Command
NSDD	National Security Decision Directive
NSPD	National Security Presidential Directive
NW	Nett Warrior
OCSIA	Office of Cyber Security and Information Assurance
ONR	Office of Naval Research
OPSEC	Operations Security
PGM	Precision-guided munition
PLA	People's Liberation Army
PLC	Programmable logic controller
POV	Point-of-view
PDD	Presidential Decision Directive
PPD	Presidential Policy Directive
Psyops	Psychological operations
R&D	Research and development
RAT	Remote access Trojan
RBN	Russian Business Network
RBSS	Raqqa is Being Slaughtered Silently
RMA	Revolution in Military Affairs
ROE	Rules of engagement
RPV	Remotely-piloted vehicle
SCADA	Supervisory control and data acquisition
SDF	Syrian Democratic Forces
SDI	Strategic Defense Initiative
SEA	Syrian Electronic Army
SFRY	Socialist Federal Republic of Yugoslavia
SIGINT	Signals intelligence
SIPRNet	Secret Internet Protocol Router Network
SOCOM	Special Operations Command

SOF	Special Operations Forces
STRATCOM	Strategic Command
SUGV	Small unmanned ground vehicle
TALOS	Tactical Assault Light Operator Suit
TAO	Tailored Access Operations
TKP	Targeted killing programme
TOR	The Onion Router
TRADOC	United States Army Training and Doctrine Command
TTP	Tehrik-e-Taliban Pakistan
UAS	Unmanned aerial system
UAV	Unmanned aerial vehicle
UCAS	Unmanned combat air system
UCAV	Unmanned combat autonomous vehicle
UGC	User-generated content
UGV	Unmanned ground vehicle
UN	United Nations
UNMOVIC	United Nations Monitoring, Verification and Inspection Commission
UNSCOM	United Nations Special Commission
URL	Uniform resource locator
USAF	United States Air Force
USCYBERCOMM	United States Cyber Command
USSR	Union of Soviet Socialist Republics
USV	Unmanned sea vehicle
UUV	Unmanned undersea vehicle
VR	Virtual reality
WMD	Weapons of mass destruction
WWW	World Wide Web
YPG	People's Protection Units

Timeline 1990–2017

This is a timeline of events covered in the book, from the 1991 Gulf War through to the present. It is inevitably selective, primarily focusing upon the western experience as well as upon those conflicts and issues discussed in the chapters. It includes key events in the War on Terror and the most significant terrorist attacks in Europe and the USA carried out in response to that campaign.

Year	Events
1990	2 August. Iraq invades **Kuwait**.
1991	17 January. The allied air campaign begins against **Iraq**. The 'Gulf War' begins.
	24 February. The allied ground campaign begins against **Iraq**.
	28 February. The allied campaign declares a ceasefire in **Iraq** and victory in the Gulf War.
	1 March – 5 April. Shia Arab and Kurdish uprisings in southern and northern **Iraq** are put down by Saddam Hussein.
	6 April. The USA establishes a no-fly-zone in northern **Iraq**, an extension of the no-fly warning given under 'Operation Provide Comfort' which had begun 3 March. From 1 January 1997 'Operation Provide Comfort' (1 and 2) was renamed 'Operation Northern Watch'. It lasted until 19 March 2003.
1992	26 August. The beginning of the US 'Operation Southern Watch' (which lasted until 19 March 2003) policing a no-fly-zone in Southern **Iraq**.
	5 December. 'Operation Restore Hope', begins: a US-led, UN-sanctioned, multinational force in **Somalia** (lasting until 4 May 1993). On 9 December 1992, the UNITAF (Unified Task Force) troops landed on the Somalian beaches to a media circus.
1993	26 February. A truck bomb explodes at the World Trade Centre in New York City, **USA**, killing 6 people and injuring over 1,000. It was an Islamist terrorist attack, masterminded by Ramzi Yousef with advice and funding from Khaled Sheikh Mohammed (who would later plan 9/11).
	12 April. NATO launch 'Operation Deny Flight', enforcing a UN no-fly-zone over **Bosnia and Herzegovina** (lasting until 20 December 2005).

(Continued)

Year	Events
	22 August. US Special Operation Forces launch 'Operation Gothic Serpent' in **Somalia** (until 13 October). It includes the 'Battle of Mogadishu', 3–4 October.
1994	11 December. A test explosion on **Philippines** Airlines Flight 434 exposes the Islamist terrorist 'Bojinka Plot' planned by Ramzi Yousef and Khaled Sheikh Mohammed.
1995	11–13 July. The 'Srebrenica massacre' of 8000 men and boys in **Bosnia and Herzegovina** by the Bosnian Serb army.
	30 August. NATO launches 'Operation Deliberate Force', the bombing of Bosnian Serbian army forces in **Bosnia and Herzegovina** (lasting until 20 September).
1996	25 June. The 'Khobar Towers bombing', targeting US servicemen in Khobar, **Saudi Arabia**, kills 20 and injures 372 people. Iran and Hezbollah are suspected of the attack.
	3 September. The USA launch 'Operation Desert Strike', with cruise missile strikes on **Iraq** in response to an Iraqi military offensive in the Kurdish civil war.
1997	17 November. 'The Luxor massacre': Egyptian Islamist terrorists kill 62 people, mostly foreign tourists, at Deir el-Bahri, **Egypt**, a major archaeological and tourist site.
1998	7 August. Al-Qaeda carries out bomb attacks on the US embassies in Nairobi, **Kenya**, and Dar es Salaam, **Tanzania**, killing 224 people and wounding over 4,000. The USA responds with 'Operation Infinite Reach', with cruise missile strikes on **Sudan** and **Afghanistan** on 20 August.
	16–19 December. The USA launches 'Operation Desert Fox', with cruise missile attacks on **Iraq** in response to its failure to comply with UN Security Council resolutions and obstruction of United Nations Special Commission weapons inspectors.
1999	24 March–11 June. NATO launches 'Operation Allied Force', a bombing campaign on **Yugoslavia** (mainly **Serbia**) in defence of **Kosovo**. On 3 June Milosevic accepts peace conditions and the Serbs withdraw from Kosovo ending the 'Kosovo War'.
	20 September. The discovery of 'Moonlight Maze' – the US term for ongoing computer attacks, most likely from **Russia**, dating back to March 1998.
2000	12 October. A terrorist attack by al-Qaeda on the USS *Cole* in Aden harbor, **Yemen**, kills 17 and injures 39 US servicemen.
2001	11 September. '9/11': Terrorist attacks on the World Trade Centre and the Pentagon in the **USA** kill 2996 people (including 19 hijackers). The attacks were organized by al-Qaeda, being masterminded by Khaled Sheikh Mohammed under Osama Bin Laden's direction.
	7 October. The USA launches 'Operation Enduring Freedom', the beginning of their global 'War on Terror', with attacks on **Afghanistan**. The 'Afghan War' begins.
	13 November. Kabul falls in **Afghanistan**.
	22 December. The al-Qaeda-trained terrorist, Richard Reid, known as the 'Shoe Bomber', is arrested after failing to detonate explosives packed into his shoe on the **USA** American Airlines flight from Paris to Miami.
2002	4 July. 'Recon', the first version of the 'America's Army' videogame is released.

(*Continued*)

Year	Events

12 October. Islamist terrorists carry out 'the Bali bombings' in the tourist district of Kuta, on the Indonesian island of **Bali**, killing 202 people and injuring 240.

2003 20 March. The allied invasion of **Iraq**. The 'Iraq War' begins.

15 April. The allies declare the invasion of **Iraq** over.

1 May. President Bush gives his 'mission accomplished' speech, declaring the end of major hostilities in the 'Iraq War'.

12 May. The 'Riyadh Compound bombings' by Islamist terrorists in **Saudi Arabia** kill 39 people and injure over 160.

5 August. An Islamist terrorist car-bomb attack on the Marriott Hotel in South Jakarta, **Indonesia**, kills 12 and injures 150 people.

15 and 20 November. Terrorist bombings in **Istanbul** kill 57 people and injure over 700.

13 December. US troops launch 'Operation Red Dawn', capturing Saddam Hussein in a farm compound near Tikrit, **Iraq**.

2004 11 March. An al-Qaeda-inspired terrorist cell carries out the 'Madrid train bombings' in **Spain**, killing 191 and injuring 1800 people.

28 April. The US TV programme *60 minutes II* broadcasts a story about systematic prisoner abuse at Abu Ghraib prison in **Iraq** by US servicemen.

29 May. Al-Qaeda terrorists carry out the 'Al-Khobar massacre', **Saudi Arabia**, killing 22 people and injuring 25.

2005 7 July. The 'London bombings': four UK Islamist suicide-bombers kill 52 people and injure 784 in blasts on the underground network and on a bus in London, **UK**.

21 July. The 'failed London bombings': five UK Islamist terrorists fail to properly ignite their explosives on the public transport system in London, **UK**. The suspects were later arrested.

22 July. Armed London Metropolitan police mistake the Brazilian Jean Charles de Menezes for a suspect in the previous day's failed bombing attack and shoot him dead on a train at Stockwell Tube Station, London, **UK**.

19 October. The trial of Saddam Hussein by the Interim Iraqi Government begins in **Iraq**. Saddam is charged with the killing of 148 Shiites from Dujail in retaliation for the failed assassination attempt on 8 July 1982.

9 November. The 'Amman bombing': Coordinated bomb attacks on three hotels in Amman, **Jordan**, by Al-Zarqawi's al-Qaeda in Iraq (AQI), kills 60 and injures 115 people.

December. The discovery of 'Titan Rain' – the US designation for attacks on its computer systems since 2003 originating in **China**.

2006 11 July. The 'Mumbai train bombings' in Mumbai, Maharashtra, **India**, kill 209 people and injure over 700.

(*Continued*)

Year	Events
	21 August. A second and separate trial of Saddam Hussein begins in **Iraq**, trying Saddam and six co-defendants for genocide during the Anfal military campaign against the Kurds of Northern Iraq.
	5 November. Saddam Hussein is found guilty in **Iraq** of the killing of 148 Shiites and sentenced to death by hanging
	30 December. Saddam Hussein is executed at 'Camp Justice', an Iraqi army base in north-eastern Baghdad, **Iraq**.
2007	11 April. Al-Qaeda in the Islamic Maghreb (AQIM) carry out the bombings in Algiers, **Algeria**, killing 33 people.
	27 April. Cyberattacks begin on **Estonia** following a dispute with Russia and ethnic Russians over the relocation of 'the Bronze Soldier of Talinn' World War II memorial. The attacks are almost certainly coordinated from Russia.
	29 June. Two car-bombs are discovered and disabled in London, **UK**. They were linked to the terrorist attack in Glasgow the next day.
	30 June. Two Islamist-inspired terrorists drive a vehicle laden with petrol and propane tanks into the main terminal of Glasgow International Airport, Scotland, **UK**, injuring 5 people.
	11 December. Al-Qaeda in the Islamic Maghreb (AQIM) explode two car bombs in Algiers, **Algeria**, killing 41 and injuring 170.
2008	2 June. Al-Qaeda bomb the Danish embassy in Islamabad, **Pakistan**, killing 6-8 people and injuring 24. It was in response to the Danish republication of cartoons of Mohammed in February.
	5 August. Cyberattacks begin against **Georgia** from **Russia** prior to the outbreak of the 'Russo-Georgia War' and continue through the ground campaign on 7–12 August.
	20 September. Terrorists carry out a suicide truck bomb attack on the Marriott Hotel in Islamabad, **Pakistan**, killing 54 and injuring 266 people.
	26 November. Members of the Pakistan Islamist terrorist group Lashkar-e-Taiba carry out coordinated terrorist attacks in Mumbai, **India**, killing 164 people and injuring over 600.
	27 December. The 'Gaza War' begins. Israel launches 'Operation Cast Lead' with airstrikes on the **Gaza** strip in response to Hamas rocket attacks. The war lasted 22 days until 18 January 2009 when Israel announced a unilateral ceasefire.
2009	28 March. The discovery of 'Ghostnet' – the US term for a series of computer cyber-espionage intrusions in 103 countries coordinated from **China**.
	30 April. UK forces end operations in **Iraq**.
	1 June. The 'Little Rock recruiting office shooting' by an Islamic convert in Little Rock, Arkansas, **USA**, kills 1 and wounds 1 other.
	29 June. US forces withdraw from Baghdad and **Iraq**.
	7 August. Cyberattacks on social networking sites aimed at one blogger in **Georgia**.

(*Continued*)

Year	Events
	5 November. A US Army Major and psychiatrist kills 13 and injures over 30 in the 'Fort Hood shootings', in Texas, **USA**. Many claim it was a terrorist attack, motivated by radical Islam.

25 December. A failed al-Qaeda 'Christmas Day bombing attempt' on the **USA** Northwest Airlines Flight 253 from Amsterdam to Detroit in which Umar Farouk Abdulmutallab ('the underwear bomber') tried to set off plastic explosives sewn into his underwear.
December. 'Operation Aurora': the US name for a series of cyber-espionage hacking attacks on Google and 20 other companies by **China**, lasting until January 2010.

2010 5 April. Wikileaks releases 'Collateral Murder' – video of the 12 July 2007 Baghdad, **Iraq**, airstrike that killed Iraqi civilians and two Reuters journalists.

25 July. Wikileaks begin the release of the 'Afghan War logs' – leaked classified US military documents relating to the war in **Afghanistan**.

September. The USA and Israel attack the Natanz nuclear facility in **Iran** with the cyber-weapon, the Stuxnet worm, to slow down the country's nuclear programme.

22 October. Wikileaks release the 'Iraq War Logs' – leaked classified US military documents relating to the war in **Iraq**.

29 October. The 'Cargo plane bomb plot' is discovered. Following Saudi intelligence, two US cargo planes from **Yemen** to the **USA** are found to contain explosives. Al-Qaeda in the Arabian Peninsula (AQAP) claim responsibility.

28 November. Wikileaks begin release of leaked American diplomatic cables.

17 December. Mohammed Bouazizi kills himself in **Tunisia**, leading to protests and a 'Tunisian revolution' that ousts President Ben Ali on 14 January 2012. This was the beginning of the 'Arab Spring'.

2011 25th January. Inspired by the Tunisian revolution, protests begin in **Egypt**, leading to the 'Egyptian revolution' that ousts President Mubarak on 11 February.

15 February. Protests begin in Benghazi, **Libya**, against Muammar Gaddafi's rule, leading to an armed uprising and the 'Libyan Civil War'.

15 March. Protests begin in **Syria** that will lead to the 'Syrian Civil War'.

17 March. UN Security Council Resolution 1973 is passed, demanding a ceasefire in **Libya**, establishing a no-fly-zone and authorizing military intervention to protect civilians.

2 May. Osama Bin Laden is shot and killed in Abbottabad, **Pakistan** by US special forces.

16 September. The **Libyan** National Transitional Council is recognized by the UN.

20 October. Muammar Gaddafi is killed whilst trying to escape from Sirte in **Libya**.

23 October. The Libyan NTC declares the liberation of **Libya** and the official end of the war.

2012 11–27 September. A series of terrorist attacks are carried out on US and European diplomatic missions worldwide, considered as a reaction to the controversial film *Innocence of the Muslims*, available on YouTube.

14–21 November. Israel launches operation 'Pillar of Defence' against **Gaza** strip militants.

(*Continued*)

Year	Events
2013	15 April. Two Chechen Islamist brothers explode two bombs at the Boston Marathon, in the **USA**, killing 3 and injuring 264.
	22 May. The British Army soldier, Fusilier Lee Rigby, is murdered by two UK Islamist extremists near the Royal Artillery Barracks in Woolwich, London, **UK**.
	6 June. Edward Snowden reveals details to the press of the **USA** Government's global, mass-surveillance programs.
	3 July. President Mohamed Morsi, elected in the June 2012 elections, is deposed by a military coup in **Egypt**.
	14th August. Security forces carry out 'the Rabaa massacre' in Cairo, **Egypt**, attacking supporters of the ousted president Mohamed Morsi, killing at least 817 and perhaps more than 1,000 people.
	21 August. The Syrian government carry out a chemical weapons attack on the rebel-held Ghouta suburbs of Damascus in **Syria**, killing between 281 and 1,729 people.
	21 September. Islamist gunmen from al-Shabaab kill 67 and wound 175 people in attack on the Westgate shopping mall in Nairobi, **Kenya**.
	16 December. UK Prime Minister David Cameron declares 'mission accomplished' in **Afghanistan**, prior to the withdrawal of UK forces by the end of 2014.
	19 December. Two British people are found guilty of the murder of Lee Rigby in the **UK**.
2014	22 February. The **Ukraine** parliament votes to remove pro-Russian President Yanukovych after demonstrations in Kiev leave about 100 dead.
	26 February. Pro-Russian unrest in **Ukraine** leads to an insurgency against the Kiev authorities.
	21 March. The Russian Federation confirms the annexation of Crimea from the **Ukraine**.
	5 June. ISIS ('The Islamic State of Iraq and the Levant') begins a major offensive in Northern **Iraq** against government forces. The city of Mosul is taken on the 10 June.
	29 June. ISIS, now renamed 'Islamic State' (IS), announces the establishment of an Islamic 'Caliphate' covering Northern **Iraq** and part of **Syria**.
	8 July. Israel launches operation 'Protective Edge' against the **Gaza** Strip, following Hamas rocket attacks and the killing of three Israeli teenagers in June.
	17 July. Malaysia Airlines Flight 17 is shot down by a missile fired from pro-Russian insurgent territory in the **Ukraine**, killing 284 people.
	8 August. The USA begins an air-campaign to halt the spread of Islamic State in Northern **Iraq**.
	20 August. IS releases a video of the beheading of US journalist James Foley in **Syria**.
	2 September. IS in **Syria** releases a video of the beheading of US journalist Steven Sotloff.

(*Continued*)

Year	Events

13 September. IS in **Syria** releases a video of the beheading of UK humanitarian aid worker David Haines.

3 October. IS in **Syria** releases a video of the beheading of UK humanitarian aid worker Allan Henning.

16 November. IS in **Syria** releases a video of the beheaded body of the US aid worker Peter Kassig.

15–16 December. The Islamist-inspired terrorist Man Haron Monis holds 18 hostages in a café in Sydney, **Australia**. Monis and 2 hostages are killed in the rescue operation.

16 December. Islamist militants attack the Army Public School in Peshawar in **Pakistan**, killing 145 people, including 132 schoolchildren.

20 December. An islamist-inspired attacker injures 3 policemen with a knife in a police-station near Tours, in **France**.

21 December. An Islamist-inspired terrorist injures 2 people after a vehicle-ramming attack in Dijon, **France**.

2015   7 January. An Islamist terrorist attack on the offices of the satirical magazine *Charlie Hebdo* kills 12 and injures 7 in Paris, **France.** The gunmen are killed on 9 January. A police officer is killed in a related shooting in Paris, on 8 January and on 9 the gunman who killed him laid siege to and killed 4 people in a kosher supermarket in Paris before being killed by police.

January. IS in **Syria** releases videos of the beheadings of Japanese hostages Haruna Yukawa and Kenji Goto.

3 February. IS in **Syria** releases a video of the execution of the captured Jordanian airforce pilot Muath al-Kasasbeh, who was killed by immolation.

10 February. The death of Kayla Mueller, a US hostage held by IS in **Syria**, is confirmed.

14–15 February. Three separate shootings by an IS-inspired gunman occur in Copenhagen, **Denmark**, killing 2 people and injuring 5.

8 March. Boko Harem in **Nigeria** pledge allegiance to IS.

18 March. A terrorist attack at the Bardo museum in **Tunisia** kills 22 and injures around 50. IS claims responsibility.

20 March. Suicide bombings at mosques in **Yemen** kill at least 137 people. IS claims responsibility.

2 April. Al-Shabaab gunmen kill at least 150 people and wound at least 79 at Garissa University College in Garissa, **Kenya**.

3 May. 'The Curtis Culwell Centre attack', Texas, **USA**: Two Islamic gunmen open fire at an exhibition building hosting cartoons of the prophet Mohammed, injuring one person before being killed. Islamic State claim responsibility.

26 June. There are three terror attacks across three continents. In **Kuwait** a suicide attack on a Shia mosque by an IS-affiliated group kills up to 25; in **France** a man is beheaded in an

Year	Events

Islamist attack on a US-owned gas factory near Lyon; and in **Tunisia** 38 people, mostly western tourists, are killed by an Islamist gunman in an attack on tourist hotels at Sousse.

16 July. 'The Chattanooga shootings', Tennessee, **USA**: an Islamist-inspired terrorist opens fire on two US military installations, killing 5 and injuring 6 people before being killed by police.

21 August. A highly-armed Islamist terrorist injures 4 people on a Thalys train in **France** on its way from Amsterdam to Paris. A massacre is averted when he is restrained by 3 passengers.

31 October. A Russian airliner carrying 217 passengers and 7 crew, travelling from **Egypt** to Russia, crashes in Northern Sinai killing everyone on board. IS's Sinai affiliate claims responsibility for downing the airliner with a bomb.

12 November. Mohammed Emwazi, the British citizen known as 'Jihadi John' who carried out executions on video for Islamic State in **Syria**, is killed by a US drone strike.

13–14 November. Eight, armed suicide-bombers and terrorists attack a number of venues overnight in Paris, **France**, killing 130 people and leaving 413 injured. Islamic State claims responsibility for the attacks.

5 December. A mass-shooting by an Islamic State-inspired married couple in the Inland Regional Center in San Bernardino, California, **USA** leaves 14 people dead and 12 injured.

5 December. A lone-attacker attempts to behead a tube-passenger during an Islamic State-inspired rampage at Leytonstone tube station in London, **UK**. He injures 3 people before being arrested.

2016 7 January. An Islamist terrorist is killed when he attacks a police-station in Paris, **France**.

22 March. Three coordinated Islamist suicide-bomb attacks are carried out in Brussels, **Belgium**, two at Zaventem airport and one on a train at Maelbeek Station, killing 32 people and injuring 340. Islamic State claim responsibility.

12 June. A gunman who had pledged allegiance to Islamic State kills 49 and injures 53 in a terrorist hate-crime at 'Pulse', a gay nightclub in Orlando, Florida, **USA**.

13 June. Two French policemen are killed in Paris, **France**, by a convicted Islamist terrorist who had pledged allegiance weeks before to IS.

28 June. A gun and bomb attack on Ataturk International Airport, **Turkey**, kills 41 and wounds 239. Islamic State are blamed.

July. There are multiple attacks across Europe. On the 14 July 84 people are killed and 303 injured when a terrorist drives a 19-tonne truck through crowds celebrating Bastille Day on the Promenade des Anglais in Nice, **France** before being killed by police. On the 18 July a teenage Afghan refugee hacks at train passengers on a train in Wuerzburg, **Germany**, wounding 5 before being shot. On 24 July, a Syrian refugee kills a woman with a machete and wounds 5 others in Reutlingen, **Germany**, before being arrested. Also on 24 July a Syrian refugee blows himself up outside a bar in Ansbach, **Germany**, wounding 15 people. On 26 July, a priest is killed in an attack on a church near Rouen, **France** by two men claiming to be from Islamic State. Most of the attacks were identified as acts of Islamist terrorism inspired by Islamic State.

Year	Events

17 September. A stabbing attack at a shopping mall in Minnesota, **USA**, injures 8 people. Islamic State claim responsibility for the attack online. Also on the 17 September, though unrelated, a pipe-bomb explodes in Seaside Park, New Jersey, **USA**, near a 5km run in support of US marines. Possibly linked to this, a bomb explodes that night in Chelsea, NYC, **USA**, injuring 29 people. A second bomb device is found nearby. On 19 September five pressure-cooker-type bomb devices are found in New Jersey. The bomber, Ahmad Khan Rahimi, was motivated by Islamist ideology.

28 November. An Islamic State-inspired terrorist carries out a vehicle-ramming and knife attack at Ohio State University, **USA**, injuring 13 before being shot and killed.

19 December. An Islamist terrorist drives a truck into a Christmas market in **Berlin**, killing 12 and injuring 56. He is discovered and shot by Italian police in Milan, **Italy** on 23 December.

2017 3 February. An Islamist terrorist carries out a machete attack at the Louvre, Paris, **France**, injuring 1 soldier.

22 March. An Islamic State-inspired terrorist drives a car into pedestrians on Westminster Bridge and stabs a policeman to death near the Houses of Parliament, London, **UK**, killing 4 people in total and injuring 49 before being shot dead.

7 April. A highjacked truck is driven into pedestrians by an Islamist terrorist in Stockholm, **Sweden**, killing 4 and injuring at least 15.

20 April. An Islamist gunman kills 1 policeman and injures 2 others and a civilian on the Champs-Élysées, Paris, **France**, before being shot dead.

22 May. An Islamist suicide-bomber kills 22 and injures 129 in an attack on a concert at the Manchester Arena in Manchester, **UK**.

3 June. Three Islamist attackers drive a van into the public on London Bridge, London, **UK**, and stab others, killing 8 and wounding 48 before being shot dead.

6 June. An Islamic State-inspired attacker is shot and arrested after attacking a police officer with a hammer outside Notre Dame Cathedral, Paris, **France**, injuring 1.

18 August. 'The Catalonia attacks': following the failure of a gas-bottle explosive attack, three Islamist terrorists use vehicles and knives to attack pedestrians on La Rambla in Barcelona and in Cambrils, **Spain**. They kill 16 and injure 152 with the perpetrators being shot dead by police.

15 September. A 'bucket bomb' on a tube train at Parsons Green station, London, **UK** fails to properly ignite, injuring 29.

31 October. 8 Killed and over a dozen injured in Manhattan, **USA**, when a truck drives into pedestrians. The driver was inspired by Islamic State videos and material.

11 December. A failed bomb attack on the Port Authority bus terminal, Manhattan, New York City, **USA**. Four people, including the bomber, are wounded.

Introduction

A new field

The book's title is 'digital war', but it's best to begin by explaining what I don't mean by that. My intention here isn't to identify a new *type* of war: I don't want to theorize a new form to stand alongside others covered in this book, such as 'virtual war', 'non-war', 'postmodern war', 'information war' or 'network-centric warfare'. The question of how the properties and biases and uses and applications of digital technology have impacted upon conflict is central to the book, but I don't want to define and defend one overarching military concept. Instead I want to do something broader and more interesting. I want to suggest that the term 'digital war' identifies and conceptualizes today, not a new form of war, but an entire, emerging research field.

The origins of this book lie in my university's novel decision around 2009 to move the media studies staff into a new department, whilst leaving the media studies degree behind for other staff to run. Finding myself adrift in my new home of 'political and cultural studies' I decided to change my teaching to better fit in. As my primary interest was the digital and as I had a long-standing interest in war I created a module called 'Digital War'. Having come up with the title, I then realized I had to think about what it might include.

Since the 1991 Gulf War there had been an explosion of work within media studies on war and media and I was especially grateful to Donald Matheson and Stuart Allan's 2009 book *Digital War Reporting* for filling a lot of my module. But my problem with media studies was that its focus on media coverage of wars, journalism and war-reporting was too limiting. The discipline was often hostile to technology, seeing any discussion of it as 'technological determinism', and it had too little interest in developments in politics or military technology and theory (for example, issues around drones, cyberwar, information war, and network-centric war etc.). Most of all, it couldn't break free from its broadcast-era origins and biases and (with exceptions such as Matheson and Allan's book) it was painfully slow to deal with the ongoing digital revolution that was making traditional mass media processes and concepts obsolete.

The politics, international relations and security studies literature offered much of value, but it had its own limitations – technological issues and media were often only cursorily treated, and the literature on digital media and many military developments were overlooked. Cultural studies and cultural theory were also useful, making important contributions to war that were, in turn, overlooked within media and politics. But to really understand what was happening in war today I had to move beyond the humanities. I had to look at commentators on the military and military theorists, at technological commentators and cybersecurity experts, at policy organizations and think-tanks, at government and military institutions, at scientific and specialist journalists, at public intellectuals, at AI and robotics researchers, and at popular culture. There was work too I didn't have the space to fit in such as the multi-media experiments of artists in response to war and work on the digitalization and archiving of war materials within library and information studies, all of which I was introduced to in the conferences I attended.

Over the following years my module expanded and its subject matter gained in popular awareness. Wikileaks brought us a new vision of our ongoing wars, popular and academic books on drones and cyberwar began to appear, lethal, autonomous robotics began to be publicly debated, and a growing awareness of the revolutionary impact of social media in conflict zones spread. Topics such as hacking, hacktivism, digital civil wars and government surveillance came to the fore; the success of Islamic State meant everyone was discussing online terrorism and propaganda; wars across the world played out now on social media platforms and people's smartphones with everyone joining in; and new developments in military AI, simulation, augmentation and weaponry made the news. Soon, everyone became conversant with the subject of cyberwar and nation state and hacking group cyberattacks, and discussions of 4Chan, trolling, the weaponization of Facebook, Twitter-bots, Troll-Farms, and Russian information war became common. By the time this book was completed, digital war had gone mainstream.

The aim of this book, therefore, is to offer a survey of this emerging field of digital war: to consider in one place a connected set of phenomena and collect the disparate literature around them into an inter-disciplinary text that will appeal to staff and students in media studies, politics, IR, security studies and cultural studies, and even, hopefully, beyond the humanities too. It is also my aim to provide a context for this knowledge. In recent years, I realized I was now teaching students who were too young to remember 9/11, whose entire lives had been lived under the 'War on Terror' and who often had a very limited or erroneous understanding of the events and developments that had marked their lifetime. I want the book to be able to function as a background primer for the topics it discusses, filling in the historical and political context for its readers. The book is intended to be broadly linear in its scope so that it builds up the reader's knowledge and so that they can follow key themes and issues as they reappear. It takes the 1991 Gulf War as its starting point as that was when developments in new military technology made their spectacular appearance as the stars of a new, live news television show, and it was also when a new wave of academic writing originated, to explain what we'd just seen.

The roots of digital technology in warfare of course predate the Gulf War. If I'd been writing a book on 'digital war' as a concept I'd have begun with the development of modern computing during and after World War II and focused especially on the Vietnam War when drones (UAVs), precision-guided munitions (PGMs, or 'smart bombs') and a smart 'electronic battlefield' were first trialed. These still make their appearance in Chapter 3, providing that longer historical context, but the book focuses on explaining developments in the conduct, operation, mediation and experience of war from 1991 to the present.

Chapter 1 considers how wars were conducted and reported in the 1990s, looking in detail at the 1991 Gulf War and the 1999 Kosovo War. The chapter explores the new model of top-down, military media management introduced in 1991 and reapplied in 1999 that was designed to win over domestic and international support and enable the US military to fight a war in the glare of the world's media. It also looks at the new, western experience of war through real-time, 24-hour, broadcast television news, showing how it palatably repackaged and presented the reality of war. The next two chapters explore the 1990s reactions to these developments. Chapter 2 considers the response of philosophers and cultural critics and their theorization of 'non-war', 'information war', 'third wave war', 'postmodern war', 'cyborg war', 'virtual war' and 'virtuous war', while Chapter 3 looks at how the US military themselves reconceptualized war through the 'RMA', 'command and control warfare', 'information war', 'information operations', 'cyberwar', 'full spectrum dominance' and 'network-centric warfare'. Most of these military ideas would be applied in Afghanistan and Iraq and throughout the 'War on Terror' and would form the background to the military research into augmentation and robotics discussed in the final two chapters.

Chapter 4 covers the global event of 9/11 and its reporting and the military conduct and media coverage of the 2001 Afghanistan War. Chapter 5 then considers how the political case was made for the invasion of Iraq as part of the 'War on Terror' and the failure of the mainstream media to challenge this narrative. It also looks at the Iraq invasion and its coverage, exploring how the system of US military media management was imposed again and how the military itself became an important informational producer during the invasion. Chapter 6 then considers how this model of military media control finally broke down in the aftermath of the Iraq War with developments in digital technology and with the 2004–05 emergence of 'Web 2.0' participative platforms and technologies. It begins with a case study of the 2004 Abu Ghraib prisoner abuse scandal, where the perpetrators themselves took the photographic images that exposed the torture, before considering how, over the following years, social media empowered soldiers to produce and share their experiences, much to the consternation of the military authorities.

Chapter 7 begins by considering how the post-invasion fate of both Afghanistan and Iraq was largely overlooked by western media and how the new leaking site, Wikileaks, exposed what the authorities knew about these wars. The chapter explores the controversy over Wikileaks and evaluates its status as a journalistic organization. Chapters 8 and 9 then turn to developments in military technology. Chapter 8 offers an overview of the history and rise of drones, their military application and the issues their use raises, while Chapter 9 traces the history of cyberwar, examining the major cyberattacks, considering the significance of cyberwar and exploring its implications for traditional concepts of warfare.

Chapter 10 considers how, today, everyone in a conflict zone, from militaries, to militias, to terror groups, to civilians, can become an informational producer. It demonstrates how the USA's 1990s' dreams of achieving military, battlefield 'full spectrum dominance' were destroyed by the rise of Web 2.0 platforms, technologies, smartphones and connectivity, leading to a new form of 'participative war' where anyone – including interested parties from around the globe – can share their experiences and images, comment, and promote their preferred cause. It takes as its case studies the 2014 Gaza War and the Syrian Civil War, from 2011 to the present. Chapter 11 explores this further through the case study of Islamic State (IS) and its online presence, propaganda and terror. It considers the operation of IS's media units, their adoption of contemporary digital technologies, the resulting online informational war and the way both the authorities and ordinary people responded to IS and tried to fight back.

The final two chapters consider emerging military technological developments. Chapter 12 looks at the rise of wearable military technologies and the technological augmentation of soldiers, simulations and experiments in brain–computer interfaces (BCI), and Chapter 13 considers developments in robotics and the emergence of Lethal Autonomous Weapons Systems (LAWS). Together they argue that these are continuing military responses to the problems of warfare. Drones have already removed the soldier's body, allowing a safe telepresence, and robotics promises to go even further and fully automate warfare, removing the human even more from the battlefield. If humans are to be deployed then wearables and augmentation systems are designed to give them a hyper-present capability and clear battlefield advantage. Digital technology is again remaking war, therefore, with new modes of non-presence, tele-presence and hyper-presence transforming military experience and combat.

This isn't an exhaustive overview of what digital war is or could be. It remains introductory, leaving out much and skimming over many technologies, issues and developments that could have been chapters in their own right. Whatever its limitations, my hope is that it's a book whose primary contribution is to raise awareness that there is an emerging field of research here; that the topics I've discussed are connected and are best examined in relation to each other; and that there is scope here for others from a range of backgrounds to contribute to this as a field and converse with each other. As ongoing developments suggest, this is a vital topic whose importance is only increasing. If we are to understand what warfare is in the twenty-first century, how it operates, its effects and where it is going, we need to consider all these elements as a whole.

Instead of filling pages with acknowledgements I'd like to briefly thank everyone who has contributed towards this project, especially my students, other academics I have met within the field, and all those who have been kind enough to listen to me talk on war. I am especially grateful to Andrew Hoskins for his friendship, for his generosity in inviting me to contribute to so many of his projects, for his confidence in my work, and for the wonderful discussions with him from which I learnt so much. I'd also like to thank Ben O'Loughlin, Marcus Leaning and Rhys Jones for their significant contributions to my work and understanding. Personally, I'd like to thank those who kept me sane, despite my workload, whilst I wrote this, especially Rob Long, Liz Wride, Heather Merrin, Rebecca Francis-Davies, Steve Vine, Leighton Evans and Ronit Knoble. The book is dedicated to Henry Merrin, Alice Merrin and Hector Merrin, who are my life.

1 Top-down war

Televising conflict in the 1990s

A new kind of war ...

The 1991 Gulf War felt different. A western public who had had little recent experience of conflict tuned in to find a new type of war. This was a war in real-time, occurring on the screens in front of them; a war carried by global 24-hour rolling news channels with live coverage seemingly from the heart of the battlefield; a clean-war of high-tech, high-precision smart weaponry; a press-release war, with daily conferences and generals talking through the day's message; and a video-game war and media spectacle consumed by the domestic audience as entertainment. The reality of the war was, of course, very different – all wars are ultimately about the violent destruction of fragile, physical bodies – but the public perception wasn't wrong. In the military management of both the actual operations and the media coverage this *was* a different kind of war.

The reason why, however, owed a lot to the past. It seemed as if during the Gulf crisis the USA was fighting as much to exorcise the ghosts of Vietnam. Ironically, a conflict that would do so much to define the future model of war was explicitly designed to put to rest the trauma of the past. The Vietnam War was a Cold War-era proxy war, fought by the USA in aid of South Vietnam against the communist North from November 1955 to April 1975. It had ended with the capture of the South by Northern forces and the humiliating withdrawal of the USA after suffering over 58,000 dead and 300,000 injured servicemen. The defeat of the greatest military power by a small, third-world guerrilla army had led to a period of soul searching in the USA and an internal crisis of confidence in American power.

Unable to accept their military defeat, by the 1980s conservative politicians and commentators had come to agree a more palatable explanation for the USA's failure. As Philip Taylor says, 'Middle America and the US establishment remained convinced that an explanation for the single remaining blemish on America's illustrious military record had been found: the enemy within had been their very own media' (Taylor, 1998:2). Vietnam was widely considered the first 'television war', with US journalists free to roam across the combat zone, sending back footage of the conflict for the nightly news. Thus television (and, implicitly, liberal journalism) was blamed for alienating public sympathy and support. The daily, televisual drip-feed of horror, US and civilian casualties, and destruction, it was argued, had turned public opinion against the war, aiding the anti-war movement and weak politicians who hadn't supported the military. In this way, the myth was created that the Vietnam War had been lost on the home front through the television set, rather than militarily. Sylvester Stallone's Vietnam vet John Rambo echoes this sentiment in the iconic 1985 Neo-Conservative action movie *Rambo: First Blood Part II*. Offered the chance of a return to 'Nam' by his commander Rambo famously asks, 'Do we get to win this time?' Thus, through the 1980s a powerful right-wing argument gained force that the USA needed to overcome the self-imposed paralysis of its defeat – the 'Vietnam syndrome' – to rediscover its pride and reassert itself on the world stage.

The problem remained of how to wage a war that wouldn't be undermined by images of death and returning bodybags. The answer lay in wars with clear outcomes, with minimized casualties, with a prepared population and with tight control of the media. As Taylor points out, the British operation in 'the Falklands War' provided a model for this. The Argentinian military invasion of the Falkland Islands on 2 April 1982 had led the UK to form a naval 'task force' to engage the Argentinian navy and airforce and attempt to retake the islands. What resulted was a short 74-day war, with simple and successful aims and a limited and controlled media coverage encompassing Ministry of Defence (MoD) briefings and reports from journalists accompanying the task force. With their control over access to and communication from the warzone the MoD could dictate terms to the media, including limiting the numbers of reporters, vetting individuals and imposing censorship agreements. The result was a highly successful propaganda campaign involving the suppression of information and the delaying of dangerous news not just to prevent any benefit to the enemy but also to manage domestic morale and opinion. In return for privileged access and caught up in the military operation they were reporting upon, the mainstream UK media proved eager to play this propaganda role, putting patriotism before objectivity.

The USA didn't immediately learn these lessons. In the October 1983 military invasion of Grenada their control of the informational environment was so tight that the press were excluded and even fired upon, whilst in the December 1989 invasion of Panama the press were allowed access but were overly restricted in their movements in order to present an image of a bloodless operation. The Gulf War, however, would see the perfection of the USA's military media management system.

Box 1.1 The Gulf War 1991

The 1991 Gulf War had its roots in existing regional conflicts. Following the 1979 Iranian Islamic revolution that deposed its ally, the Shah, and its humiliation in the Iranian hostage crisis from 1979 to 1981, the USA's regional policy shifted. America looked now to Saddam Hussein's Iraq as a counterweight to both Islamic fundamentalism and to Soviet expansionism (following the latter's invasion of Afghanistan in December 1979). Hence the USA's support for Saddam Hussein when he launched the Iran–Iraq War on 22 September 1980. By 1982 this support included money, intelligence, weapons, equipment and training for Iraqi forces. Saddam's actions were motivated by a history of border disputes with Iran and a fear that their revolution might inspire the suppressed Shia majority in Iraq, but his hopes of an easy victory and territorial gains proved naïve. The war ended after eight years on 20 August 1988 with a strategic stalemate and claimed loss of up to half a million soldiers and countless civilian lives. The war left Iraq with economic problems and debts. Suspecting that Kuwait was over-producing oil to depress Iraqi's much needed oil revenues, it sought to rectify these problems by claiming disputed oil-rich territories on the Kuwaiti border. Having been told by

US ambassador April Glaspie that America had 'no opinion' on Arab–Arab conflicts, Iraq invaded Kuwait on 2 August 1990, seizing control of the country by the next day and deposing its monarch, the Emir. The international community criticized the invasion. The UN Security Council passed resolution 660 on 3 August condemning the invasion and demanding the withdrawal of Iraqi forces, as well as resolution 661 on 6 August, imposing economic sanctions on Iraq, and resolution 678 on 29 November, which gave Iraq a deadline of 15 January to withdraw, authorizing member states to use all necessary means to force compliance after this date.

Alongside international diplomatic and political pressure on Iraq and international sanctions introduced on 6 August 1990 that would last until 2003, the USA launched 'Operation Desert Shield', forming a coalition of 34 countries and building up military forces in the region to defend Saudi Arabia and prepare for war. By the time the deadline passed on 15 January 1991 there were 956,000 coalition troops in the area, 543,000 of them US. When Iraq failed to withdraw, the USA launched 'Operation Desert Storm' on 17 January. The air campaign lasted until 24 February, quickly achieving air supremacy, flying over 100,000 sorties and dropping 88,500 tons of bombs on the Iraqi military and civilian infrastructure. The ground campaign, launched on 24 February was astonishingly successful, being called off after a PR-perfect 100 hours on the 28 February, following the destruction, mass retreat or surrender of the Iraqi army and the liberation of Kuwait.

Although the Iraqi army had been comprehensively defeated, the USA had no intention of pursuing it into Iraq to depose Saddam Hussein. The war aims had been achieved and the USA recognized it needed a strong regional counterweight to Iran. It was also wary of taking responsibility for the long and difficult process of state building that would follow his overthrow. Long-term troop deployment would be costly, unpopular and risk turning into another Vietnam. Instead the USA encouraged uprisings by the Shia in the south and the Kurds in the north but the Iraqi military brutally suppressed these with helicopter gunships. The USA was forced to implement 'no-fly-zones' in northern and southern Iraq which its airforce had to enforce for the next decade. The survival of Saddam, his Republican guard and his weapons programmes, together with Iraqi activity in the no-fly-zones necessitated continued US military action and major missile strikes on Iraq in 1993, 1996 and 1998. These unresolved issues and ongoing US antagonism towards Saddam would lead to Iraq being targeted again in 2003 in the Iraq War.

The Gulf War

Following the Iraqi invasion of Kuwait on the 2 August 1990 and the passage of UN Security Council resolutions demanding Iraqi withdrawal and authorizing the use of force to achieve this, the USA began to prepare for military action. As well as the physical build-up of troops in the region and international diplomacy and coalition building, this required the selling of the war on the domestic front. The US public and politicians needed to be convinced that the war was necessary and that the USA could overcome the 'Vietnam syndrome' to successfully fight it. Although there was considerable public and media support for war, the cause was helped by President Bush's demonization of Saddam Hussein as equivalent to Hitler (a trope that would become common in the international coverage) as well as by highly publicized stories of Iraqi atrocities.

The most famous of these was the 10 October 1990 testimony of 'Nayirah', a volunteer nurse at the al-Addan hospital in Kuwait who told the US Congressional Human Rights Caucus that she had seen armed Iraqi soldiers enter the hospital and remove equipment and incubators to be taken back to Iraq, leaving the babies to die on the floor. Her story was widely reported, appearing on ABC and NBC TV news, being cited by senators in the Senate debate to approve military action and being quoted repeatedly by President Bush in the following weeks. It was only revealed in 1992 that 'Nayirah' was the daughter of the Kuwait ambassador to the USA and that her participation had been organized by the 'Citizens for a Free Kuwait' campaign run by the US public relations company Hill & Knowlton and funded by the Kuwaiti government. She had not been a volunteer at the hospital and although Iraqi soldiers had been involved in looting and violence there was no evidence to support the incubator story. Such was its emotive power, however, that, as Knightley says, it proved to be 'the definitive moment in the campaign to prepare the American public for the need to go to war' (Knightley, 2000:488).

By the time the UN deadline had passed the US-led coalition was prepared for war. Crucially the USA had realized it would be fighting two wars simultaneously: a physical, military campaign in the middle east and a global informational campaign. This media campaign had three key targets: first, it was aimed at domestic populations to aid morale and retain support for war; second it was aimed at an international audience and especially the broader coalition members to ensure their continued support and to demonstrate the legitimacy of the action; and third it was aimed at the enemy as propaganda, hoping to demoralize the Iraqi leadership by demonstrating the coalition power.

The coalition media campaign had three elements: official briefings by political leaders in Washington and London; closely controlled military briefings from the command centres in Dhahran and Riyadh, and reports from journalists who had been selected for combat zone access. The military developed a 'pool system' whereby a selected number of predominantly Anglo-American journalists were accredited by the military and allowed to operate alongside troops. They were organized in 'media reporting units' overseen by censors, with reports sent to 'forward transmission units' who also had the right to censor and who relayed reports home for copy to be freely distributed among news outlets. In the event, apart from the invasion plans, censorship was rare, partially as the pool system blurred the line between the military and journalists leading to a self-censorship caused by the close identification with the troops and the operations, and partially because there was little to report prior to the ground invasion and little chance to file copy once it had begun.

The Gulf War, Taylor says, 'was the first major conflict fought against the background of accessible global telecommunications' (Taylor, 1998:x). Hudson and Stanier describe it as 'the most widely and swiftly reported war in history' and 'arguably the greatest media event in history' (Hudson and Stanier, 1997:209). This is largely due to the television coverage, with the world's public tuning into a near-continuous feed of 24-hour rolling news mixing studio commentary, expert opinion, official briefings, live coverage from reporters across the region and even, for the first time, broadcasts from the enemy capital itself. As Sturken argues, it was the Gulf War, not Vietnam, that was a 'television war', as the latter was shot almost exclusively on film and was subject to the delays of the developing process: 'There was always at least a twenty-four-hour delay before images of the Vietnam War reached the United States. The Persian Gulf War, by contrast, took place in the era of satellite technology and highly portable video equipment. It was technologically possible for the world to watch the Persian Gulf War as it happened' (Sturken, 1997:125–26). Cumings goes further: the reason, he says, why this was the *real* television war was not just because of the live reporting or saturation coverage but because of the way in which television itself imploded with the military operations, through its 'radically distanced, technically controlled, eminently "cool" postmodern optic which, in the doing, became an instrument of the war itself' (Cumings, 1992:103).

Indeed, television was almost co-substantial with the war: as Philip Taylor says, 'the Gulf War broke out on television' (Taylor, 1998:31), and viewers followed it that night in real-time. In the USA, ABC captured its outbreak, cutting into their 6.30pm news programme to go live to Gary Shepard in Baghdad who announced, 'something is definitely underway here … Obviously an attack is underway of some sort'. However, it was CNN's coverage of the first night of the bombing that would become one of the most famous moments in media history. Their rental of a 'four-wire' communications system enabled CNN to keep broadcasting after other news organizations were affected by the destruction of Baghdad's communications tower and disruption of power supplies. Over a billion people worldwide watched CNN's through-the-night telephone commentary from the Al-Rashid Hotel by Bernard Shaw, Peter Arnett and John Holliman, including, remarkably, the political leaders in Washington, London and Baghdad, who used it for on-the-spot intelligence as to the progress of the campaign.

Quickly, the daily military briefings from US Central Command in Riyadh became one of the defining elements of the Gulf coverage, representing the most obvious example of the military's control of the global perception of the war. It was here that the military decided upon, produced and disseminated the day's message, narrativizing the conflict and its events for the global media. General Norman Schwarzkopf's explanation of what was happening and the coalition's military operations were aided by a powerful new tool: 'smart-bomb' videos. Laser-guided bombs ('precision-guided munitions', or PGM) had been developed and tested in Vietnam but by 1991, with the development of cheap, miniature computers and guidance systems, a new generation of 'smart' munitions was available. Most used a plane fitted with a nose-cone 'forward-looking infrared radar' (FLIR), which sent a laser signal to pick out a target for a bomb whose light-sensing nose-cone could follow this laser to the target. The aircraft's computer could also send signals to adjust the bomb's control fins in flight to increase its accuracy. The FLIR information could be converted into visible images shown on the computer console in the cockpit whilst some bombs also included nose cameras sending back a record of their fall. The primary aim of these videos was to aid damage assessment but the key development of the Gulf War was to employ them as part of the military briefings.

In this way, the smart-bomb became a dual weapon. As a military explosive it had a localized, precise, destructive effect, but as an image weapon it had a global, resonating, *productive* effect, carrying a message to the world about the US operation. Cumings argues the bomb was a 'video press release', 'simultaneously image, warfare, news, spectacle, and advertisement for the Pentagon' (Cumings, 1992:122). It functioned, therefore, as an advert for US power in the post-Cold War era and for its military and defence industries, whilst also providing imagery and news for the media and a spectacle for the watching population. Most importantly, the smart-bomb demonstrated a new ideal of a pinpoint, hi-tech, 'clean' war of 'surgical strikes' that avoided the civilian 'collateral damage' of dumb-munitions. As such it played a significant propaganda role by helping transform the image and idea of war itself. As Philip Knightley wrote:

> Ever since the British invented military censorship in 1856 ... wartime news management has had two main purposes: to deny information and comfort to the enemy and maintain public support. In the Gulf War the new element has been an effort to change public perception of the nature of war itself, to convince us that new technology has removed a lot of war's horrors.
>
> Taylor (1998:262)

With the smart-bomb, therefore, western violence was presented to its domestic audience as a moral force. As Aksoy and Robins argue

> The clear message was that 'smart' was good, and brilliant was virtuous. 'Smart' weapons, it was being claimed, could actually save the lives of soldiers and civilians alike in the Gulf. To reduce error, to be so deadly accurate and efficient, was a reflection of the virtuous triumph of western technology'.
>
> Aksoy and Robins (1991:331)

Broughton similarly sees the videos as aimed at the home front, 'recruiting participation at the hearth of virtually every American living room', with their violence 'soliciting the perceptual complicity of the viewing citizenry' (Broughton, 1996:140). The aim, he says, was to promote the 'New World Order' – the claimed dominance of western values in the post-Soviet era and the belief that these values could be globally policed. With its avoidance of civilian casualties, its precision and its vision of justice being carried out, the smart-bomb clip thus became 'the primary signifier' of this new world, seizing 'the ethical territory' of the moral high-ground. 'Accuracy was transformed into a sign of noble intention', Broughton says, with the bombardment being presented as the 'performative juridical founding of the New World Order' (Broughton, 1996:141).

The video feeds also functioned as a means of personalization and identification for their audiences. Within the anonymous prosecution of the war, Broughton says, 'it was left largely to the missile to provide a model of individuation' (Broughton, 1996:151). The bomb appeared in coalition briefings as an active moral agent and individual, honing in on an unseen enemy edited out of the videos or disintegrated in the explosions. Thus it was the bomb with which we identified, Broughton argues: 'The viewer, falling under the thrall of the smart-bomb video, took up a specific, symbolic position, not as abstract, transcendental subject but as concrete, material body'; one fusing with the projectile and enjoying the scopic pleasure of the descent (Broughton, 1996:150–52). These falling, vertiginous, point-of-view shots simultaneously positioned the spectator physically on the side of the bomb (giving as Broughton says, 'a bomb's-eye view'), whilst also positioning them epistemologically, as all they knew was what the bomb saw, and morally, as the spectator identified with the bomb itself. Hence the McLuhanist electronic implosion into the real offered by television was extended here in an implosion into the military technology itself. Watching these videos, the domestic audience *became* the bomb.

The other major propaganda success for the west was the 'Patriot' missile. Saddam hoped to broaden the conflict by drawing Israel in, hence his launching of ageing Scud missiles from mobile launchers against Israeli cities. Under intense pressure to prevent an Israeli response that would split the coalition, the USA provided a new anti-missile system to stop the threat. The Patriot's fame began on 18 January 1991 when it achieved, Knightley says, 'an historic knockout' as 'the first defensive missile to destroy an incoming offensive missile' (Knightley, 2000:496). What followed was an intense media focus on the continuing US attempts to destroy Scud missiles and launchers. Taylor argues:

> The success of the American patriot missiles in intercepting the Scuds provided, in microcosm, a televisual symbol of the conflict as a whole. It was a technological duel representing good against evil: the defensive Patriots against the offensive Scuds, the one protecting innocent women and children against indiscriminate attack, the other terrifying in their unpredictable and brutal nature. The very resonance of their names implied it all. Here was beneficial high-technology, a spin-off of the American SDI ('Star Wars') programme, being utilized against comparatively primitive weapons of mass destruction from the old Cold-war era: the Patriot was the 'Saviour of the Skies' and the 'Darling of the US Arsenal'.
>
> Taylor (1998:70)

Here too, the 'liveness' of the media coverage was used to increase audience identification and excitement. On the night of 17–18 January, for example, western reporters in Israel broadcast updates on the Scud attacks. These included CNN reports from Jerusalem of reporters wearing gas masks, discussing the explosions they could hear and worrying about chemical attacks. As Taylor points out, 'CNN's cameras were pointing at the wrong place. In fact, it was all, in a sense a non-event; Jerusalem was not attacked. Some 25 miles away Tel Aviv was, but not with chemicals' (Taylor, 1998:69). The reality, however, was less important than the media image and its domestic impact.

In contrast to the west's use of the media, Saddam's propaganda efforts appeared clumsy and ineffective. His televisual appearance on 23 August 1990 with western hostages from British Airways Flight 149, captured at Kuwait airport, was intended to demoralize the west whilst reassuring it of the safety of these 'human shields' provided no attacks were launched, but Saddam's avuncular attempt to pose with a young British boy appeared threatening and horrified the global audience. Similarly, Iraqi television's parading of captured coalition pilots on 20 January 1991 had the opposite effect to that intended. Their blank, bruised and beaten faces mouthing the opinions of the Iraqi state convinced no one and only hardened public opinion about the necessity and justice of the bombing.

Iraq also misplayed western journalists. The intelligence value of CNN's live Baghdad reports for the west led to an Iraqi order to cease transmitting on 17 January whilst fears of journalists aiding western damage assessment together with far fewer casualties to display due to the unexpected precision of the coalition weapons led to Iraq expelling all but two western journalists on the 19th. This might have been a mistake as reports from Baghdad-based journalists were controversial in the west, splitting domestic unity, with many critics accusing reporters of becoming mouthpieces of the Iraqi regime. CNN's Peter Arnett, for example, was described by US House of Representatives as 'the voice of Baghdad', whilst Conservative MPs in the UK called the BBC 'the Baghdad Broadcasting Corporation'.

One of the most criticized events took place on 23 January when Peter Arnett reported from a 'baby milk factory' he was taken to, which the Iraqis claimed had been destroyed by coalition bombing a few days earlier. The USA denounced Arnett's report as Iraqi propaganda and insisted the site was 'associated with biological warfare production' (Taylor, 1998:113). All later evidence points to Arnett's report being correct, but Iraq's propaganda coup was fatally holed by two important errors. One was CNN's shots of an Iraqi working inside the factory earlier with a white lab-coat with the words 'baby-milk factory' stitched on the back, and the other was a crude sign shown propped against railings outside the building with the handwritten English words 'Baby Milk Plant'. The interior shots were genuine and were shot in August 1990 and the sign was simply an attempt to draw attention to the site, but both were widely ridiculed in the western media and seen as evidence of Iraqi lies. Here, coalition misinformation trumped truthful Iraqi propaganda.

Saddam was also blamed for things he hadn't done. At the end of January, he was accused of 'environmental terrorism' when Iraqi forces opened valves at the Sea Island oil terminal, dumping oil into the Persian Gulf to prevent a possible sea-borne invasion. Iraq's culpability for this is certain but the images that appeared in the western media on 25–27 January of dying, oil-drenched sea-birds that provoked such sympathy among its animal-loving audiences were misleading. The Iraqi oil-spill hadn't yet reached land and it was only one of several oil-slicks of disputed provenance. The oil-covered cormorants desperately trying to breathe were not killed by the Iraqi oil but by a slick from tankers that the coalition had bombed. Whatever the cause, the claims of an 'environmental disaster' only helped galvanize more support for the war.

Western journalists were allowed back into Iraq by the end of January, with the authorities hoping to make political capital from the destruction by escorting them to sites of civilian casualties. There were still fewer of these than expected, however, and Iraq had little to counter the western narrative of events until 4.30am on 13 February when two 2,000lb laser-guided bombs hit an installation at Amiriyah that was being used as a civilian bomb shelter, killing 408 people, the majority women and children. Iraq immediately lifted all reporting restrictions and the earliest media reports by western journalists were honest and graphic about the deaths.

For once the coalition military authorities were on the back foot. By the time of the evening's Riyadh briefing the line had been worked out, with Brigadier-General Richard Neal declaring 'I'm here to tell you that it was a military bunker. It was a command and control facility'. He even suggested that it was 'plausible' that Saddam had deliberately placed civilians in the bunker for a propaganda coup (Taylor, 1998:194–95). A White House press conference by Marlin Fitzwater later that evening repeated the claim that this was 'a military target' and the intimation that Saddam was responsible for the casualties, commenting: 'We don't know why civilians were at this location but we do know that Saddam Hussein does not share our value in the sanctity of life' (Taylor, 1998:196–97). Soon after, Defence Secretary Dick Cheney confirmed that this was a military facility and similarly suggested Saddam had planned the deaths.

Though many papers repeated these claims as fact, the lack of regret and the repeated insistence of the infallibility of coalition intelligence and precision weapons caused significant damage to the image of the military. Its officials soon admitted privately that this was a simple intelligence mistake, but publicly the American military held the official line. What should have been a significant propaganda coup for Iraq, however, was limited by the nature of the news coverage. The images that were shown were shocking, but sensibilities necessitated a self-censorship of the real horror. Few channels showed images of the dead. Taylor argues that CNN's brief broadcast of the rows of charred bodies in a long shot proved to be 'the first and only time audiences could get a glimpse on this station of the real carnage caused to human life' (Taylor, 1998:193). There was no evidence, however, that the public wanted more than this. The clean, bloodless war the military and media crafted for their audience was the only one that they wanted to see.

For a war that appeared to get you closer to the real than ever before – that appeared to be broadcast right from the heart of the enemy capital, in real time, with live footage of Scud attacks, reports from troops stationed with the soldiers and with military briefings that exposed the inner workings of the operation and even gave you a bomb's-eye view of what it was like to actually *be* the missile – what we saw was only a constructed reality. The truth about the western military action was very different from that shown on the television screens. Focused as it was on the air strikes on Baghdad and the Patriot–Scud battle, for example, the media completely missed the real centre of the conflict, what Paul Rogers calls 'the systematic destruction of the Iraqi forces in Kuwait and south-eastern Iraq'. These forces, he says, 'were mainly peasant conscripts and reservists as the elite forces were generally kept away from the most dangerous zones. They were exposed to the latest-generation of area-impact munitions: weapons designed to kill and maim over the widest area possible' (Rogers, 1991). Indeed, despite the media focus on precision weapons the dumb reality was that they comprised only 6,250 of the 88,500 tons of bombs used during the war. One source suggested that 70 percent of the bombs dropped missed their target and although precision weapons had a greater success rate of around 80 percent the problem of damage assessment remained and many targets required repeated attacks to make sure they'd been destroyed.

The Patriot–Scud battle was especially misleading. A US Gulf War air power survey reported later that despite 1,500 missions against Iraq's mobile Scud launchers not a single one was destroyed, whilst the number claimed destroyed at the time was four times greater than the total launchers deployed. Video 'proof' of launchers being destroyed by laser-guided bombs shown in a press conference on 30 January was later disproved as the launchers were only fuel trucks. Despite Bush claiming that Patriots had knocked out 41 out of 42 Scuds and Schwarzkopf's claim of a 'one hundred percent' success rate, after the war an Israeli examination concluded that only one or none of the Scuds had been destroyed, whilst a 1993 US Armed Services Committee report concluded, 'A post-war review of photographs cannot produce even a single confirmed kill of a Scud missile'. In fact, the Patriot missiles may have increased the danger for Israelis as they often came down themselves, causing more damage. A 1993 US Congressional report commented that, 'US forces greatly overestimated the Iraqi equipment destroyed during the air-phase of the war – and couldn't do any better today' (Merrin, 1994:448; Knightley, 2000:496).

The reality of the ground invasion was also unreported, mostly because it moved too fast for the pooled journalists. Few photographs and little footage has emerged since of these operations. Frontline resistance to the ground force was minimal and ineffective. Along 70 miles of trenches, earthmovers and bulldozers went in first, simply filling in the Iraqi earth and sand defences, burying the troops alive. Out of 8,000 troops in this sector, 2,000 managed to surrender. Colonel Anthony Moreno later reported, 'For all I know we could have killed thousands. I came through right after the lead company … What you saw was a bunch of buried trenches with people's arms and things sticking out of them' (Merrin, 1994:446). With little chance to resist, the Iraqi forces fled, surrendered or were killed or captured en masse. By 26 February coalition forces were struggling to keep up with the frontline and to process the 20,000 prisoners of war. The war that Saddam Hussein had famously claimed in a speech on 17 January would be 'the mother of all battles', had quickly turned into the mother of all defeats.

With the Iraqis' rapid withdrawal from Kuwait, the final story was the liberation of Kuwait City on the 26 February – appropriately enough for this 'postmodern war' in which the media had played such a central role, by a unilateral CBS news crew. Kuwaiti freedom, Kuwaiti celebrations and Iraqi atrocities now formed the core of the news: as Taylor says

> While the allied forces were attempting to "cut off and kill" the fleeing Iraqi army to the north, the journalists who had liberated Kuwait City were having a field day with reports which, to some, appeared like Paris 1944 with television ... The cameras, in other words, were pointed away from the scene of the actual fighting, and at a liberated city which was no longer part of the military action.
>
> Taylor (1998:249)

The most famous scene of conflict on the final day was on 'the highway of death' – Highway 80 on the outskirts of Al Jahra – where on the night of 26–27 February a massive column of Iraqi military and other vehicles fleeing back to Iraq was attacked and destroyed by coalition aircraft. In fact, there were two roads, with a similar column on Highway 8 also coming under attack, though this went unreported for almost two weeks. The aim was to cut off the retreating Iraqi forces, to inflict as much damage as possible on their military and even, for the pilots, to get on the score sheet before the war ended. One commander said, 'it was a turkey shoot for several hours', whilst a pilot referred to it as 'like shooting fish in a barrel' (Taylor, 1998:250). Estimates of the casualties vary wildly, with claims many troops fled into the desert, although a figure of between 800 and 1,000 dead seems most likely. It was the PR dream of a 100-hour war with the successful completion of all objectives, combined with a fear that too many Arabic casualties would delegitimize the military action and turn opinion against the west that convinced President Bush to call a ceasefire from midnight EST on 28 February.

Footage of the aftermath of 'the highway of death' didn't emerge until 1 March. The spectacular photos of the carnage would become some of the most iconic images of the war, but here still there was little sign of death. Few bodies remained at the scene and most outlets presented only the sight of blown-out vehicles. ITN reporter Alex Thomson complained about the lack of coverage:

> Seldom had the job of the correspondent looked so clear cut as it did on Mutla Ridge: show people, tell people. The whole point about the Gulf War was that it was censored to such a degree that the images went in precisely the opposite direction. They gave the casual sense that no-one had died or been hurt. Mutla Ridge was a rare, golden opportunity to try to put that right ... it was crucially important to report, to show at last that this war like any other was about killing people.
>
> Knightley (2000:499)

In the UK only *The Observer* printed a photograph of a body, publishing on 10 March an image by Kenneth Jarecke of a charred, grimacing face burnt into position at the empty windscreen of a vehicle. Causing outrage, it was almost the only dead body seen in mainstream coverage: the west had seen more dead birds than humans in this war. The real catastrophe, the public were told, was ecological, with a clear-up operation in place to stem the 600 burning oil wells the Iraqis had ignited as they left. A later clash after the ceasefire – 'the battle of Rumaila' on 2 March, when coalition forces engaged and annihilated a large column of retreating Iraqi Republican Guard armoured forces and killed an estimated 700 troops – went almost unreported.

As Sturken points out, the treatment of the dead of 'the highway of death' was consistent with the whole coalition and media approach to the war. 'Though the image of the war on CNN may have made it appear that the television screen was the war's primary location, this illusion effaced the war that took place among human bodies and communities'. Discussions of the 'virtual war', she says, eclipse 'the fact that it was still a conventional war, fought with conventional weaponry, in which the body of the other was obliterated' (Sturken, 1997:127). The predominant military and media narrative of the war was instead the battle of technologies: 'Instead of images of human beings at war, the media presented images of a war of machines: tanks, bombs, helicopters and planes' (Sturken, 1997:133). This emphasis on the spectacle of the non-human – 'of weapons against weapons' – represented, Sturken argues, a deliberate process of 'dis-remembering bodies' in order to present a 'bloodless' war (Sturken, 1997:132). Hence the images from Highway 80 and Highway 8. At the end of the war we only saw broken vehicles.

The Gulf War was a remarkable military success. Taylor describes it as 'one of the most clear-cut and one-sided military victories in the history of warfare' (Taylor, 1998:265), although perhaps this should not have been such a surprise, given that an Iraqi military exhausted by an eight-year war with Iran faced the greatest conventional military force assembled to that date. The war aims were achieved quickly and easily, with few casualties. The USA lost 146 soldiers, 35 of them to friendly fire whilst Britain lost 47 soldiers, 9 of them to friendly fire. Of the total 379 coalition force deaths only 190 were killed by the Iraqis, with the rest coming from friendly fire or accidents. In contrast, some estimate Iraqi military losses at 20–35,000 with perhaps 3,500 civilians killed during the war.

Western euphoria was quickly tempered, however, by the war's aftermath. The USA needed a strong Iraq to balance Iran regionally and was reluctant to depose Saddam and engage in any nation building that would commit troops to an indefinite and dangerous stay. There were confused hopes of a popular, democratic, internal revolt but the CIA-provoked uprisings among the southern Shia and northern Kurds from March to April 1991 were put down with brutal force. The USA refused to intervene in the Iraqi military action, with President Bush declaring 'I am not going to have a single soldier shoved into a civil war that has been going on for ages' (Merrin, 1994:454). Television coverage of the Kurdish suffering, however, exposed the humanitarian crisis and pressured the coalition to send in troops and to impose southern and northern 'no-fly-zones' for the Iraqi airforce to prevent further massacres. The survival of Saddam and his military and weapons programme necessitated a continued policing of the no-fly zones and repeated military action including major missile strikes in 1993, 1996 and 1998.

The media operation of the Gulf War was more successful, as the military's media management system effectively secured public support for the war and ensured it remained on side throughout its operation. With it, Taylor says:

> The coalition therefore demonstrated that modern democracies could fight wars … in the television age without allowing too much of war's "visible brutality" to appear in the front rooms of their publics. Only that which was deemed acceptable by the warring partners was permitted but, thanks to the presence of western journalists in Baghdad, the illusion

was created that war was being fought out in full view of a global audience. However, the absence of cameras in Kuwait or at the Iraqi front-line meant that neither the main reason for the war, nor the battlefields where it was mainly won and lost, were being seen.

Taylor (1998:278)

Beyond the success of presenting and packaging a palatable 'war' for the domestic audience, perhaps the greatest coalition achievement was in the unification of the military and media operations to create a new model of warfare. Ironically what began as a response to western weakness – to the fear of employing force and the need to wage a war without casualties and with public support – produced, in its astonishing success, an ideal model of war and its military and informational management that could be applied now to future conflicts. It was a model that would be deployed next in Kosovo, though by then western optimism had been shaken by the events of the decade and the realization that its power could not always be so easily used.

'The New World Order' 1991–1999

Initially, western confidence in military intervention and international alliances was high. The 1991 Gulf War had showcased US military superiority and with President Gorbachev's resignation on 25 December 1991 and the dissolution of the USSR the USA became the sole remaining superpower. President Bush had already set out his vision for the future in a speech given on 29 January 1991 in which he suggested US action in the Gulf was part of a broader project, arguing: 'What is at stake is more than one small country; it is a big idea: a new world order, where diverse nations are drawn together in common cause to achieve the universal aspirations of mankind – peace and security, freedom, and the rule of law' (Bush, 1991). Bush saw the USA as having a 'leadership' role in this new era, but as a pragmatic politician he saw the 'New World Order' as defined by international cooperation to end acts of national aggression, rather than by unilateral US military action. This explains the USA's intervention in Somalia in August 1992 with 'Operation Provide Relief', a well-intentioned attempt to join the multinational UN relief effort.

An outgoing President Bush extended this aid in December with 'Operation Restore Hope', which authorized the deployment of troops to prevent the looting and extortion that was hampering food distribution. Following the model of the Gulf, the administration understood success on the airwaves mattered as much as on the ground so they informed the media in advance about their plans, leading to remarkable scenes on 9 December 1992 as the US military nighttime beach assault was greeted not by armed resistance but by lights, cameras and live TV coverage. Interviewed about the mission as he unloaded, Lt. Kirk Coker of the US Marines was asked, 'Don't you think it's rather bizarre that all these journalists are standing out here …?'. 'Yes', he admitted: 'It really was and you guys really spoiled our nice little raid that we wanted to come in without anyone knowing it' (YouTube, 2014). As Thomas Keenan notes, today 'images and publicity have become military operations themselves, and the military outcome cannot be easily distinguished from the images of that operation' (Carruthers, 2000:221). As a deliberately constructed, globally broadcast military and imagic operation advertising the US presence, the beach landing epitomized this postmodern implosion of both media and event and the media and military.

As Hudson and Stanier note, the US Somalian intervention turned out to be 'an almost unmitigated disaster' as a 'mission creep' moved the objectives from aiding relief to disarming the warlords, capturing General Aidid and restoring order (Hudson and Stanier,

1997:245). Humiliation in 'the Battle of Mogadishu' on 3–4 October 1993 – where a raid by Delta Force rangers ended with the shooting down of two Black Hawk helicopters and television images of the body of a US soldier being dragged through the streets – confirmed the Clinton Presidency's desire to reappraise its role. Scaling back its relief efforts it finally withdrew in March 1994. The USA's failure in Somalia largely explains its reluctance to intervene in the Rwandan genocide that broke out that April. Over the next 100 days up to 1 million Tutsis and moderate Hutus were slaughtered whilst the UN and countries such as Belgium and the UK refused to intervene. The graphic television images of the killings, many carried out by machete, and of the aftermath of people burnt alive in buildings were broadcast to an impotent western audience, exposing the naiveté of post-Gulf War optimism.

The international community had also been slow to get involved in the conflicts surrounding the break-up of the Socialist Federal Republic of Yugoslavia (SFRY). Both Croatia and Bosnia and Herzegovina had attempted to follow Slovenia in declaring independence from the Federal Republic, leading to conflict with their own Serb populations and with the Serbian Federal Army. The Croatian War began in March 1991 and the Bosnian War began in April 1992, both being marked by fierce fighting, the shelling of towns, mass rape, executions and 'ethnic cleansing', especially by Serb forces. Media coverage of the conflicts was widespread but there was little desire for intervention among western governments. Ultimately, however, as Hudson and Stanier note, 'the combined force of the world's press and television' and their graphic accounts of the violence and atrocities forced the west's hand (Hudson and Stanier, 1997:278), and in August 1992 the UK and France agreed to send troops under UN auspices to protect aid supplies.

Media pressure led to other largely tokenistic gestures, such as 'Operation Irma', a series of staged airlifts from Sarajevo organized after the wounding of five-year-old Irma Hadžimuratović. BBC television pictures of 'little Irma' in August 1993 led to considerable public interest and her case became a cause célèbre in the UK, with tabloids competing to evacuate her for specialized medical treatment in London. With the UK government reluctant to meaningfully intervene, the desire to be seen 'doing something' and the photogenic human-interest value of the story led to a media-created 'pseudo-event' (Carruthers, 2000:217) that had little effect or long-term consequence. Workers at the UN headquarters confirmed this, joking that 'Irma' stood for 'instant response to media attention' (Moorcraft and Taylor, 2008:131).

The greatest embarrassment for the west was the UN's failure to protect the UN Bosnian 'safe haven' of Srebrenica, where the deliberate withdrawal of the Dutch Protection Force allowed the Bosnian Serb army to massacre 8,000 men and boys from 11–13 July 1995. NATO had been policing no-fly zones since 1993 and had engaged in its first ever combat activities in clashes with the Serb airforce on 28 February 1994, but the Srebrenica massacre and the Serbian shelling of Sarajevo's marketplace on 28 August now prompted a decisive UN and NATO response. From 30 August to 20 September 1995 NATO launched 'Operation Deliberate Force' which included a bombing campaign against Bosnian Serb army forces. Its success pressured the Federal government to accede to negotiations that would lead to the Dayton agreement in November 1995, formally signed in Paris on 14 December, bringing an end to the Bosnian War.

Despite this late effort, overall western intervention in the Yugoslav conflicts had been reluctant, inadequate, and a reaction to Serbian actions or media coverage. Combined with the catastrophes in Somalia and Rwanda it seemed that the post-Gulf War dream of an enforceable 'New World Order' had been hasty and naïve. These failures almost certainly influenced the decision to intervene decisively in Kosovo a few years later.

Box 1.2 Kosovo 1999

Kosovo was a region within Serbia, in the post World War II Socialist Federal Republic of Yugoslavia with a population that was, by 1999, 88 percent Albanian and 7 percent Serbian. Ethnic Albanians had always been in the majority but it was the Serbs that dominated the government, security forces and industrial employment. By the late 1960s Ethnic Albanian Protests led to greater decentralization, reforms recognizing a muslim Yugoslav identity and a shift towards Albanian dominance in public authorities and employment. By 1974 Kosovo had been granted major autonomy but Albanian nationalism, demands for full provincial status and protests led to a 1981 crackdown and the reassertion of Serbian rights. Inter-ethnic tensions worsened through the 1980s and in 1989 the President of the Serbian Republic, Slobodan Milošević, reduced Kosovo's semi-autonomous status and instituted a cultural repression of the Albanian population. A non-violent resistance movement attempted to promote Albanian rights and, before its dissolution by Serbia, in 1990 the Kosovo assembly passed a resolution claiming Kosovo to be a Republic. In September 1992, they also declared it to be an independent state and elected a president, although only Albania recognized it. The 1995 agreements that ended the Bosnian War ignored the situation in Kosovo and by 1996 the Kosovo Liberation Army (KLA) had emerged to fight for the rights of ethnic Albanians and for full independence. The situation deteriorated into all-out war by 1998 with a series of attacks, massacres and reprisals. A NATO-brokered ceasefire failed in December 1998 and in January 1999 Serb forces carried out a massacre of 45 Kosovo Albanians at Račak that attracted international opprobrium. A multinational international conference was called in February, drafting the 'Rambouillet Accords' by 18 March, but the Serbian government refused to accept its demands for Kosovo to become an independent province administered by NATO troops. Their refusal to sign was taken as sufficient reason for the commencement of military action by NATO.

NATO's 'Operation Allied Force' was a bombing campaign against Serbia by an 11-nation coalition from 24 March 1999 to 10 June. In this time 1,000 aircraft flew 38,000 missions against the Serb capital Belgrade and against Serb Army forces in Kosovo. Although it was launched for 'humanitarian' reasons it proved more controversial than the Gulf War due to the absence of UN authorization and many on both the political right and left questioned its purpose, relevance and legitimacy. NATO was reluctant to commit ground troops and the air campaign had only a limited success against poor weather conditions and experienced ground troops in a forested, mountainous region that offered none of the attacker benefits of the Gulf deserts. The commencement of the air campaign also allowed the Serbs to step up their ethnic cleansing of the region, with up to a million ethnic Albanians being displaced by Serb aggression or fleeing the conflict, leading to a significant refugee crisis on the borders. Ultimately the plans for a ground invasion finally convinced Milošević that NATO was serious and he agreed a ceasefire on 3 June. UN 'KFOR' troops entered Kosovo on 12 June to oversee its administration and its movement towards self-government and ultimately to an independence that was announced on 17 February 2008. Serbia refuses to recognize Kosovo's independence and UN forces remain in place.

The Kosovo conflict

In office since May 1997, the new UK Prime Minister Tony Blair was emerging as a vocal supporter of military action for 'humanitarian' reasons and there was a sense that a decisive response was now needed in the emerging conflict in Kosovo – not only to prevent more massacres but also to make up for the lack of intervention in Bosnia. As Knightley argues, the Kosovo campaign against the Serbs was intended in part to retrospectively punish Serbia for its earlier violence and its 'humiliation' of the USA (Knightley, 2000:518). Thus NATO's 'Operation Allied Force' was launched on 24 March 1999 in support of the Kosovo people and the Kosovo Liberation Army (KLA) against the Serbian army and state. An 11-nation coalition, dominated by the USA, participated in an air campaign involving 1000 aircraft and 38,000 missions that lasted until 10 June, ending with the capitulation and withdrawal of the Serb army.

The military and media operations in Kosovo were explicitly modelled on those of the Gulf. As soon as the operation began, Knightley says, 'a meticulously prepared system of propaganda and media control – especially in the United States and Britain – swung into action', with NATO bringing to it, 'all the skills for managing the media and arousing public support that its member countries … had polished during the Gulf War' (Knightley, 2000:501–02). Militarily the campaign followed the Gulf War in using air strikes to hit the enemy capital, remove their command, control and communications infrastructure and 'degrade' forces in the field. NATO also replicated the Gulf War media campaign, combining coordinated political briefings in Washington and London with daily military briefings from NATO's Brussels HQ. Here, following Schwarzkopf's example, NATO's spokesman Jamie Shea presented the day's events and activities from his podium in the Media Operations Centre (MOC), interspersing his discussion with maps and film from cockpit cameras and gun and bomb-sights. Here again the information flow was tightly-controlled with details released to the media, Matheson and Allan argue, 'managed, massaged and manipulated by NATO and the Pentagon … with an eye to filling up the airtime with the alliance's message of the day, rather than filling in the blanks for the thirsty media and the uninformed public' (Matheson and Allan, 2009:37). Just as in 1991, for all the media coverage, Knightley says, 'the public drowned in wave after wave of images that added up to nothing' (Knightley, 2000:504).

As in the Gulf, military action needed public support and this was achieved again by the demonization of the enemy. Like Saddam, Milošević was compared to Hitler, whilst the Serbs were Nazis, involved, in the Bosnian War, in land grabs, mass executions, 'genocidal' ethnic cleansing and the establishment of concentration camps for their enemies. This 'Nazification' campaign (Hammond and Herman, 2000:70–78) was widely embraced by both western politicians and the media, playing an important part in justifying the intervention and mobilizing domestic opinion in its favour. For Tony Blair Kosovo became a moral crusade: it was, he said, 'no longer just a military conflict. It is a battle between good and evil; between civilization and barbarity' (Hammond and Herman, 2000:70). The media, Knightley argues, simply repeated this line, painting the conflict 'in black-and-white terms with simple "goodies and baddies"' (Knightley, 2000:501).

Matheson and Allan point out that even the opening of the conflict echoed the Gulf War. When the airstrikes began on 24 March reporters in Belgrade attempted 'rooftop journalism' from their hotels, with CNN successfully capturing bright green 'nightscope' footage of the bombing (Matheson and Allan, 2009:32). Serb authorities, however, quickly arrested western journalists, smashed their equipment and escorted them to the border, treating them, as CNN's Brent Sadler said, as if 'we were part of the whole attack structure' (Matheson and Allan, 2009:32) – an assumption that was reasonable, given the use of live television for intelligence in the Gulf War. The few reporters who remained faced strict reporting restrictions, although new technological developments such as satellite phones and cell phones empowered individual reporters to produce and send material from wherever they were. With a lack of access to Serbia or the Kosovo warzone, reporters gathered instead at the borders to report the growing refugee crisis and interview those fleeing the country. As Matheson and Allan point out, 'Under such circumstances journalists could do little more than report what the refugees were claiming, comparing and contrasting their accounts with official statements. Independent verification of what was happening on the ground remained frustratingly elusive' (Matheson and Allan, 2009:38).

There were, however, important new developments in the Kosovo War. Through the 1990s the USA had been theorizing concepts such as 'cyberwar', 'information war' (IW) and 'information operations' (IO) and some of these were tested in Kosovo. The simplest information operations included psychological operations (PSYOPS) such as the mass dropping of leaflets, broadcasting pro-allied messages and creating online propaganda. Western governments used the internet to disseminate official updates, press briefings and speeches, as well as maps, images, audio and videos, whilst the Serb government countered this with its own online accusations of violence against Serbs and encouraged critical emails to western news organizations. As *Newsweek* noted: 'Now the web is a vivid mirror of the struggle for Kosovo, a first in war' (Matheson and Allan, 2009:35).

The Pentagon held back from employing full-scale cyberwar in Kosovo due to legal and political constraints, although it formally established an 'information warfare' group supporting the air campaign that claimed to have 'great success' (Brewin, 1999). If US cyber-operations were covert and secretive, pro-Serbian hacking was very visible, enabling it to score a considerable propaganda victory online. The Serbian hacking group 'The Black Hand', and possibly Russian hackers too, were successful in defacing NATO and western government websites and in launching simple denial-of-service (DOS) attacks and virus-infected email attacks. Hackers calling themselves the 'Hong Kong Danger Duo' deleted the whitehouse.gov website and replaced it with the message 'Protest USA's Nazi action! Protest NATO's brutal action!', whilst other hackers took down NATO's Public Affairs website and their email system, forcing Jamie Shea to apologize for the service in a press briefing. Other pro-NATO hackers responded against Yugoslav systems, with 'Dutchthreat' replacing an anti-NATO page with their own 'Help Kosovo' page, whilst US hackers 'Teamspl0it' left anti-war, anti-NATO and anti-Milošević messages across a number of websites (Messmer, 1999; Nuttall, 1999; Geers, 2008).

The NATO airstrike on the Chinese embassy in Belgrade on 8 May that killed three Chinese reporters was explained at the time as an 'accident' due to a map error but it was retrospectively revealed as a deliberate act of information warfare as the embassy was being used to transmit Yugoslav army communications. Although NATO apologized, China condemned the bombing as a 'barbarian act', leading to Chinese hackers joining the online attack on NATO and US websites. Other physical attacks such as those upon roads, railway lines and bridges and the airstrike on the head office of Radio Television Serbia on 23 April that killed 16 journalists were all implicitly part of NATO's information operations in targeting Serbia's command, control and communications (C3) systems. Information was also, of course, central to the entire bombing campaign, with NATO's Supreme Allied Commander Europe, General Wesley K. Clark developing, Ignatieff says, an effective 'computerized, real-time target development and review process' (Ignatieff, 2000:99–103), whilst the Kosovo War also marked the first time the USA's Predator unmanned aerial vehicle (UAV) system was deployed, being used for surveillance.

Another new development was the use of the internet by people in the warzone. Faced with propaganda from both Serb and NATO sources, the internet became an important site of conversation and debate for those wanting to understand the events happening around them. Personal messaging, chat-rooms and bulletin boards helped people exchange information, discuss official claims and share opinions. Thus, the internet emerged as an important news source, with individuals, activists and amateur journalists in the warzone and human-rights groups and refugee organizations all using it as a means to report on the crisis. As Lasica says, the internet 'provided an alternative channel that offered deeper coverage, more interactivity, and, most significantly, greater diversity of voices and viewpoints' (Matheson and Allan, 2009:44). The Kosovo conflict took place in a country whose populations were participants in the informational environment, adding to it and propelling it forward: in the words of Ellen Goodman in the *Boston Globe*,

> the conflict in Kosovo is a chat-room war, an email war, a Web site war, a war in which anyone with a PC and a phone line can literally become a correspondent … A war in which anyone with a netserver can log on to the war zone.
>
> Matheson and Allan (2009:45–46).

The new news producers included the 'cyber-monk', Father Sava Janjic, a Serbian Orthodox monk who collected accounts of the conflict and news to present a more complex reality than the simplistic picture of good vs. evil seen in the international press. Using daily emails, real-time chat and web updates, the monastery website, Matheson and Allan argue, 'was effectively transformed into a site for independent reporting, attracting attention around the globe' (2009:42).

Another figure who captured the imagination was Kujtesa Bejtullahu, a 16-year-old Albanian Muslim in Pristina, Kosovo whose emails as 'Adona' to her US 'electronic pen-pal', Finnegan Hammill, were read out on National Public Radio's 'Morning Edition', before being picked up by CBS and CNN. Her first-person account fostered a personal connection to the events in Europe, putting a human face on the conflict for those in the USA. Hammill's description of her as 'Anne Frank with a laptop' soon caught on, with the media calling her the 'Anne Frank of Kosovo', and by 27 March she was being quoted by President Clinton as he explained his reasons for launching airstrikes. Bejtullahu was a real person and her emails were genuine but, framed with emotive references to the Nazi genocide, they were used to support military intervention and became an important weapon in the propaganda battle.

This early 'citizen journalism' had an obvious appeal for the media. In the absence of other news sources it provided accessible, cheap copy, with the added value of representing authentic experiences and voices. Its focus on personal events and emotions provided 'human interest' stories that had a greater emotional resonance and populist appeal than the packaged reports of professional journalists whilst also helping to simplify a highly complex cultural and political reality for a western audience. The problem was many of these accounts were anonymous, few could be independently verified, and there remained an uncertainty as to who was posting information and their agenda. As Brooke Shelby Biggs, news editor of Mojo.wire commented,

> in place of real reporting we're offered a flood of unmediated dispatches from non-journalists often with a personal interest in how the war is fought and how it ends … These people have no ethical mandate to be unbiased any more than the government of Yugoslavia or the US Defense Department does.
>
> Matheson and Allan (2009:53)

This new space of information earned Kosovo the popular title of 'the first internet war' (Matheson and Allan, 2009:41), but this needs some qualification. Only about 1 percent of the population in Serbia and Kosovo was able to get online and those that could were overwhelmingly urban and educated. Net access in Kosovo was far scarcer than in Belgrade and in the rural areas where the Serb forces were pursuing their ethnic cleansing it was almost non-existent. Given also the language barriers for an international audience and the limitations of pre-Web 2.0 online culture, the internet was not yet a force able to significantly challenge the dominance of the mainstream broadcast media.

With broadcasters reliant on official channels, NATO's military media management system achieved considerable success in setting and controlling the narrative of the conflict. Both its justification for the war and its account of its progress were accepted and reported by the mainstream media. In particular, stories about Serbian atrocities were common in the press. In the UK, for example, even left-of-centre newspapers such as *The Observer* and *The Guardian* accepted the humanitarian case for the war and published front page stories of Serb violence, massacres and 'death squads'. KLA claims about the systematic use of rape and even of Serb 'rape farms' and 'rape camps' were also uncritically accepted and widely reported, even though later investigations found little evidence to support them. In contrast, violence against Kosovo Serbs and the ethnic cleansing of the Serb population, both during and after the Kosovo War, was largely ignored by the western media.

NATO's control of the debate wasn't perfect. Many questioned the legality of a war without UN authorization and the status of a 'war' that had occurred without a declaration, without Parliamentary support and which had no other aim than to 'degrade' the Serbs. In the UK there was also criticism from both the right, who saw the region's conflicts as irrelevant, and from the left, who saw NATO action as anti-Serbian imperialism. Others saw the intervention as linked to NATO's need to prove its relevance and efficacy on its 50 anniversary or pointed to a US President who, the month before, had faced an impeachment trial for the Monica Lewinsky scandal and needed to regain popular support. As in the Gulf, however, the main threat to the alliance's message was the issue of casualties. At the time, most journalists accepted NATO's line that civilian casualties (such as those in the Grdelica bombing, which killed 14 people on a train) were a 'mistake'. The truth, however, as Knightley says, 'was that NATO not only bombed civilian targets accidentally but that it also bombed them deliberately', as part of its war on the communications infrastructure (Knightley, 2000:509–10).

It was an accident, however, that nearly derailed NATO's bombing campaign on 14 April when NATO planes mistook a convoy of Albanian refugees near Gjakova for Serbian military vehicles, killing 73 and injuring 36. Like the denunciations of the deaths in Amiriyah in the Gulf War, the allied response was a public relations disaster. With CNN showing graphic images of civilian bodies and vehicles, a stream of explanations came from NATO all denying responsibility and blaming Serbia. It was the own goal of these denials that led to the reorganization of the Brussels Media Operations Centre by Tony Blair's press secretary and 'spin doctor' Alistair Campbell. His new policies involved greater honesty and a faster acknowledgement of NATO's mistakes – before ultimately assigning blame to Milošević. The latter, Ignatieff says, had gambled on 'the tenderness of western hearts', hoping to exploit civilian casualties for propaganda value and although this strategy failed civilian deaths (such as those at Koriša, a village where NATO killed at least 87 Albanians) all 'sowed doubt' in a western public that had initially been in favour of the bombing (Ignatieff, 2000:193).

Where the Gulf War model came closest to failure was militarily, in the differences between Iraq and the former Yugoslavia. Unable to achieve air superiority and facing the continued threat of the Serb airforce and anti-aircraft systems, NATO was forced into higher altitude bombing which made it harder to hit targets and confirm damage. Kosovo also lacked the perfect conditions of the Gulf deserts, with poor weather and forested, mountainous regions hampering the success rate of precision technologies and allowing Serbian ground forces to continue to operate. The USA and NATO were also reluctant to follow the air campaign with a ground invasion, fearing the long, bloody war that could result against experienced troops who knew the region well and who had survived the air campaign. In the end the destruction of the Serbian power grid, the alliance's forced plans for ground troops and the loss of support from its key ally Russia following USA–Russian negotiations all pushed Milošević into agreeing a ceasefire.

Like the Gulf War, what seemed like a simple military victory was complicated by its aftermath and later reevaluation. Many blamed the air campaign for accelerating, or even causing the ethnic cleansing, whilst the effectiveness of the air campaign and its role in forcing Milošević's surrender were also questioned. Moreover, Milošević and his regime had survived the war and the Serb army hadn't been defeated. It had retained its air capacity during the war, resisted NATO for 78 days and employed a range of decoys that led NATO to overestimate Serb military losses. Military historian John Keegan's claim that Kosovo represented a historical 'turning point' in proving 'a war can be won by air power alone' proved premature: after the war NATO claimed it had destroyed 120 Serbian tanks, 220 armoured personnel carriers and 450 mortar pieces but a suppressed US Air Force report later suggested NATO verifiably destroyed just 14 tanks, 18 armoured personnel carriers and 20 artillery pieces (Barry, 2000). Serbia even won a victory of sorts, in stalling Kosovo's independence and forcing the deployment of a UN Protection Force that remains there to this day.

Claims of Serbian massacres, mass graves and rape camps similarly proved to be over-inflated, with little evidence for organized 'systematic killing'. The failed post-war search by western 'mass graves correspondents' seemed to confirm Audrey Gillan's comment during the war that 'The story being seen at home is different from the one that appeared to be happening on the ground' (Knightley, 2000:521). US State Department claims on 19 April of 500,000 Kosovar Albanians missing, feared dead, were revised down to 100,000 in May, and by June the UK government was claiming that 10,000 ethnic Albanians had died in the war, a figure that is now broadly accepted. Where the coalition military was successful, however, was in avoiding alliance casualties, with NATO achieving the remarkable statistic of zero combat losses (with two killed in a non-combat accident).

Conclusion

What was created at the beginning of the decade in the Gulf War, therefore, and redeployed in Kosovo, was a new model of conflict: a 'top-down' war, involving the close management and control from above of both the military and media theatre of operations. Militarily this model aims for complete superiority in both the physical and non-physical domains, ideally combining overwhelming conventional force and the ability to strike precisely, as desired, with a complete knowledge of the battlefield and its targets. Control of the theatre extends to the exclusion of enemy contact, conducting operations with clear, achievable aims and spatial and temporal limits and the possibility of both rapidly removing oneself from or returning to the warzone as expedient.

This military operation is accompanied by a media campaign that similarly aims for a complete control of the narrative of the conflict, imposing its will, images, explanations and vision upon the broadcast outlets that carry the message down to the public, with the aim of mobilizing support for its actions. Air supremacy combines, therefore, with *airwave supremacy* to ensure the most successful possible application of military forces, the minimization of its own and civilian casualties and the most favourable domestic and international interpretation of the events. If, as the Gulf and Kosovo campaigns demonstrated, the actual political reality, the conduct of the military operations and their aftermath diverge from this ideal, this may matter less than the successful global management of conflict perception at the time.

The result of these changes was a transformed experience of war for the west. Now, going to war meant little danger for the military combatants, becoming, as the critics suggested, a 'Nintendo' war or 'deadly video game' (Taylor, 1998:75; Knightley, 2000:483), with targets on a screen being eliminated at a distance. Meanwhile, war for the domestic audience became something to be watched on television, requiring no risk or mobilization and with a scopic pleasure in the special effects that was disconnected from the reality of the violence. A Steve Bell cartoon in *The Guardian* after the Gulf War captured this domestic consumption of war best. It showed a penguin asked by his penguin grandchild what he'd done during the war replying, 'Me? I watched television and I'm proud of it!' (Merrin, 1994:452). It was this model of war that would come to define western military experience over the next 25 years, across various theatres of operation, in particular in Afghanistan in 2001 and in Iraq in 2003. But it was this model too – with its ease of use and safe western conduct and consumption – that would provoke considerable anti-western sentiment, radicalizing many to take action back against both its military operations and its disconnected and blasé domestic populations.

Key reading

The best book on the media coverage of the Gulf War is Taylor (1998), although I would also recommend Kellner (1992), Knightley (2000), Hudson and Stanier (1997), Cumings (1992) and Sturken (1997). Freedman and Karsh (1994) provides a good factual overview of the conflict. Aksoy and Robins (1991) and Broughton (1996) provide important discussions of smart-bomb technology. George Bush's 'New World Order' speech is online (1991) and good discussions of intervention in the 1990s can be found in Hudson and Stanier (1997) and Carruthers (2000). Footage of the US beach landing in Somalia is available on YouTube (2014). The best discussions of the media coverage of Kosovo are found in Matheson and Allan (2009) and Knightley (2000). Other important texts on Kosovo include Ignatieff (2000) and Hammond and Herman (2000).

References

Aksoy, A. and Robins, K. (1991) 'Exterminating Angels: Morality, Violence and Technology in the Gulf War', *Science as Culture*, 2(3), pp. 322–36.

Barry, J. (2000) 'The Kosovo Cover-Up', *Newsweek*, 15 May, http://www.newsweek.com/kosovo-cover-160273.

Brewin, B. (1999) 'Kosovo ushered in cyberwar', 27 September, FCW, http://fcw.com/Articles/1999/09/27/Kosovo-ushered-in-cyberwar.aspx.

Broughton, J. (1996) 'The Bomb's-Eye View: Smart Weapons and Military TV'. In S. Aronowitz, ed., *Technoscience and Cyberculture*, London: Routledge, pp. 139–65.

Bush, G. H. W. (1991) 'Address Before a Joint Session of the Congress on the State of the Union', 29 January, http://www.presidency.ucsb.edu/ws/?pid=19253.

Carruthers, S. L. (2000) *The Media at War*. London: Macmillan.

Cumings, B. (1992) *War and Television*. London: Verso.

Freedman, L. and Karsh, E. (1994) *The Gulf Conflict*. London: Faber and Faber.

Geers, K. (2008) 'Cyberspace and the changing nature of warfare', *SC Magazine*, 27 August, http://www.scmagazine.com/cyberspace-and-the-changing-nature-of-warfare/article/115929/.

Hammond, P. and Herman, E. S. (2000) *Degraded Capability: The Media and the Kosovo Crisis*, London: Pluto Press.

Hudson, M. and Stanier, J. (1997) *War and the Media*. Stroud, Gloucs: Sutton Publishing.

Ignatieff, M. (2000) *Virtual War*, London: Chatto & Windus.

Kellner, D. (1992) *The Persian Gulf TV War*, Oxford: Westview Press.

Knightley, P. (2000) *The First Casualty*. London: Prion Books.

Matheson, D. and Allan, S. (2009) *Digital War Reporting*. Cambridge: Polity.

Merrin, W. (1994) 'Uncritical Criticism? Norris, Baudrillard and the Gulf War', *Economy and Society*, 23(4), pp. 433–58.

Messmer, E. (1999) 'Kosovo cyber-war intensifies: Chinese hackers targeting U.S. sites, government says', CNN.com, 12 May, http://edition.cnn.com/TECH/computing/9905/12/cyberwar.idg/

Moorcraft, P. L. and Taylor. P. M. (2008) *Shooting the Messenger: The Political Impact of War Reporting*, Washington, DC: Potomac Books.

Nuttall, C. (1999) 'Kosovo info-warfare spreads', BBC News, 1 April, http://news.bbc.co.uk/1/hi/sci/tech/308788.stm.

Rogers, P. (1991) 'Myth of a Clean War Buried in the Sand', *The Guardian*, 19 September.

Sturken, M. (1997) *Tangled Memories*, London: University of California Press.

Taylor, P. M. (1998) *War and the Media*. Manchester: Manchester University Press.

YouTube. (2014) 'Operation Restore Hope Beach Landing, Mogadishu Somalia', posted 2 January, https://www.youtube.com/watch?v=Xj9Fn3qG-Cw.

2 Non-war and virtual war

Theorizing conflict in the 1990s

Re-thinking war

The Gulf War faded fast. Despite its occupation of every newspaper and television channel and its spectacular, global dissemination as a media event, the war itself and its bubble of media coverage was quickly forgotten. By the following year Bruce Cumings' *War and Television* only half jokingly asked, 'Remember the Gulf War? Or was that last season's hit show?' (Cumings, 1992:103). If the media and the public had moved on, academically, at least, the Gulf War proved more enduring. The conflict spurred interest in the relationship between war and the media and the problems of war reporting and a minor academic industry developed through the 1990s exploring its contribution to these debates.

In addition to these commentaries there also emerged during the decade a series of important reflections on the nature of warfare itself. Inspired by the Gulf and Kosovo Wars, philosophers such as Baudrillard, Virilio and the Tofflers and critics such as Hables Gray, Ignatieff and Der Derian took the opportunity to re-theorize conflict, developing original analyses of the new forms of 'non-war', 'information war', 'third wave war', 'postmodern war' and 'virtual war' that were emerging. As I'll argue, inspired by the Gulf War and by ongoing developments in technology the US military was simultaneously rethinking and remaking warfare to foreground the concept of information and impact of information technology, but the philosophical and critical analyses of war in this era remain significant in capturing important changes both in what war had become and how it was experienced in the west.

Jean Baudrillard and 'non-war'

The most original and provocative analysis of the Gulf War was offered by the French philosopher Jean Baudrillard (1929–2007). In a remarkable series of essays published in *Libération*, he personally matched the war's build up, outbreak and successful denouement with his own acerbic, escalating denial of the entire conflict. His first essay, published on 4 January argued simply that 'the Gulf War will not take place'. When this claim proved premature, he was unrepentant, asking in an essay published on 6 February, at the height of the air campaign, 'is the Gulf War really happening?'. Finally, on 29 March, a month after the war's conclusion, he published a third essay confidently entitled 'The Gulf War did not take place'.[1] Baudrillard's essays received a lot of attention at the time, but little of it dealt in detail with his arguments. Popularly identified as the leading thinker and 'high priest' of 'postmodernism' (Baudrillard, 1989), his essays were largely read in the light of this movement and its ideas rather than for their contribution to how we could understand contemporary war.

Heavily influenced by post-1968 French theory, 'postmodernism' emerged in the 1980s and early 1990s as a broad philosophical, sociological and cultural movement, developing a critique of both post-Enlightenment industrial 'modernity' and its processes, values and truth claims and of twentieth century, 'modernist' forms of art, literature, music and architecture. 'Postmodernism', therefore, was an amorphous movement, encompassing different claims across several fields of inquiry. With his analysis of our society of 'simulation', apparent pleasure in its mass media fictions and provocative writing style and arguments, Baudrillard was co-opted as the quintessential 'postmodern' philosopher, even though he rarely used the term and was resolutely opposed to the world it described.

The Left were especially critical of postmodernism, seeing it as politically reactionary in its denial of the relevance of industrial capitalism; as philosophically charlatanistic in its style and mode of thought; and as fundamentally nihilistic in its relativistic denial of reality and truth. As the movement's figurehead, Baudrillard became a target for their hostility. Building on Douglas Kellner's earlier critique of Baudrillard's 'postmodernism', Christopher Norris attacked his Gulf essays as final proof of the bankruptcy of this movement. His 1992 book *Uncritical Theory* lambasted Baudrillard for the 'moral and political nihilism' of his 'Berkeleian transcendental idealism' in seeing the Gulf War merely as 'a figment of mass-media simulation' taking place, Norris said, 'in the minds of a TV audience' (Norris, 1992:194, 196, 11). Except this critique had no relationship to anything Baudrillard was actually arguing, completely missing the significance of his analysis of the war.

When Baudrillard wrote in his first essay that 'the Gulf War will not take place', he wasn't alone in this view. It wasn't certain that the west would use force and Baudrillard's discussion of the 'crisis' of war in the post-Cold War era and the west's 'self-deterrence' was defensible (Baudrillard, 1995:24–5). Indeed, some of his points are prescient, especially his warning that replacing the declaration of war with a UN-mandated 'right to war' is dangerous as 'the disappearance of the symbolic passage to the act' presages, he says, 'the disappearance of the end of hostilities, then of the distinction between winners and loser'. His conclusion, that 'since it never began, war becomes interminable' (Baudrillard, 1995:26), certainly anticipates the postwar fall-out wherein a coalition 'victory' required ongoing military activity for the next decade. Faced with the outbreak of war Baudrillard refused to back down. Instead he decided to escalate his position to challenge the greatest, globally mediated event of the age, developing, over his subsequent two essays, a coherent and important theory of contemporary 'non-war'.

Crucially, in these essays, Baudrillard does not deny that something occurred, but for him it was a 'non-war', not a war. If, as Clausewitz says, war is the 'collision of two living forces', each attempting to defeat the other, then this didn't occur. The coalition's military superiority meant they were able to follow their own strategy with minimal resistance. This was, therefore, a programmed, modelled 'simulation' of war; one whose success excluded all contact, communication and enemy participation. Instead of a conflict there was the unilateral *imposition of war* upon the opposition, with the coalition refusing to engage the enemy and merely 'annihilating him at a distance' (Baudrillard, 1995:43). 'Simulation', therefore, isn't about unreality; rather it refers to the efficacious production and materialization of the real in the world. Here the ideal model of war – of a rapid, successful war with limited casualties – was so successfully realized by the military that the enemy didn't even have a chance to respond. War reversed into non-war.

This was a war, therefore, that followed its model so completely that the enemy's only role was as a target. 'Everything unfolded according to a programmatic order, in the absence of passional disorder', Baudrillard says: 'nothing occurred which would have metamorphosed events into a duel' (Baudrillard, 1995:73). Because if war is a contest – a dramatic, agonistic struggle for supremacy – then where there is *no contest* then there is *no war*. 'War is born of an antagonistic, destructive relation between two adversaries', Baudrillard says, whereas 'this war is an a-sexual, surgical war, a matter of war-processing in which the enemy only appears as a computerized target' (Baudrillard, 1995:62).

Baudrillard's critical position here is derived from late nineteenth- to early twentieth-century Durkheimian social anthropology and the work especially of Mauss, Durkheim, Bataille and Caillois on 'primitive societies' and their lived production of social relationships and meaning. These ideas would become the basis for Baudrillard's own theory of 'symbolic exchange': of a mode of being and reciprocal communication that he would oppose to the lifeless semiotic forces of western consumer societies and their replacement of human contact with technological mediations. Baudrillard was inspired especially by Marcel Mauss's theory of 'the gift' and the dramaturgy of giving and receiving in tribal societies. Mauss argued that this was about more than the circulation of property but was rather a dualistic battle for personal recognition following strict rules and creating clear social obligations and relationships. Thus, non-war is an explicit refusal of this relationship – 'any dual or personal relation is altogether absent', Baudrillard says, being replaced by the indifferent 'clean relation' of technology (Baudrillard, 1995:44–5). It is a refusal too of the recognition of the other that characterizes the face-to-face scene of the gift. Here there is no interest in the other, except for removing them: 'the Americans inflict a particular insult by not making war on the other but simply eliminating him' (Baudrillard, 1995:40).

Baudrillard's critique takes aim, however, not just at the idea of 'war' but also at the idea of occurrence – of what it means for something to actually take place. Given the discrepancy between the forces and the coalition programming, Baudrillard says, this war was 'won in advance': 'We will never know what it would have been like had it existed. We will never know what an Iraqi taking part with a chance of fighting would have been like. We will never know what an American taking part with a chance of being beaten would have been like'. It was closer, he argues, to 'an ultra-modern process of electrocution … with no possibility of reaction' (Baudrillard, 1995:61). This calls into question the *eventness* of the event for if this victory was inevitable, can we really say such an event 'happened', Baudrillard asks? 'Is there still a chance that something which has been meticulously programmed will occur? Does a truth which has been meticulously demonstrated still have a chance of being true?' (Baudrillard, 1995:35). For Baudrillard the acting-out in the real of something that has been so completely planned is an empty formality, the shadow of an event whose substance has already been used up and whose experience proves or adds nothing extra.

The two adversaries didn't even confront each other, Baudrillard argues, 'the one lost in its virtual war, won in advance, the other buried in its traditional war, lost in advance' (Baudrillard, 1995:62). This lack of relationship was central to the coalition military strategy – from the unchallenged air campaign, to the frontline bulldozers that treated the enemy as already dead and only worth burying, to the unstoppable sweep of the ground forces, to the bombed-out vehicles on the 'highway of death'. The war Iraq thought it was going to fight – a war of trenches, territory and men – didn't take place: the first world's Third World War left no room for the third world's First World War. Just as western media is based, Baudrillard argues, upon a 'non-communication' (Merrin, 2005:20–22) – upon the abolition and replacement of the lived and actualized human relationship – so this war followed a similar path of unilateral technological mediation, with 'emitter and receiver on opposite sides of the screen'. Except 'instead of messages it is missiles and bombs which fly from one side to the other, but any dual or personal relation is absent' (Baudrillard, 1995:44).

For Baudrillard, therefore, this was not a war but a massacre: 'It is as though the Iraqis were electrocuted, lobotomized, running toward the television journalists in order to surrender or immobilized besides their tanks, not even demoralized: de-cerebralised, stupefied rather than defeated – can this be called a war?' (Baudrillard, 1995:67–8). He emphasizes especially the question of casualties and the paradox that a war that is too successful at limiting the winner's casualties undermines its own claims as a war. 'The minimal losses of the coalition pose a serious problem that never arose in any earlier war', he says. Though superficially a cause for 'self-congratulation', they actually expose the war for what it is: 'the prefiguration of an experimental, blank war … a war even more inhuman because it is without human losses'. Contrast that with the unknown numbers of Iraqi dead, he says, abandoned as 'sacrificed extras' in the desert (Baudrillard, 1995:73).

If, as Clausewitz says, war is the pursuit of politics by other means, then the Gulf non-war represents, Baudrillard says, 'the absence of politics pursued by other means' (Baudrillard, 1995:30). This is due to both the lack of planning for its aftermath and because the real aims, Baudrillard says, were 'transpolitical'. This was 'a preventative, deterrent, punitive war' to prove the west's power and to domesticate and eliminate the symbolic 'alterity' of the Arab 'other' (Baudrillard, 1995:30; 56; 36–7). 'The victory of the model is more important than victory on the ground', Baudrillard argues (Baudrillard, 1995:55), hence the outcome of the war was less important than its global functioning as a demonstration of US and western power and deterrence of resistance to its order and values.

Baudrillard's critique of the military 'non-war' is accompanied by a critique of the media 'non-event'. As he argued in *The Consumer Society* in 1970, television doesn't transport us into the heart of the event, but instead into its simulation. Mass media operate, he says, by 'the disarticulation of the real into successive and equivalent signs', neutralizing 'the lived, unique, eventual character of the world' and replacing it with its own semiotic combination. What the media give us, therefore, is 'a filtered, fragmented world', one 'industrially processed' into 'sign-material' combined into a 'neo-reality' that comes to assume the force of reality (Baudrillard, 1998:122–24.). Thus, what the Gulf War viewer experienced was a simulated 'war' produced for them through rolling news coverage, real-time reports, live commentary, studio speculation, military briefings and video footage and graphics and maps. This is 'war' as a breathless, live media event brought to us in close-up with a remarkable hyperreality, but it is also a 'war' founded for us on a fundamental distance from the real, as something consumed at home, involving no risk of death for the audience. This was a 'war' with the violence removed, replaced and eclipsed by its semiotic spectacle.

In his 1978 book *In the Shadow of the Silent Majorities* Baudrillard described again this process of mediation, arguing that 'there is the implosion of the medium itself in the real, *the implosion of the medium and the real* in a sort of nebulous hyperreality where even the definition and the distinct action of the medium are no longer distinguishable' (Baudrillard, 1983:101). The movement to live reporting exacerbates this process. 'War implodes in real-time', Baudrillard argues. 'The instantaneity of the event and its diffusion' produces, he says, 'an involution of the event' – the implosion of the event and its mediation – that precipitates us not into the reality but into a state of virtuality (Baudrillard, 1995:47–9). What we experience is an event whose 'reality' cannot be disentangled from its mediation. Hence this 'virtual war' is marked for us by uncertainty: the uncertainty of live events, of the absence of images, of the lack of hard information, and of the continuous analysis and talking heads. Thus, it becomes a 'symptomatic' war: one requiring a constant interpretation of information by the military, the media and the audience. In such a situation 'truth' is replaced by credibility in the moment. 'This uncertainty invades our screens like a real oil slick', Baudrillard argues, the famous image of the blind, oil-covered seabird stranded on the beach becoming 'the symbol-image of what we all are in front of our screens, in front of that sticky and unintelligible event' (Baudrillard, 1995:32).

Baudrillard is especially critical of the vacuity of the war reporting. Real-time coverage furnishes us with 'images of pure, useless, instantaneous television' whose function is merely to fill the vacuum and block the TV screen (Baudrillard, 1995:31). This is best illustrated, Baudrillard says, by 'those who pontificate in perpetual commentary on the event', whose analysis makes us feel 'the emptiness of television' as never before. But no one is held to account for such 'idiocies', Baudrillard says, since they are erased by those of the next day as 'everyone is amnestied by the ultra-rapid succession of phony events and phony discourses' (Baudrillard, 1995:51).

Baudrillard's conclusion exhibits his wry humour. The war ended, he says, with the Americans inflicting 'a perfect semblance of military defeat' upon Saddam, in exchange for 'a perfect semblance of victory' (Baudrillard, 1995:71). But everything here is in order, he adds. This war is not a war but that is compensated for by the fact that the information is not information either: indeed, 'everything becomes coherent if we suppose that, given this victory was not a victory, the defeat of Saddam was not a defeat either. Everything evens out and everything is in order: the war, the victory, and the defeat are all equally unreal, equally non-existent' (Baudrillard, 1995:82).

Though Baudrillard would become a key reference point in popular discussions of 'postmodern war', his essays are better understood apart from that movement. His aim wasn't to promote postmodernism but instead to develop a critique of the western military and media processes. Far from being nihilistic, therefore, his essays were an immediate, highly moral response to the conflict, puncturing the west's claims of a clean, surgical and moral war. Whereas his critics merely took the opportunity to attack him, Baudrillard himself took aim at the entire global event and its media narrative. In contrast to both left and right who accepted the 'truth' of this war, Baudrillard rejected its occurrence and its historical status and position.

Baudrillard's philosophical discussion of 'war' remains, of course, open to criticism from within international relations and military studies. Significant differentials of force and one-sided victories aren't new in conflict and don't necessarily justify the denial of a war's status, but Baudrillard's paradoxical concept of 'non-war' is important in recognizing that western technological superiority had transformed this ideal of war into a realizable model that could be unilaterally, globally imposed without resistance. Equally important is Baudrillard's analysis of the media's role. His theory of simulation and the media event already anticipated our experience of the war, but his essays extend this critique, skewering especially media coverage of the event and its banality and emptiness. By 2003 many had caught up with this, with Baudrillard's Gulf essays becoming a key reference for critics of the media's overly excited coverage of the Iraq War.

Also significant is Baudrillard's recognition of the media's role in mobilizing entire populations as consumers and supporters of the war. If the aim of the war was global social control, Baudrillard says, then the media coverage represented a similar process as the domestic level: 'the complement of the unconditional simulacrum in the field is to train everyone in the unconditional reception of broadcast simulacra' (Baudrillard, 1995:68). In particular Baudrillard emphasizes our spectatorial complicity in this model of war, in our lack of taste for 'real drama or real war' and the 'hallucinogenic pleasure' we derive from its lethal spectacle. His comment that our 'indifference' to the violence represents 'our definitive retreat from the world' (Baudrillard, 1995:75) presciently identifies in the Gulf War what would become one of the most important features of our mediatized non-war.

Because what Baudrillard highlights is how easy it is to support wars in which the reality of violence has been removed and which mobilize us only as consumers of television news. This would attain a greater significance in the new millennium where, with the 'War on Terror', domestic audiences would get used to neverending military operations, becoming blasé about activities carried out in their name. These operations and the apparent disinterest of western publics in the destruction and casualties they caused would fuel anger around the world and give rise to terrorist responses aimed at re-injecting the reality of those wars back to their domestic audiences. Thus, rather than being, as Elliott argues, 'the worst contribution to the Gulf controversy' (Elliott, 1992:13), Baudrillard's essays constitute one of the most important in their critique not only of the war and its mediation but also for their identification of a model of 'non-war' and its experience to which the west would return to in the following decades.

Paul Virilio and 'information warfare'

Fellow French philosopher and technological critic Paul Virilio (1932–) rejected Baudrillard's reading of the Gulf War. He offered instead a very different analysis of contemporary war, emphasizing the impact of electronic tele-technologies, the dominance of real-time systems over real space, and their creation of a global, panoptical system of military and civil control. War has always been central to Virilio's life and work. He has described himself as a 'war-baby' in living through World War II, claiming that 'war was my university' (Virilio, 1997b); his first book, *Bunker Archaeology* (1975) was a study of the German North Atlantic bunker defences, and discussions of weaponry and conflict have reappeared throughout his oeuvre, most notably in his analysis of the relationship between visuality, vision technologies and conflict, *War and Cinema* (1984). Virilio is best-known as the theorist of 'dromology', or the science of speed. In particular his work traces the movement from 'metabolic speed' – the speed of the animal body – to 'relative speed' – the transportation revolution of trains, cars and planes – to 'absolute speed' – the transmissions revolution built on the speed of electromagnetic waves. His books on the Gulf and Kosovo wars, *Desert Screen* (1991) and *Strategy of Deception* (1999) explore the impact of this absolute speed upon contemporary conflict.

For Virilio the Gulf War is 'the first total electronic war. Broadcast live via satellite' (Virilio, 2002:44). He follows Marshall McLuhan's vision of electronic technologies as 'imploding' space and time in their instant speed, but for him these systems do more than abolish real-space and lived time, they supersede it, *replacing* the physical theatre of conflict with their own electronic environment. They form a *separate* electronic reality that is privileged above and is more important than the physical one. Hence, Virilio says, 'since the night of 17th January, we have entered into the excess of a new war: the war of real time, of omnivoyance, of omnipresence, that supplants the ground war, the war of real space that made up the history of nations'. This is a conflict, therefore, that no longer plays out only on the geographic horizon, 'but first of all on the monitors, the control screens of televisions of the entire world' (Virilio, 2002:46).

The very first act of the war, a laser-guided bomb on the Iraqi army's communications centre in Baghdad, highlighted the importance for the coalition of informational dominance and their aim, Virilio says, of 'a total control of the electromagnetic environment above Iraq' (Virilio, 2002:95). This is the first 'absolute' war, where 'the control of the geophysical environment of the adversary and that of the armed forces will have given way to the control of the microphysical and electromagnetic environment of the hostile milieu' (Virilio, 2002:113). For Virilio, therefore, the physical space of deployment is now secondary to informational space: 'Real time, that is to say the *absolute speed* of electromagnetic exchanges, dominates real space, in other words the *relative speed* of exchanges of position, occasioned now by offensive and defensive manoeuvres' (Virilio, 2002:121). Thus, Virilio argues today '*arms of communication* prevail for the first time in the history of combat over the traditional supremacy of arms of destruction' (Virilio, 2002:113–14).

For Virilio what we are seeing is the emergence and supremacy of a 'fourth front' – after land, sea and air – of 'information', and a new fourth space, 'the orbital', dominated by military and media satellites. 'Hence the great metamorphosis of the "postmodern" war', he says, which 'denies both the offensive and the defensive, solely to the benefit of the control and interdiction of the battlefield' (Virilio, 2002:121). Conflict is now dominated by an electronic 'logistics of perception', a generalized, orbital tele-surveillance whose aim is 'to make an identification, to interpret as quickly as possible the signs, images and trajectories' of the battlefield in order to control the enemy and foreclose their activities. Thus, 'henceforth the fourth front becomes the principle front and comes to supplement, indeed supplant, the strategies of land, sea and air actions' (Virilio, 2002:68).

Success, therefore, is based now around the control not of 'immediate perception' on the ground, but of 'mediated perception', of 'an electro-optic in real time' (Virilio, 2002:54). As the aim of this electronic war is total battlefield perception, then imperceptibility becomes a central strategy for success or survival. As Virilio says, '*if that which is seen is already lost*' then 'it is necessary to block at all costs long-range tele-detection or, at the very least to reveal one's presence as late as possible' (Virilio, 2002:109). Where once it was necessary to conceal one's forces and camouflage equipment, now one must also 'camouflage the trajectories' of one's forces and activities, using decoys and countermeasures to lure the enemy's attention towards false movements and illusory trajectories (Virilio, 2002:54–5). One's aim must be to reduce '*the dimensions of the envelope of detection* of the weapons systems engaged in conflict'. Hence, again, destructive systems are secondary today to the communicational systems:

> Here lies the technological revolution of this war in the Gulf: the question of the *stealth of the materiel tends to supplant that of the speed* of the weaponry with the greatest velocity, be it missile or combat aircraft … the absolute speed of the waves of electromagnetic detection prevails henceforth over the relative speed of the supersonic or hypersonic flying aircraft. To no longer *lose sight* of the enemy is thus to *gain* the upper hand or indeed even to win the conflict, this war in which the disappearance from sight tends to prevail over the power of conventional or non-conventional explosives.
>
> Virilio (2002:110)

Real time, therefore, trumps real space: the absolute speed of electromagnetic waves has a clear advantage 'over the materiel of war that only moves at the relative speed of some hundreds or thousands of kilometers an hour' (Virilio, 2002:55).

Just as the military optic is doubled between the optic of real space and the electro-optic of the real time of military or civil tele-spectators, 'there is simultaneously a doubling of the front, a commutation between the place of action – the Near East – and the place of its immediate reception – the entire world' (Virilio, 2002:56). Thus, we cannot speak of just a localized battlefield. Dominated by its instantaneous, global, tele-topical control and experience, 'the "electronic battlefield" does not stop at the Iraqi–Saudi border; it now extends, through the mediation of satellites, to the entire world' (Virilio, 2002:120; 70).

The perceptual control of the battlefield is matched, therefore, by 'the total control of its public representation, on the global level' (Virilio, 2002:108). Again, real time is central to this: for Virilio, 'the most important thing is the impact, the very immediacy of the televised news ... the important thing, as in a publicity clip, is to affect people's minds, to tele-affect emotionally the tele-spectators before their screens' (Virilio, 2002:51–2). Hence, again, the electromagnetic instant dominates: 'Space and the informative content of the image count infinitely less than time, the reality of a scene that plays out live' (Virilio, 2002:52). It is the real-time experience of the images that capture and hold the audience and form their experience.

'The message of this mediated war', Virilio argues, is actually 'the promotion of the virtuality of wars to come'. This will be 'war of the radically unthinkable, where the role of people in war would be only that of impotent tele-spectators and victims of "intelligent" weapons' (Virilio, 2002:66). This is implicitly a divided future where the most advanced countries will provide the tele-spectators and the global targets of its weaponry will provide the civilian victims. But it isn't only western civilians that will benefit from distance. Virilio notes the use of surveillance drones in the Gulf War and their anticipation of the future:

> For the first time in history ... aerial and orbital war prevailed over classical terrestrial combat, so much so that we witnessed, in the last hours of this war, some forty Iraqi soldiers surrendering to an airplane without a pilot, an aerial reconnaissance drone with a three-metre wingspan, equipped with a simple video camera – the future vision of this *logistic of electro-optic perception* that allowed soldiers under shelter at their consoles, to take enemy prisoners without having to move, solely by the panic produced by the over-flight of a scaled-down model!
>
> Virilio (2002:107)

In his 1995 book *Open Sky* Virilio presents a dystopian caricature of the present – of the development of 'terminal citizens': sedentary, plugged-in, hyper-equipped and connected individuals unable to intervene in their proximate environment but using an array of sensors for 'tele-present' interactivity (Virilio, 1997a:9–21). It would prove to be a prescient vision too of the drone operators who would fly the weaponized descendants of that surveillance drone in the War on Terror, launching missiles from a safe distance, abolishing the real space and bodies of the theatre of conflict with their lethal, real-time technologies.

By the decade's end, war had advanced again. For Virilio the Kosovo War represented a movement beyond the 'electronic warfare' of the Gulf to a full 'information warfare' (Virilio, 2000:18). This has as its goal, Virilio says, the realization of 'global information dominance', following the US Air Force Chief of Staff's stated aim of being 'capable of finding, tracking and targeting virtually in real-time any significant element moving on the surface of the earth' (Virilio, 2000:17–18). This 'truly panoptical vision' rests, Virilio argues, on three fundamental principles: '*the permanent presence of satellites* over territories, the *real-time transmission* of the information gathered and, lastly, the *ability to perform rapid analysis* of the data transmitted to the various general staffs' (Virilio, 2000:18). It is these 'cybernetic monitoring and surveillance systems', Virilio suggests, that allow the US 'to dominate across the full spectrum of military operations' and establish their 'globalitarian' hegemony (Virilio, 2000:41–2).

This information war also involves a clash 'in Hertzian space' (Virilio, 2000:31), Virilio says, broadcasting one's own propaganda into the theatre of conflict whilst simultaneously degrading or controlling the enemy's informational capacities. Hence the bombing of the Serbian media offices was part of '*a nodal war*' whose aim was '*the policing of images*' and a control of the enemy's 'information flows', cutting them off from their own populations (Virilio, 2000:24; 22; 24–5). The graphite bombs that cut off Serbia's electricity systems were also part of this strategy, as a distinct 'energy intervention' affecting also the media and the civilian population. 'When you know the strategic importance of this primary energy source in these days of "information revolution"', Virilio argues, 'you are better placed to understand the logic of this act of war interrupting all communications' (Virilio, 2000:26–7).

But in Kosovo this logic of electronic perception was not perfect. Whereas the desert glacis of the Gulf enabled a 'war at the speed of light', in Yugoslavia real space was able to reassert itself. 'The nature of the terrain in the Balkans seems to have been totally left out of account in the reasoning of NATO's leaders', Virilio notes (Virilio, 2000:1). Not only was there no hope of a 'lightning campaign', but the Serbs countered the monitoring ability of the USA by finding cover, dispersing or remaining static, resisting electronic speed 'by apparent inertia', merely awaiting the ground assault NATO were reluctant to launch (Virilio, 2000:3, 19).

For Virilio, the west's 'humanitarian' justification for the war linked closely with the information systems used to prosecute it to create a new hybrid: 'After the eye of God pursuing Cain into the tomb, we now have *the eye of Humanity* skimming over the oceans and continents in search of criminals', Virilio argues. The result is a 'global tele-surveillance of social or asocial behaviour'. Hence information war becomes a juridical project as the 'ethical dimension of the global information dominance programme' opens up 'the possibility of *ethical cleansings*' to oppose the 'ethnic cleansings' on the ground (Virilio, 2000:21–2). Virilio's analysis refers to Kosovo in 1999 but this idea of a global, panoptical system hunting down 'evil' in the name of western, liberal values would find its fullest expression in the years after 9/11.

The Tofflers and 'third wave war'

Another vision of the transformation of contemporary conflict came from Alvin and Heidi Toffler. In his bestselling 1980 book *The Third Wave* Alvin Toffler (1928–2016) had claimed that human civilization could be divided into three epochs, arguing that after the 'first wave' Neolithic, agricultural revolution and the 'second wave' nineteenth-century industrial revolution we were witnessing the emergence of a new, 'third wave', post-industrial society in which 'knowledge' was the defining element (Toffler, 1981). Now, in 1993, the Tofflers explored the impact of this third wave upon warfare. For the Tofflers the emergence of the new wave not only creates internal tensions, but also impacts on global relationships, creating a new, 'trisected world': 'one sharply divided into three contrasting and competing civilizations – the first still symbolized by the hoe; the second by the assembly line; and the third by the computer' (Toffler and Toffler, 1994:21). Thus, instead of the post-Cold War bringing a new era of peace and stability, the end of bipolarity and second wave industrialism 'could well trigger the deepest power struggles on the planet as each country tries to position itself in the emerging three-tiered power-structure' (Toffler and Toffler, 1994:25).

For the Tofflers 'the way men and women make war' reflects 'the way they work' and create wealth (Toffler and Toffler, 1994:33). Each new wave of civilization, therefore, produces its own mode of warfare, transforming every element of the armed forces 'from technology and culture, to organization, strategy, tactics, training, doctrine and logistics' (Toffler and Toffler, 1994:32). First wave wars, for example, 'bore the unmistakable stamp of the First Wave agrarian economies that gave rise to them, not in technological terms alone, but in organization, communication, logistics, administration, reward structures, leadership styles and cultural assumptions' (Toffler and Toffler, 1994:37). Similarly, second wave warfare reflected back its own world, with the industrialization of conflict, the mass mobilization of populations as military forces or workers, standardized, mass-produced weaponry, the development of the machine gun and mechanized warfare and the industrialization of death in the trenches of World War I and death-camps of World War II. Thus, the aim of second wave warfare was the mass production of materiel and the mass destruction of opposing forces. In contrast third wave warfare reflects its own de-massified, accelerated, innovative, knowledge-based economy.

The Gulf War represented a moment of transition, the Tofflers argue, as a 'dual war', combining second wave attrition methods – heavy bombardment and 'industrialized slaughter' – with third wave weapons and strategies based upon knowledge and information (Toffler and Toffler, 1994:66–9). Although knowledge has always been vital in combat, nevertheless, they argue, a revolution is occurring that places it today 'at the core of military power' (Toffler and Toffler, 1994:69), as seen in the Gulf War in the role of computing in the coordination of the campaign and in the command and control warfare against Iraq. The Tofflers see smart weapons as the best example of third wave warfare, in using digital technology to achieve a precise goal, with de-massification replacing the mass effect of 'dumb' bombs. Thus, just as businesses today use computers to cut expenditure and deliver more efficient and individual results, so too does the military. Weapons based on information thus replace weapons based on firepower: as the Tofflers argue, 'de-massified destruction, custom-tailored to minimize collateral damage, will increasingly dominate the zones of battle' (Toffler and Toffler, 1994:73).

The second wave, set-piece attritional warfare of the Gulf is not the future, the Tofflers argue. Instead the trisected world will see a 'radical diversification' of warfare in which all the methods of first, second and third wave conflict co-exist and confront each other (Toffler and Toffler, 1994:81). This means that today's military needs to plan for multiple threats and 'niche wars'. Instead of a superpower-based balance of power, today 'we see a bewildering diversity of separatist wars, ethnic and religious violence, coups d'état, border disputes, civil upheavals, and terrorist attacks, pushing waves of poverty-stricken, war-ridden immigrants (and hordes of drug traffickers as well) across national boundaries' (Toffler and Toffler, 1994:90). Thus, the future of war will be marked by 'distributed threats', 'small wars' and 'low-intensity conflict'. In response third wave nations are developing a range of new technologies, including the militarization of space, robotic and drone warfare, cyber-intrusion, smart weapons, exo-skeletons and wearable technology and non-lethal weaponry. In 'a world seething with potential violence', today's military, they conclude, needs advanced 'knowledge warriors' and 'knowledge strategies', dedicated to using informational capacities, systems and advantages both to win, or to forestall conflict (Toffler and Toffler, 1994:139–52).

There are obvious issues with the Tofflers' vision. Their privileging of technology and reduction of human civilization to a simplistic tripartite structure is highly problematic and helps explain their marginalization within mainstream academia. Outside it, however, they were widely read. The public, US politicians and policy makers all read *The Third Wave*, and *War and Anti-War* proved especially influential upon a military who were attempting at the time to theorize 'information war'. The Tofflers' vision of the global disruption caused by new technology, of the complexity and unpredictability of the post-Cold War era and the need to update the industrial-era military 'machine' with third wave forces, technologies and strategies all chimed with military strategists trying to confront the future of warfare.

Chris Hables Gray: 'Postmodern' and 'cyborg' war

With its broadcast mindset, privileging of television and print, and focus on news coverage and journalistic reporting, mainstream media studies paid little attention to these ideas. They found a better reception, however, within cultural studies. More interested in technology, more open to theories drawn from literary criticism critical theory, sociology and continental philosophy, and more aware of developments in 'new media', it was cultural studies – and 'cyberculture' especially – that pioneered the critical analysis of digital technology in the 1990s. In particular cultural studies gave rise to new ideas of 'postmodern war' and 'cyborg war'.

Debates around 'postmodernism' were central to cultural studies in the 1990s, with its existence, definition and central elements all attracting considerable commentary and criticism. As I've suggested, developments in media were central to these debates, hence the common co-option of Baudrillard's work. The Gulf War, therefore, was a gift for those interested in postmodernism. The media spectacle of smart-bomb videos, military conferences and live, 24-hour television from the event itself appeared as powerful evidence of our 'postmodern culture' and discussions of 'postmodern war' subsequently proliferated in the academic and popular literature. Appropriately, this concept was itself a postmodern bricolage, drawing from a variety of arguments and sources, most obviously Baudrillard's work. The term wasn't, however, used by Baudrillard and though in common use it lacked a systematic analysis.

Chris Hables Gray's 1997 book *Postmodern War* was one of the few attempts to develop the concept. He identifies here 'the key tropes of postmodern war', which, he says, include the increased lethality, speed and scope of the battlefield, the fragmentation of battlefields, the deployment of high technology, the substitution of men with technology, the aim of minimized casualties and apparently bloodless deaths, the erasure of nature as a category, and the role of discourse as a weapon through the media (Hables Gray, 1997:168–84). Ultimately, Hables Gray's book is only partially successful. Though strong on the historical roots of 'postmodern war' in World War II, the Cold War and Vietnam, this detail comes at the price of a loss of specificity of the key concept and a failure to elucidate its character in depth. Where the book is more successful, however, is in its discussion of 'cyborg soldiers' and warfare.

The question of 'the body' was popular in cultural studies in the 1990s and the linked field of cyberculture explored real and metaphorical developments in the coupling of humanity and technology. Hables Gray was at the centre of these debates, publishing on cyborg soldiers in 1989 and editing *The Cyborg Handbook* in 1995, which contained an important essay by Robins and Levidow on the cyborg relationship of modern war (Hables Gray, 1989; Hables Gray, 1995:119–25). For Robins and Levidow, 'vision and image technologies mediate the construction of a cyborg self' (Hables Gray, 1995:120). The military operator, integrated with their weapons and surveillance systems, experiences the enemy only as a screen target, developing a 'technological psychosis' in which a phantasy of total visibility and control combines with a paranoid anxiety over potential threats, leading to the hunting-down of 'thing-like' objects devoid of any humanity or moral value. Those watching television at home suffer a similar cyborg identification and psychosis, they argue, experiencing the fear, anxiety and pleasures of the hunt in a close coupling with the screen and the cameras.

In *Postmodern War* Hables Gray returns to this idea of a 'cyborg soldier' (Hables Gray 1997:195–211, 2002:55–65, 2003). Whilst soldiers have always used weapons, today, he says, 'it is not that the solder is influenced by the weapons used; now he or she is (re)constructed and (re)programmed to fit integrally into weapons *systems*' (Hables Gray, 1997:195). The US military especially is moving towards a man–machine integration: 'a cybernetic organism (cyborg) model of the soldier that combines machine-like endurance with a redefined human intellect subordinated to the overall weapons system', with the aim of producing 'a postmodern army of war machines, war managers and robotized warriors'. The ideal soldier is now either a machine or one who has been made to act like one through drugs, discipline and management and who exists in a closer interface with computer and weapons systems (Hables Gray, 1997:196).

Hables Gray explores contemporary developments in this interface, arguing that one of the key areas for the implementation of this cyborg soldier is the aircraft and helicopter pilot who is being brought into an increasingly intimate connection with increasingly intelligent technological systems. This was already seen in 1991 in the Apache helicopter pilot's pioneering connection to their weapons systems:

> Attached to the pilot's helmet was a two-inch-square semi-transparent monocle that extended about an inch or two in front of the pilot's right eye. Projected onto the monocle was the targeting information that came from the Apache's infrared targeting systems. There was also a cross hair-type targeting device. All of the Apache's weapons systems were linked electronically to the monocle. All a weapons officer had to do was look at a target, lay the cross hairs on it, and fire his weapon of choice.
>
> Frontline (2015)

Although discussions of the 'cyborg' would become unfashionable and fade in the new millennium, real-world military research into its possibilities would lead to real advances, such as in drone systems, brain–computer interfaces, closer man–machine integration and wearable computing and exo-skeletons. 'Cyborg' warfare, therefore, would prove to play a key role in twenty-first-century conflict.

Michael Ignatieff and 'virtual war'

Alongside 'postmodern war', another term that gained currency in the 1990s was 'virtual war', a concept that again owed much to Baudrillard and Virilio. Their ideas were a significant, and uncredited, influence, for example, on *Virtual War*, Michael Ignatieff's

1999 book on the Kosovo War and contemporary conflict. At the heart of Ignatieff's critique is again the question of experience:

> The Kosovo conflict looked and sounded like a war: jets took off, buildings were destroyed and people died. For the civilians and soldiers killed in airstrikes and the Kosovar Albanians murdered by Serbian police and paramilitaries the war was as real – and as fraught with horror – as war can be. For the Citizens of the NATO countries, on the other hand, the war was virtual. They were mobilized, not as combatants but as spectators. The war was a spectacle: it aroused emotions in the intense but shallow way that sports do.
>
> Ignatieff (2000:3)

In this, Kosovo shares much with the Gulf War, but Ignatieff follows the Tofflers and Virilio in seeing the Gulf War as a war of the past. Instead of being the first in a new stage of wars, Ignatieff says, the Gulf War was

> the last of the old wars: it mobilized a huge land force and the vast logistical support required to sustain it, and it was fought for a classic end, to reverse a straightforward case of territorial aggression against a member state of the United Nations. Soldiers were committed in full expectation of casualties.
>
> Ignatieff (2000:5)

Kosovo was different. It was the first 'virtual war': it was fought for more complex political reasons, had no official UN or legislative sanction, lacked any face-to-face, direct physical engagement and achieved its aim of zero casualties for the military.

This new, 'risk-free' virtual war owes its existence to developments in technology, Ignatieff says, and especially to the 'Revolution in Military Affairs' (RMA). This originated in the 1970s as a way to break the nuclear stalemate. To make the use of conventional weapons morally and politically acceptable, 'it was essential to increase the precision of their targeting; to minimize the collateral or unintended consequences of their use; and to reduce, if not eliminate, the risk to those who fired them'. In order to make war viable again, therefore, it had to become 'bloodless, risk-free and precise' (Ignatieff, 2000:164).

The first breakthrough was in precision-guided munitions (PGM), which were introduced in May 1972 when, after hundreds of conventional bombs and the loss of many aircraft and men, the US Airforce succeeded in destroying the Thanh Hoa bridge in Vietnam with the first laser-guided bombs. By 1991 PGM had advanced sufficiently to play a major part in the Gulf War, transforming in the process the public expectation of war: 'Having been told to prepare for as many as 25,000 casualties, the electorate discovered the intoxicating reality of risk-free warfare' (Ignatieff, 2000:168). By the time of the Kosovo War the new, precise lethality-at-a-distance transformed war from attritional destruction into information warfare and the targeted destruction of 'the nerve centres – command posts, computer networks – which direct the war-machine' (Ignatieff, 2000:169).

The second element of the RMA, Ignatieff says, was the role of computers, which increased the ability to collect, process and act upon information, allowing the commander to 'coordinate the actions of all his forces in near-perfect knowledge of the battle-space' (Ignatieff, 2000:171). Echoing the Tofflers, Ignatieff says that these two developments call into question 'the heavy industrial armies created to fight WWII', and end 'a century of total war' (Ignatieff, 2000:171). The new systems allowed politicians to cut defence spending whilst retaining their preparedness and encouraged the development of new theories of warfare based upon smaller, lighter, more rapid and coordinated forces, although, Ignatieff notes, these weren't ready by 1999, hence the dependence on a purely virtual, aerial war.

In addition to 'risk-free' warfare, Ignatieff identifies other elements of today's 'virtual war'. This includes the rise of 'virtual consent', or the avoidance of legislative consent for war together with the use of linguistic subterfuge to redefine events and actions to avoid democratic checks and balances upon the executive. It also includes the central role of 'media war' and the control of imagery and use of 'spin' to ensure public support, as well as the rise of military lawyers, who now play a key role in targeting decisions. Linked with a system of PGM their presence provides a false belief, Ignatieff argues, that military actions are legal, justified, 'clean and mistake-free' (Ignatieff, 2000:198). Yet the legal coverage they bring doesn't make the actions moral or morally justifiable and the same west that appeals to international law, Ignatieff points out, refuses to allow its own troops to be bound by it.

Ignatieff also points to the 'virtual values' of war today and the clash between our rhetoric and our actual commitments; to the 'virtual rhetoric' used to mobilize civilian support; and to the use of 'virtual alliances' to build legitimacy and support for war, whilst giving 'the illusion of unity' in covering up differences in tactics, coordination and aims. Appropriately enough this 'virtual war proceeds to virtual victory' (Ignatieff, 2000:208). Whereas 'total wars' were fought to the death, today's virtual wars are self-limiting, Ignatieff says, leaving demonized regimes in place due to the west's fear of chaos or of nation building. Such outcomes in Iraq and Serbia, are, however, far from satisfying: as Ignatieff argues, 'virtual victory is a poor substitute for the real thing' (Ignatieff, 2000:210). And with the inevitable loss of western monopoly over smart weapons, there is no reason to believe virtual, risk-free war will result in a safer world. As Ignatieff concludes, 'virtual war … is a dangerous illusion' (Ignatieff, 2000:212).

Ignatieff's analysis of 'virtual war' re-treads many ideas better presented earlier in the decade by Baudrillard and Virilio and aspects of his classification have also dated, but the book's lasting value lies in the questions it asks about our experience of war and the future implications for democracy. In contrast to the mass-mobilizations of the past that swept up populations as war workers or soldiers, today's wars are marked, Ignatieff says, by a 'virtual mobilization' that requires little risk or involvement: 'We now wage wars and few notice or care. War no longer demands the type of physical involvement or moral attention it required over the past two centuries' (Ignatieff, 2000:184). The decline of mobilization and of 'the martial ideal' of sacrifice for the state impacts, therefore, upon our relationship with war. 'War thus becomes virtual, not simply because it appears to take place on a screen but because it enlists societies only in virtual ways'. This is war, Ignatieff argues, as 'a spectator sport': 'war affords the pleasures of a spectacle, with the added thrill that it is real for someone, but not, happily, for the spectator' (Ignatieff, 2000:191).

The ability to wage wars without combat fatalities is significant here. This is a huge military achievement but it is an ethically questionable one, Ignatieff says, as the contest of war loses its legitimacy when one side can kill with impunity. In Kosovo, the west avoided these issues by emphasizing their rules of engagement and humanitarian aims, but the ethical questions remain for the future, Ignatieff says: for 'if war becomes unreal to the citizens of modern democracies, will they care enough to restrain and control the violence exercised in their name?' (Ignatieff, 2000:4; see also 2000:163). If war becomes cost free, then what democratic controls will remain over it?

> Democracies may well remain peace loving only so long as the risks of war remain real to their citizens. If war becomes virtual – and without risk – democratic electorates may be more willing to fight, especially if the cause is justified in the language of human rights and even democracy itself
>
> Ignatieff (2000:179–80).

This, of course, assumes the western public will be informed about, be able to find out about, or even care about its governments' wars. Developments in cyberwar, for example, may be beyond the scrutiny of journalists and commentators, whilst the general demobilization of the population means 'the conflicts of the future may take place without anyone even realizing they are occurring' (Ignatieff, 2000:190). Like Baudrillard and Virilio before him, Ignatieff recognizes the problems this new model of war may present for the future. His warning about the public's declining relationship to conflict would become especially important in the new millennium in an era dominated by ongoing, global military operations, special forces actions, electronic warfare and drone strikes by the west, much of which would go unreported.

James Der Derian and 'virtuous war'

By the end of the decade, therefore, the once-controversial ideas of Baudrillard and Virilio had become more acceptable. Another critic heavily influenced by their work was James Der Derian, who in 2001 published *Virtuous War*, his 'travelogue' exploration of war in the 1990s. Der Derian identifies here the post-Vietnam, post-Cold War emergence of a 'postmodern' mode of conflict. Developing 'from the battlespace of the Gulf War and through the aerial campaigns of Bosnia and Kosovo', it combines the technological superiority of precision munitions (the *virtuality* of violence-at-a-distance) with an ethical superiority derived from Just War and Holy War theory (the *virtue* of its humanitarian or liberatory intervention) in a new 'virtuous war'. 'Ethically intentioned and virtually applied', Der Derian argues, this virtuous 'infowar' uses 'computer simulation, military dissimulation, global surveillance and networked warfare', 'to deter, discipline and if necessary destroy the enemy', preferably 'with minimal casualties where possible' (Der Derian, 2009:xx, 232–33).

Thus, Der Derian says, 'technology in the service of virtue has given rise to a global form of virtual violence, *virtuous war*' (Der Derian, 2009:xxvii). Like Ignatieff, he recognizes here the contradiction of a virtuous intervention that is based on the removal of oneself from any danger and the valorization of one's own lives above those of others in the combat zone in its aim for 'no or minimal casualties' (Der Derian, 2009:xxxi). Indeed, despite its claimed humanitarianism, 'something fundamentally human is being lost' in new infowar attacks on the 'vital life-support systems' of communication and energy infrastructures (Der Derian, 2009:146). There is no ethical accountability here either for the civilian victims of the long-lasting effects of these attacks. As Der Derian says, 'This is the darker side of virtuous war that goes largely unreported, or is belatedly revealed, long after the first images of technological wizardry yielding political success have been burned into the public consciousness' (Der Derian, 2009:146).

Following Baudrillard and Ignatieff, Der Derian develops, therefore, another critique of 'the disappearance of the body, the aestheticizing of violence, the sanitization of war' (Der Derian, 2009:165), but Der Derian also goes further in his exploration of this mode of warfare. In particular he investigates the 1990s 'Revolution in Military Affairs' (RMA), the contemporary military re-theorization of conflict and the increasing military significance of new technological partnerships with the media and entertainment industries. Hence, building on Eisenhower's 1961 identification of a 'military-industrial complex', Der Derian identifies a new 'virtual revolution in military and diplomatic affairs', caused by the 'virtual alliance' of what he calls the 'military-industrial-media-entertainment network', or 'MIMEnet' (Eisenhower, 1961; Der Derian, 2009:xxx; 2009:xxvii).

Thus, MIMEnet is the result of the expansion of the military-industrial complex through close relationships with media and entertainment industries. Their primary contribution lies in their development of new simulation training technologies that allow the military to hone their strategies, tactics and skills for future forms of combat. Hollywood, Disney and new technological start-ups all, therefore, now feed into the RMA and new military theories, enabling the detailed digital simulation of combat experiences, situations and even entire battle zones. What emerges, Der Derian argues, is a new 'simulation triangle', or a convergence of 'the military, new media and Mickey' (Der Derian, 2009:81–2).

Der Derian's book sets out to explore this convergence. His tour takes in visits to Fort Irwin's 'National Training Centre' (NTC), the Hohenfels 'Combat and Manouver Training Centre' (CMTC) and the 'Institute of Creative Technologies' (ICT) at the University of Southern California where he watches digitized wargame battlefields and virtual simulations in action. He also investigates 'Simulation, Training and Instrumentation Command' (STRICOM) and its 'Joint Simulation System' (JSIMS), experiences various virtual environment training systems and vehicle simulators, and participates in the domestic US military exercise 'Urban Warrior'. One reason for the turn to simulation technologies, he suggests, is their cost-effectiveness at a time of budgetary limitations, but beyond this there is an entire philosophical project: that of the implosion of the real and virtual and the latter's ascendency.

Der Derian quotes Jorge Luis Borges' fable of an empire that created a 1:1 scale map that would cover the territory precisely. This is an idea we have now left behind, Baudrillard says. Today's abstraction, Baudrillard argues, is not that of the map or double that comes after and reflects the real, rather

> It is the generation by models of a real without origin or reality: a hyperreal. The territory no longer precedes the map, nor does it survive it. It is nevertheless the map that precedes the territory – *precession of simulacra* – that engenders the territory.
>
> Baudrillard (1994:1)

This is how the US military are developing simulation, Der Derian suggests: as a way not simply to model but to precede, eclipse and replace the real:

> What is qualitatively new is the power of the MIMEnet to seamlessly merge the production, representation and execution of war. The result is not merely the copy of a copy, or the creation of something new: it represents a convergence of the means by which we distinguish the original and the new, the real from the reproduced.
>
> Der Derian (2009:xxxvi)

This is the new 'danger' of these simulation technologies, Der Derian suggests: 'I felt as if reality itself, like light being sucked into a black hole, was disappearing in the Simulation Triangle'. Developments in military simulation, he argues, are leading to 'a brave new world that threatens to breach the last fire walls between reality and virtuality' (Der Derian, 2009:167). As Baudrillard warns, the simulations Der Derian sees on his visits display 'a capability to precede and replace reality itself' (Der Derian, 2009:96). In an increasingly uncertain world, der Derian warns, it is these maps that will be used to predetermine policy intentions and even military interventions.

Der Derian's *Virtuous War* was published in 2001 at a time when the US military was ready to unleash its new ideas of lighter, mobile, networked forces but lacked an explicit enemy. In his 2009 revised edition he reflects back on eight years of the War on Terror, finding in the US response the explicit realization of the idea of 'virtuous war' – of high-tech virtual wars based upon 'a rhetoric of total victory over absolute evil'. In the War on Terror, he concludes, 'virtuous war became the ultimate means by which the United States intended to re-secure its borders and assert its global suzerainty' (Der Derian, 2009:xx).

Conclusion

We can see, therefore, that through the 1990s, in response to the new developments of the Gulf War and Kosovo War, philosophers, critics and commentators developed a series of related reflections on the changing nature of warfare in the west. In part this was a reflection on the changing nature of war – the new centrality of knowledge, information and information systems; the role of information technology in precision-guided munitions and in the real-time panoptical surveillance and control of the battlefield; the development of a 'risk-free' hi-tech 'clean war' of 'surgical' strikes and minimized casualties; the emphasis on ethical warfare and just or humanitarian intervention in the name of values; and the close cyborg coupling of man and machine in IT and weapons systems. In part, too, it was a reflection on the media coverage – the new experience of 24 hour, real-time and live coverage seemingly from the heart of the 'event'; the implosion of media and military; the mobilization of domestic support for war, the breathless awe at the technological 'spectacle' and the identification of the audience with weapons systems.

At the heart of many of these critiques, however, was a warning about the changing experience of warfare in the west – the absence of casualties, the aesthetization and sanitization of violence, the virtuality of a war experienced only through the television screen and the implications of these both for western democracy and for global relations in the future. Many of these concerns would prove to be valid in the following decades.

Key reading

For the key ideas covered in the chapter see Baudrillard (1995), Virilio (2000; 2002), Toffler and Toffler (1994), Hables Gray (1997), Ignatieff (2000) and Der Derian (2009). For an introduction to Baudrillard's philosophy and his discussion of the Gulf War see Merrin (1994, 2005). The debate around Baudrillard's 'postmodernism' can be followed through his supporters such as Gane (1991a, 1991b) and Merrin (2005) and critics such as Callinicos (1989), Kellner (1989), Best and Kellner (1991), Elliott (1992) and Norris (1990, 1992). Baudrillard's own earlier discussion of media (1983, 1994, 1998) is essential to understanding his view of 'non-war'. Virilio's other writings on warfare are also worth reading (1997c, 2009) and further information about his work and ideas can be found in Armitage (2000, 2012) and James (2007). The best background reading for the Tofflers' ideas of war is the theory of 'three waves' found in Toffler (1981). Chris Hables Gray's theory of cyborg warfare is also developed in his other texts (1989, 1995, 2002, 2003).

Note

1 The three essays were published together in 1991 as *The Gulf War Did Not Take Place*. See Baudrillard (1995).

References

Armitage, J. (ed.) (2000) *Paul Virilio. From Modernism to Postmodernism and Beyond*, London: Sage.

Armitage, J. (ed.) (2012) *Virilio and the Media*, Cambridge: Polity.

Baudrillard, J. (1983) *In the Shadow of the Silent Majorities*, New York: Semiotext(e).

Baudrillard, J. (1989) 'Politics of Seduction', Interview with Baudrillard, *Marxism Today*, January, pp. 54–55.

Baudrillard, J. (1994) *Simulacra and Simulation*, Ann Arbor: The University of Michigan Press.

Baudrillard, J. (1995) *The Gulf War Did Not Take Place*, Sydney: Power Publications.

Baudrillard, J. (1998) *The Consumer Society*, London: Sage.

Best, S. and Kellner, D. (1991) *Postmodern Theory: Critical Interrogations*, London: Macmillan.

Callinicos, A. (1989) *Against Postmodernism: A Marxist Critique*, Cambridge: Polity Press.

Cumings, B. (1992) *War and Television*, London: Verso.

Der Derian, J. (2009) *Virtuous War*, Abingdon: Routledge.

Eisenhower, D. D. (1961) 'Military-Industrial Complex Speech, Dwight D. Eisenhower, 1961', http://coursesa.matrix.msu.edu/~hst306/documents/indust.html.

Elliott, G. (1992) 'A Just War? The Left and the Moral Gulf', *Radical Philosophy*, 61 (Summer), pp. 10–13.

Frontline. (2015) 'Weapons: Ah-64 Apache', PBS, http://www.pbs.org/wgbh/pages/frontline/gulf/weapons/apache.html.

Gane, M. (1991a) *Baudrillard's Bestiary*, London: Routledge.

Gane, M. (1991b) *Baudrillard: Critical and Fatal Theory*, London: Routledge.

Hables Gray, C. (1989) 'The Cyborg Solider'. In L. Levidow and K. Robins, eds., *Cyborg Worlds: The Military Information Society*, London: Free Association Books, pp. 43–72.

Hables Gray, C. (ed.) (1995) *The Cyborg Handbook*, London: Routledge.

Hables Gray, C. (1997) *Postmodern War*, London: Routledge.

Hables Gray, C. (2002) *Cyborg Citizen*, London: Routledge.

Hables Gray, C. (2003) 'Posthuman Soldiers in Postmodern War', *Body & Society*, 9(4), pp. 215–26.

Ignatieff, M. (2000) *Virtual War*, London: Chatto & Windus.

James, I. (2007) *Paul Virilio*, Abingdon: Routledge.

Kellner, D. (1989) *Jean Baudrillard: From Marxism to Postmodernism and Beyond*, Cambridge: Polity.

Merrin, W. (1994) 'Uncritical Criticism? Norris, Baudrillard and the Gulf War', *Economy and Society*, 23(4), pp. 433–58.

Merrin, W. (2005) *Baudrillard and the Media: A Critical Introduction*, Cambridge: Polity.

Norris, C. (1990) *What's Wrong With Postmodernism*, London: Harvester Wheatsheaf.

Norris, C. (1992) *Uncritical Theory: Postmodernism, Intellectuals and the Gulf War*, London: Lawrence and Wishart.

Toffler, A. (1981) *The Third Wave*, London: Pan Books.

Toffler, A. and Toffler, H. (1994) *War and Anti-War*, London: Little, Brown.

Virilio, P. (1997a) *Open Sky*, London: Verso.

Virilio, P. (1997b) 'Interview with Paul Virilio', by Der Derian, J., *Speed*, 1.4, http://nideffer.net/proj/
   SPEED/1.4/articles/derderian.html.
Virilio, P. (1997c) *Pure War*, New York: Semiotext(e).
Virilio, P. (2000) *Strategy of Deception*, London: Verso.
Virilio, P. (2002) *Desert Screen*, London: Continuum.
Virilio, P. (2009) *War and Cinema*, London: Verso.

3   Informational and networked war

Remaking conflict in the 1990s

The US military

The US military was also rethinking war in the 1990s. Influenced by the emerging concept of a 'Revolution in Military Affairs' (RMA), by the key role of new technology in the Gulf War and by rapid developments in information technology through the decade, the US armed services and military theorists began to develop their own understanding of how conflict had changed and where it was going. The increasing recognition of the importance of computers and the possibility of their penetration, together with the invention of the World Wide Web and its take-off by the middle of the decade led to new discussions of 'C2W', 'C3I', 'C4I', 'information warfare', 'cyberwar', 'information operations', 'full spectrum dominance' and 'network-centric warfare'. Through the decade, therefore, the US military published a series of policy documents and theoretical essays that reflected and built on ongoing changes in computing in order to construct a twenty-first-century vision of military strategy and operations. These conceptions would find favour with politicians and prove highly influential in the new millennium, establishing the basis not only for the military operations of the War on Terror but also for continuing research into future combat capabilities.

Computing and warfare

'Knowledge' and 'information' have, of course, always been central to warfare. Intelligence about the enemy's forces, strategies, abilities, plans and about the battlefield have all been essential to military success. There has also always been a close relationship between developments in communication and the military. Historically, military signalling relied upon couriers on foot or horseback, waterborne vessels, and hill-top signals using fire and mirrors. In the late eighteenth century the mechanical telegraph transformed communication, allowing the rapid passage of complex information through signaling towers. It was quickly adopted by the French and British governments for military use and its successor, the electric telegraph, similarly played an important role in military coordination during the Crimean war (1854–56), American Civil War (1861–65) and Franco-Prussian War (1870–71). The telephone would be used by the military on the Indian North-West Frontier from 1877 and during the Second Afghan War (1879) and the Boer War (1899–1902). 'Wireless telegraphy' was important in World War One, with both the UK in 1914 and the USA in 1917 banning amateur broadcasting and reserving it for military use, and even television had an indirect role to play in World War II with many of its technologies and inventors contributing towards the war effort. In the postwar period nation states funded satellite research. The USSR's launch of Sputnik on 4 October 1957 sparked a 'space race' and in the following decades military satellites were developed for reconnaissance and intelligence, as part of navigation systems and early warning defences and for military communications.

The key contemporary development in military information and communications, however, has been that of computing and networked computing technology. This too has strong military roots. Computing has its origins in the production of technical aids for calculation. In the nineteenth century Charles Babbage and Herman Hollerith each attempted to produce calculating devices and by the 1930s large, mechanical and electro-mechanical sorting and calculating devices were in use in the accountancy, engineering, banking, manufacturing and insurance industries. It was the military demands of World War II and government funding, however, that provided the major spur to computing as anti-aircraft guns required complex calculations to hit fast-moving targets. The US Navy and IBM helped Howard H. Aiken build the Harvard Mark 1, in use at Harvard by February 1944, and the US Army funded John W. Mauchly and J. Presper Eckert's ENIAC (Electronic Numerical Integrator and Calculator) at the University of Pennsylvania, although it wasn't operational until July 1947.

In the UK computing was developed for code-breaking purposes. At Bletchley Park, electromechanical 'bombes' proved unable to break the German Geheimschreiber machine so a team headed by M. H. A. Newman and T. H. Flowers built a computer called Colossus, operational by December 1944, for that task. Whereas the UK broke up its Colossus machines after the war, computing technology was central to postwar US military research and defence. ENIAC was used to design the hydrogen bomb, EDVAC (Electronic Discrete Variable Automatic Computer) was used by the US Army Ordnance, and MIT's Whirlwind became the basis for the US Air Force's real-time SAGE (Semi-Automatic Ground Environment) early warning air defence system.

The US military was also behind the development of networked computing. The origins of the internet are usually traced to the launch of ARPA, the Advanced Research Projects Agency which was created in 1958 as a response to the Soviet Union's launch of Sputnik, with a mission to fund a community of the best scientific minds and promote 'blue-sky thinking' and research with possible government or military applications. Building on J. C. R. Licklider's 1962 'memo' advocating connected computers (an 'intergalactic computer network'), in 1966 Bob Taylor, the head of ARPA's IPTO (the Information Processing Technologies Office) realized such a network would allow researchers to better share their work and ideas. In 1968 Bolt, Beranek and Newman (BBN) of Cambridge, Massachusetts, won the contract to construct the 'ARPAnet' and at 22.30pm on 29 October 1969 the first link was established between the University of California, Los Angeles (UCLA) and the Stanford Research Institute (SRI). By 5 December, a four-node network was established with the addition of the University of Utah and University of California, Santa Barbara. What began as a means to improve ARPA's research, therefore, became its most important innovation. Contrary to popular myth, the internet itself wasn't invented to create a communications system that could survive a nuclear war, although one important element of its structure – 'packet-switching' within a 'distributed network' – *was* derived from Paul Baran's research for the RAND Corporation into 'survivable' communications, reinforcing the military connections.

During the Cold War the US military became increasingly reliant on information technologies. Computer-based processing played an essential role in secure military telecommunications, in surveillance (with the launch and operation of spy satellites) and in intelligence (with the encryption, interception and decryption of classified communications). Military needs spurred computing research: as Rattray says, 'the rush to improve the designs of nuclear warheads and delivery systems, along with the space race, pushed the development of advanced information technologies in both computing and communications during the 1950s and 1960s' (Rattray, 2001:312). New technologies such as solid-state transistors and digital telecommunications switches, for example, were created in commercial laboratories by AT&T and IBM for use by the Department of Defense.

Computing also became central to weapons systems. Eventually information technology would underlie everything from the operation of nuclear weapons, nuclear submarines, aircraft carriers, advanced strike and air superiority aircraft and missile technologies through to the electronic countermeasures and signals processing technologies required of antisubmarine warfare. 'Even the design of tank armor and fire control relied on the use of advanced computer processing and simulation' (Rattray, 2001:312). The reason for the US drive for information technology was to offset the problem of the greater size of the Soviet army. As Rattray comments, developments in computing were 'crucial to sustaining a U.S. high-technology edge for its numerically inferior conventional forces' (Rattray, 2001:312).

Early on, the US military had a vision of a fully computerized battlefield. In a speech in Washington on 14 October 1969 to the Association of the United States Army, General William Westmoreland, claimed, 'On the battlefield of the future, enemy forces will be located, tracked and targeted almost instantaneously through the use of data links, computer assisted intelligence evaluation, and automated fire control'. The American people, he concluded, would 'welcome and applaud the developments that will replace wherever possible the man with the machine' (Dickson, 2012:220–21). His confidence was based on real developments. In September 1966, the US military had begun the development of what became 'Operation Igloo White': the creation of an 'automated battlefield' to prevent North Vietnamese supplies passing along the Ho Chi Minh Trail on the Laotian border to forces in the south.

Operational from January 1968 to February 1973, the system comprised four elements – (1) a string of dropped or planted electronic, acoustic or seismic sensors able to transmit real-time information to (2) transmission relay aircraft who sent information to (3) the command and control base, the Infiltration Surveillance Centre (ISC) at Nakhon Phanom in Thailand, which processed the data through the most powerful computers then available, two IBM 360/Model 65 computers. Data was then displayed electronically for analysts who transmitted the coordinates to (4) strike aircraft who could attack the target. The system was able to provide a target location within 100 square meters, with an average five-minute time between ISC target acquisition and ordnance delivery. Most assessments of Igloo White are critical of its success rate, its technical limitations and its cost-effectiveness against small targets: 'The average annual operating costs of "Igloo White" were around one billion dollars. It was estimated that it cost around $100,000 to destroy a single truck during its operation' (Uziel, 2015). Its real failure, however, lay in the continued use of the Ho Chi Minh Trail and the success of the supply route.

The Vietnam era saw many innovations in military information technology. The microchip, for example, aided the development of early laser-guided precision-guided munitions. Texas Instruments began research into PGMs in 1964, testing their first 'Paveway' bomb in April 1965 and sending prototype weapons to Vietnam for combat testing. In May 1968, the F-4D became the first aircraft to use laser-guided munitions and on 13 May 1972 they were famously used to destroy the Thanh Hoa bridge, which had survived innumerable conventional attacks. Less than two weeks later, on 25 May 1972, NASA tested one of the first digital ('fly-by-wire') aircraft systems using an experimental F-8 Crusader, opening the era of computer-controlled military aircraft.

These developments, however, had only a limited effect on the Vietnam War, which was dominated by an industrial model of destructive, attritional warfare. It was the failure of this model and defeat in Vietnam that galvanized the USA into rethinking its military operations. The US Army established Training and Doctrine Command (TRADOC) on 1 July 1973 under the command of General William E. DePuy specifically to explore new war-fighting strategies and in the following years advanced technologies would come to play an increasingly important role in its vision of future combat. In 1976 DePuy's TRADOC had produced a new theory of 'Active Defence' that aimed to combat Soviet forces on the European battlefield using 'forward defence' and firepower, but DePuy's successor in July 1977, General Donn A. Starry, had a more radical vision of how a war could be fought. Building on Colonel John Boyd's 1976 advocacy in 'Patterns of Conflict' of rapid, smaller, more maneuverable forces, his discussions with Don Morelli (the head of TRADOC's 'Office of Doctrine Formation'), and his own reading of Toffler's *The Third Wave*, on 25 March 1981 Starry published *AirLand Battle and Corps 86, TRADOC Pamphlet 525-5*.

Though still focused on the European Soviet threat, this new 'AirLand Battle' strategy emphasized rapid movement, close air and land coordination and the use of new technologies to strike key targets in the 'extended battlefield' behind the front. One of its central features was its emphasis on information technology to compensate for the Soviets' numerical superiority and to overcome a larger enemy. Following directly on from Igloo White's concept of an electronic battlefield, the aim was to secure victory through an 'integrated battlefield' that comprised 'sensor/surveillance systems' to provide targeting and intelligence information, 'accurate', lethal delivery systems and a 'command control' integrating 'all source intelligence in near real-time', allowing real-time responses (Department of the Army, 1981:7). These ideas were embodied in the US Army's *Field Manual (FM) 100-5 (Operations)* on 20 August 1982 and in 1984 AirLand Battle became the primary battle plan of US NATO forces.

Alongside this increasing emphasis on computers was a growing realization of their own security risks. The national security implications of connected networks had already been recognized by the late 1960s, the US Defense Science Board group had issued a report on computer security in 1970 and the US military was conducting exercises looking for its own vulnerabilities only a few years later. Yet it wasn't until the 1980s that the issue came to prominence. By then cheap home computers had taken off and there was the growing worry of both an emerging hacker culture and a generational shift in which children knew more about computing than adults. John Badham's 1983 film *WarGames* exploited these fears, showing a teenage hacker entering military systems to play a 'thermonuclear war' simulation game and nearly setting off World War III.

WarGames was taken seriously by President Reagan, who had a private viewing of the film and discussed it after with members of Congress, but two other real events also made the US government take notice of 'cyberspace'. The first was the 'Cuckoo's Egg' case in 1986 where it was discovered that a team of German hackers had broken into computers at Lawrence Berkeley National Laboratory and other research institutions and military commands looking for information on the US Strategic Defense Initiative (SDI) and satellites to sell to the KGB. The second was the release of the 'Morris Worm', which spread round internet connections in November 1988, demonstrating their vulnerability. Over the next decade would come the gradual realization that computing and information networks represented a new 'space' that required both defensive and offensive capabilities.

The 'Revolution in Military Affairs'

Technological developments also spurred on new claims of a 'Revolution in Military Affairs' (RMA). The origins of RMA theory are attributed to Soviet military theorists who, since the 1960s, had argued that scientific-technical progress drove successive revolutions in military capacities. Marshal V.D. Sokolovskiy's 1962 book *Soviet Military Strategy*, for example, traced a series of key innovations, arguing that 'modern nuclear weapons' had led to a new military revolution. Recognizing the US technological developments in Vietnam, by the 1970s Soviet theorists were suggesting that 'automated reconnaissance-and-strike complexes, long-range high-accuracy terminally guided combat systems ... and qualitative new electronic control systems' could constitute another such revolution (Krepinevich and Watts, 2015:194).

In the 1980s Marshal Nikolai Ogarkov in particular was outspoken in his belief in the obsolescence of the unwieldy Soviet army and the need to remake it into a smaller strike force. In papers and lectures he described a 'military technical revolution' (MTR) involving the synthesis of new technology, military systems and organizational and operational adaptation to create a new way of waging war. In a candid 1983 remark to a US journalist (unreported until 1992) Ogarkov admitted the USA's technological superiority:

> We cannot equal the quality of U.S. arms for a generation or two. Modern military power is based on technology, and technology is based on computers. In the US ... small children – even before they start school – play with computers. Computers are everywhere in America. Here we don't even have computers in every office of the Defence Ministry. And for reasons you know well, we cannot even make computers widely available in our society. We will never be able to catch up with you in modern arms until we have an economic revolution. And the question is whether we can have an economic revolution without a political revolution.
>
> Gelb (1992)

These ideas attracted the attention of Andrew W. Marshall, the Head of the US Department of Defense 'Office of Net Assessment' (ONA). Created by President Nixon in 1973, the ONA was an internal 'think-tank' for the Pentagon, tasked with looking into the future and assessing transformations in military strength. Marshall became the key US proponent of the idea of the MTR which he renamed a 'Revolution in Military Affairs', pushing for ONA to produce an assessment of the concept.

For those who hadn't been following these developments in weapons systems and military theory, the 1991 Gulf War appeared revelatory. It was the first time that the USA's post-Vietnam high-tech weaponry and new military strategies had been fully and publicly unleashed. In particular, 'it was the first large-scale use of stealth aircraft against a significant military power, and the first intense application of PGMs to achieve the operational and strategic objectives of a campaign' (Krepinevich and Watts, 2015:198).

For many, it demonstrated how technological advances had definitively changed warfare, providing 'strong evidence of the boost in military effectiveness made possible by the use of stealth aircraft, precision-guided munitions, advanced sensors, and the global positioning system (GPS) constellation of satellites the United States had deployed in the 1980s' (Krepinevich and Watts, 2015:198). The Gulf War, therefore, appeared to be clear evidence of the RMA that was beginning to be discussed.

Except things weren't so simple. Although the military had successfully deployed its 'AirLand Battle' strategy in the Gulf as well as the latest precision technologies, the war remained, in many ways, a more traditional conflict. Wanting to avoid casualties and finally end the 'Vietnam syndrome', the US Army had insisted upon overwhelming numbers and force. This became known as the 'Powell doctrine', a term named after commander General Colin Powell's vision of military action, which was itself based upon the earlier 'Weinberger doctrine'. This was a set of conditions for the deployment of military power introduced by US Secretary of Defense Caspar Weinberger in his speech 'The Uses of Military Power' to the National Press Club in Washington, DC, on 28 November 1984. To avoid the 'quagmire' of Vietnam, Weinberger said troops should only be deployed: as a last resort, with the support of the public and congress, when national security interests were at stake, with the intention of winning, and with clear political and military objectives in place and the means to accomplish them. The 'Powell doctrine' echoed this, insisting upon clear objectives, a risk and cost analysis, the need for domestic and international support, the use of overwhelming military force when force had to be applied and a clear 'exit strategy'. Thus, in its size and massive use of force, the Gulf War remained an attritional conflict.

Marshall also had reservations about the Gulf War's evidence of an RMA. He believed the precision weapons were at an early stage of development, that the military systems hadn't yet been integrated into a fully functioning reconnaissance-strike complex and that the RMA involved more than technology, requiring new operational concepts and new modes of fighting. Marshall argued, therefore, that an RMA was still needed to produce a 'mature precision-strike regime' with 'information dominance' but the success of the Gulf War meant the military themselves saw little reason to change. With a longer vision of transformation in mind and wary of how rapidly advantages were lost, Marshall wanted the USA to complete the RMA to ensure its 'long-term competitive position' (Krepinevich and Watts, 2015:199). In July 1992 Andrew Krepinevich published his preliminary ONA MTR assessment that Marshall had tasked him to undertake, explaining the concept of the 'military technical revolution' and attempting to assess its future development. By then, however, the debate on the future of warfare had widened and RMA theorists found themselves competing with new concepts, in particular the idea of 'information warfare'.

'Information warfare' and 'C2W'

The term 'information war' was used by the scientist Thomas P. Rona, in a 1976 report for Boeing entitled 'Weapons systems and information war', whilst in October 1992 Alan D. Campen published a book describing the Gulf War as 'the first information war' due to its reliance on computing systems and satellites. The first official appearance of the term, however, came on 21 December 1992 when the Department of Defense published DOD directive TS3600.1 entitled 'Information Warfare' (IW). Its focus, however, was upon computer network attacks (CNA) and its classified nature limited its influence and take-up. The concept also appeared in the March 1993 Chairman of the Joint Chiefs of Staff Memorandum of Policy (MOP) 30, which discussed the idea of 'Command and Control Warfare (C2W)', defined as: 'the military strategy that implements information warfare on the battlefield and integrates physical destruction. Its objective is to decapitate the enemy's command structure from its body of combat forces' (Rattray, 2001:315). C2W involves, it explains,

> The integrated use of operations security (OPSEC), military deception, psychological operations (PSYOP), electronic warfare (EW) and physical destruction mutually supported by intelligence to deny information to, influence, degrade, or destroy adversary command and control facilities, while protecting command and control capabilities against such actions.
>
> Fredericks (1997:98)

C2W therefore had links with information warfare (understood primarily as a mode of EW) but saw itself as a separate and broader concept. MOP 30 was taken up by the armed forces who were tasked with developing C2W programs and integrating the concept into their operational war plans.

'Cyberwar'

One of the most important re-theorizations of war in this era was by the RAND researchers John Arquilla and David Ronfeldt, whose essay 'Cyberwar Is Coming!' was published in *Comparative Strategy* in Spring 1993. Its roots lay in the ideas of mobility emphasized in John Boyd's 'Patterns of Conflict' and Starry's 'AirLand Battle'. Its opening set the scene:

> Suppose that war looked like this: Small numbers of your light, highly-mobile forces defeat and compel the surrender of large masses of heavily armed, dug-in enemy forces, with little loss of life on either side. Your forces can do this because they are well-prepared, make room for manoeuvre, concentrate their firepower rapidly in unexpected places, and have superior command, control, and information systems that are decentralized to allow tactical initiatives, yet provide the central commanders with unparalleled intelligence and 'topsight' for strategic purposes.
>
> Arquilla and Ronfeldt (1993:23)

In such a situation, they argue, warfare is no longer about who has the most materiel or forces but about 'who has the best information about the battlefield' (Arquilla and Ronfeldt, 1993:23). For Arquilla and Ronfeldt, the key contemporary technological innovation is 'the information revolution' (Arquilla and Ronfeldt, 1993:25) and this has huge implications for the battlefield.

Writing at a time when the World Wide Web was less than two years old and the first graphical browser was just being released, they have a clear vision of its significance. They recognize 'the advance of computerized information and communication technologies', the 'sea-changes' in 'how information is collected, stored, processed, communicated and presented', the value of information to organizations and the disruptive effect of the network:

> The information revolution ... sets in motion forces that challenge the design of many institutions. It disrupts and erodes the hierarchies around which institutions are normally designed. It diffuses and redistributes power, often to the benefit of what may be considered weaker, smaller actors. It crosses borders and redraws the boundaries of offices and responsibilities. It expands the spatial and temporal horizons that actors should take into account. And thus, it generally compels closed systems to open up.
>
> Arquilla and Ronfeldt (1993:26)

By 'making it possible for diverse, dispersed actors to communicate, consult, coordinate and operate together across greater distances and on the basis of more and better information than ever before' the information revolution favours a move from hierarchical institutions to 'multi-organizational networks', consisting of many parts linked together to act as one (Arquilla and Ronfeldt, 1993:27). These changes, Arquilla and Ronfeldt argue, impact upon 'how societies may come into conflict, and how their armed forces may wage war'.

Arquilla and Ronfeldt's main thesis is that there is a distinction now between 'netwar' and 'cyberwar'. Netwars are 'societal-level ideational conflicts waged in part through internetted modes of communication' (Arquilla and Ronfeldt, 1993:27). They are a non-military 'information-related conflict' between nations or societies that tries to 'disrupt, damage, or modify what a target population "knows" or thinks it knows about itself and the world around it'. As such netwars encompass diplomacy, propaganda, psyops, political and cultural subversion, computer infiltration and media interference etc. being distinguished by 'their targeting of information and communications' (Arquilla and Ronfeldt, 1993:28). In contrast, 'cyberwar refers to conducting, and preparing to conduct, military operations according to information-related principles'. They constitute physical, military actions against an enemy force, attacking their information and communications systems to give oneself a knowledge advantage (Arquilla and Ronfeldt, 1993:30).

Cyberwar, however, involves much more than these C2 (command and control) or C3i (command, communications, control, intelligence) attacks on the opposition's information structures; rather, it is a mode of physical strategy. At its heart are military operations based around a 'decentralization of command and control', with multiple, independently acting, networked agents able to rapidly manoeuvre and strike at larger, more unwieldy enemy forces, using information technology to give 'topsight', or an understanding of the big picture' to commanders who can therefore manage the battlefield complexity (Arquilla and Ronfeldt, 1993:30–1). Using these strategic advantages, Arquilla and Ronfeldt argue, 'modern cyberwarriors should be able routinely to defeat much larger forces in the field' (Arquilla and Ronfeldt, 1993:52). Thus, they conclude: 'cyberwar may be to the twenty-first century what blitzkrieg was to the twentieth century' (Arquilla and Ronfeldt, 1993:31).

This idea of cyberwar is not new, they admit, pointing especially to Mongol battlefield tactics, an historical example. Thus, cyberwar is not primarily technological, but constitutes rather a networked form, with decentralized command in the field and a communication system allowing continual top-level oversight (Arquilla and Ronfeldt, 1993:36). The biggest challenge in implementing this form, Arquilla and Ronfeldt argue, is the need for an 'institutional redesign' of the entire US military, moving from a large, hierarchical institution designed to fight attritional, set-piece battles against similarly organized opponents to a network structure that necessarily erodes all of the older hierarchies (Arquilla and Ronfeldt,

1993:30; 40). In the post-Cold War era, however, the threats the military face have changed and this the military needs to reflect this. 'Most adversaries that the United States and its allies face in the realm of low-intensity conflict … are all organized like networks' (Arquilla and Ronfeldt, 1993:40), they argue:

> The revolutionary forces of the future may consist increasingly of widespread multi-organizational networks that have no particular national identity, claim to arise from civil society, and include some aggressive groups and individuals who are keenly adept at using advanced technology, for communications as well as munitions. How will we deal with that?
>
> Arquilla and Ronfeldt (1993:49)

The current military system will struggle with these opponents, they say, offering an important conclusion: 'Institutions can be defeated by networks. It may take networks to counter networks. The future may belong to whoever masters the network form' (Arquilla and Ronfeldt, 1993:40). The US military are beginning to become aware of the information revolution, Arquilla and Ronfeldt claim, but they are still trying to make these new tools fit into older ways. Instead, they suggest, the organization 'should be restructured, even transformed, in order to realize the full potential of the technology' (Arquilla and Ronfeldt, 1993:41). Cyberwar, they conclude, is not merely a new set of techniques, rather it is 'a new mode of warfare that will call for new approaches to plans and strategies, and new forms of doctrine and organization' (Arquilla and Ronfeldt, 1993:43).

Arquilla and Ronfeldt were among the first to use the term 'cyberwar', though what they mean by it is very different from its contemporary use to describe computer-based denial of service and penetrative attacks. For them it is a mode of real-world combat, though there is some blurring here as they also suggest online attacks may be part of their 'cyberwar' (Arquilla and Ronfeldt, 1993:44–5). Although Arquilla and Ronfeldt's use of the term would be eclipsed, their ideas of warfare would prove influential on other military theorists through the decade, especially upon the concept of 'network-centric warfare'.

'Information warfare', 'C4I' and 'information operations'

By 1994 more attention was being paid to the information revolution and its impact. The first significant public government explanation of information warfare was published in January 1994 in the annual report of the Secretary of Defense, which stated that Information Warfare

> consists of the actions taken to preserve the integrity of one's own information systems from exploitation, corruption or destruction, whilst at the same time exploiting, corrupting, or destroying an adversary's information systems and, in the process, achieving an information advantage in the application of force.
>
> Fredericks (1997:98)

Meanwhile Alvin and Heidi Toffler's *War and Anti-War* was published in 1993, becoming an important text for national security officials, whilst new digital intrusions brought the question of computer security back to prominence. Hackers from the Netherlands accessed US Department of Defense sites from April 1990 to May 1991, whilst between April and

May 1994 there was a series of over 150 intrusions against the Air Force command and control research facility at Rome Laboratory at Griffiss Air Force Base, New York, eventually tracked back to hackers in the United Kingdom. Both a 1994 Defense Science Board (DSB) Summer Study Task Force report, *Information Architecture for the Battlefield*, and a November 1996 DSB report subsequently urged the increasing protection of vulnerable information systems as 'a national strategic concern' (Rattray, 2001:318–19). Winn Schwartau's influential book *Information Warfare*, which focused on computer security and the threat of electronic attacks, was also published in 1994.

The US military was also developing its own concept of 'information war'. Following the initial 1992 DOD directive, the individual services each began to formulate their own IW doctrines. The Air Force was the first to engage with the concept, identifying it as a priority in April 1993 and renaming their Electronic Warfare Centre the Air Force Information Warfare Centre in September, with the stated aim of focusing on 'battlefield information dominance'. On 15 August 1995, the USAF created the 609th Air Information Warfare Squadron, located at Shaw AFB, in South Carolina. Its first doctrinal formulation of this IW was its August 1995 white paper, *Cornerstones of Information Warfare*, which took as its point of departure today's 'information age' (Department of the Air Force, 1995). The paper distinguishes between 'information age warfare', which uses IT as part of its combat operations, and true 'information warfare', which 'views information itself as a separate realm, potent weapon and lucrative target'. It defines IW as 'any action to deny, exploit, corrupt, or destroy the enemy's information and its functions; protecting ourselves against those actions; and exploiting our own military information functions', including within this definition psychological operations, electronic warfare, military deception, physical destruction, security measures and information attacks on data. IW, the paper argues, is important to the Air Force as a means to accomplish its missions and as a target set for its missions and thus needs integrating into Air Force doctrine. 'The ultimate aim?', the paper asks: 'Incorporating information warfare into the way the Air Force organizes, trains, equips and employs'.

The Air Force followed this in October 1996 with another white paper entitled *Information Warfare*, and the November 1996 report *Global Engagement: A Vision for the 21st Century Air Force*. These papers and their ideas were all integrated into the revised *Air Force Basic Doctrine*, published in September 1997, which foregrounded 'information superiority' (understood as 'the ability to collect, control, exploit and defend information, while denying an adversary the ability to do the same') and claimed that 'dominating the information spectrum is as critical to conflict now as controlling air and space, or as occupying land was in the past' (Department of the Air Force, 1997:31). On 5 August 1998, the Air Force codified its vision of IW with the publication of Air Force Doctrine Document (AFDD) 2–5, *Information Operations* (IO: a term introduced by the Army in 1995). The document makes it clear that IO includes and subsumes what was previously called C2W.

AFDD 2–5 collects and summarizes the Air Force's understanding of information operations to date. It begins by asserting the importance of the new realm of information and the need for information superiority. Although not new, thanks to the growth of information systems and the speed of transmission and processing this may 'become central to the outcome of conflicts in the future'. This IO consists of two elements, the authors argue: 'information-in-warfare' (what was understood above as the use of information systems as part of war) and 'information warfare' proper, which involves the defence of or attack upon information systems (Department of the Air Force, 1998:2). IW depends upon 'counter-information' (CI), or the establishment of information superiority 'by neutralizing or influencing adversary information activities' and preventing them taking the advantage. This CI can be 'offensive' (OCI), being designed 'to limit, degrade, disrupt or destroy adversary information capabilities' (through psychological operations, electronic warfare, military deception, and physical or information attacks) or 'defensive' (DCI), in protecting one's own 'information, information systems and information operations from any potential adversary' (Department of the Air Force, 1998:9–10).

The US Army also tried to claim information warfare. In Spring 1994 the Chief of Staff declared 'the Army's institutional response to the demands of the information age is Force XXI, a structured effort to redesign the Army – units, processes and organizations – from those of the industrial age to those of the information age' (Rattray, 2001:323). The ideas were developed in TRADOC pamphlet 525-5 *Force XXI Operations*, published on 1 August 1994, and the position paper *Force XXI, America's Army of the 21st Century*, published on 15 January 1995. The papers recognize the more unpredictable 'dynamic world' of the post-Cold War era and the ongoing advances in information technology and systems. Their response is a Force XXI Army that is characterized by 'doctrinal flexibility', 'strategic mobility' (with lighter, more mobile forces using new information systems), and 'flatter' and 'less rigidly hierarchical' organizations whose main imperative will be 'to gain information and continued accurate and timely shared perceptions of the battlespace' (Department of the Army, 1994:3-1-3-3). Thus, the Army envisage a complete real-time informational system. The Advanced Battlefield Command System (ABCS) and software

> will broadcast battlefield information, as well as information from other sources, and integrate that information, including real-time friendly and enemy situations, into a digitized image that can be displayed graphically in increasingly mobile and heads-up displays. These images will, in essence, depict a unit's actual battlespace. Collective unit images will form a battlespace framework based on shared, real-time awareness of the arrangement of forces in the battlespace …
>
> Department of the Army (1994:3–4)

In this vision, top-level commanders 'will share a common, relevant picture of the battlefield', whilst individual units and soldiers 'will be empowered for independent action because of enhanced situational awareness, digital control, and a common vision of what needs to be done' (Department of the Army, 1994:3–4). Friendly-fire situational awareness will be achieved by 'the digitization of each weapons platform and potentially each soldier' so that commanders will know the position of all forces. All of this will be based upon 'spectrum supremacy', or the complete control of one's own and the adversary's information and information systems (Department of the Army, 1994:3–6).

TRADOC next published pamphlet 525–69, *Concept for Information Operations* on 1 August 1995, formalizing these ideas on 27 August 1996 in FM (field manual) 100–6 *Information Operations*. This paper summarizes again the new challenges of the 'information age' and the new military significance of information and information systems. Information is 'an essential foundation of knowledge-based warfare', it argues: when transformed into capabilities, 'information is the currency of victory' (Department of the Army, 1996:iv). Information, again, allows commanders to 'coordinate, integrate and synchronize combat functions on the battlefield' whilst the achievement of 'information dominance' enables them to define 'how the adversary sees the battlespace, creating the opportunity to seize the initiative and set the tempo of operations' (Department of the Army, 1996:iv).

Where 525–69 and FM 100-6 are important is in the broader vision they bring to IW. FM 100-6 emphasizes the 'global information environment' (GIE) which includes 'all individuals, organizations or systems, most of which are outside the control of the military or National Command Authorities, that collect, process, and disseminate information to national and international audiences' (Department of the Army, 1996:1–2). All military operations take place within the GIE 'which is both interactive and pervasive in its presence and influence'. Many informational agents operate within the GIE, including other governments and their services, NGOs, private voluntary organizations, the global news media, international agencies, social and cultural elements such as religious movements, and even 'individuals with the appropriate hardware and software to communicate with a worldwide audience' (Department of the Army, 1996:1–3). They all do so beyond military control and with the potential to impact upon global perceptions and military operations.

For the military, therefore, the battlefield is not the only space they need to consider. The achievement of 'information dominance' also demands the control of the expanded 'military information environment' (MIE) (Department of the Army, 1996:1–1). This is that part of the GIE that concerns the military operation: 'the environment contained within the GIE, consisting of information systems (INFOSYS) and organizations – friendly and adversary, military and nonmilitary, that support, enable or significantly influence a specific military operation' (Department of the Army, 1996:1–4). Thus, the paper defines information operations as

> continuous military operations within the MIE that enable, enhance, and protect the friendly force's ability to collect, process and act on information to achieve an advantage across the full range of military operations; IO include interacting with the GIE and exploiting or denying an adversary's information and decision capacities.
>
> Department of the Army (1996:2–3)

Hence the Army developed the concept of IO as an alternative to and expansion of IW: as 'a broader approach' recognizing 'that information issues permeate the full range of military operations (beyond just the traditional context of warfare) from peace through global war' (Department of the Army, 1996:2–2). Over the next few years this concept of IO would gradually supersede IW.

The Navy were also considering the issue of information war in the 1990s. They initiated their 'Copernicus' project on 1 October 1990, an evolving C4I (command, control, communications, computers and intelligence) architecture and doctrine. This inspired the Joint Chiefs of Staff's (JCS) 'C4I For the Warrior' project in November 1992, which they described in a June 1993 pamphlet as aiming for 'a fused, real-time, true representation of the battlespace – an ability to order, respond and coordinate horizontally and vertically to the degree necessary to prosecute his mission in that battlespace' (Joint Chiefs of Staff, 1995:1–1). This led to projects by each of the services – the Navy and Marine Corps' 'Copernicus … Forward', the Army's 'Enterprise Strategy' and the Air Force's 'Horizon' programme – aimed at developing integrated information systems and C4I programmes.

Thus, by the mid-1990s, C4I had become an important part of US doctrinal thinking. In 1995 the JCS published *C4I For the Warrior – a 1995 Progress Report*, the Air Force published *Horizons '95. C4I, A Vision for the Future* and the Navy published their report *Copernicus ... Forward C4I for the 21st Century*, each of which explained the current thinking on C4I, the central role of coordinated information in the battlespace to aid the warfighter, the need for interoperability across the services, and the aim of developing a 'common operational picture' of the combat for commanders. Each also referred to real developments in computer-based 'situational awareness' systems trying to implement this informational battlespace. These programmes led to the development of the overall Global Command and Control System (GCCS), which the JCS announced as operational on 30 August 1996. A DOD press release on 30 September made it clear that GCCS was part of a larger project:

> Implementation of GCCS is a major accomplishment. Nevertheless, it is still only an early phase of a larger, more ambitious effort known as C4I (Command, Control, Communications, Computers, and Intelligence) for the Warrior. The C4I for the Warrior vision is to bring together all command, control, communications and intelligence elements in order to speed information flow throughout the battlefield. The ultimate objective is to link U.S. forces, allowing them to communicate and trade information more rapidly than ever before. By putting in place the basic infrastructure to support such a network, GCCS represents a critical step in achieving this objective.
>
> Department of Defense (1996)

Although the relationship between C4I and IW isn't always addressed in these policy documents, as *Copernicus ... Forward C4I for the 21st Century* makes clear, C4I capabilities are the foundation for 'information dominance' and the basis for information warfare (Department of the Navy, 1995a: 12–13; 20).

The Navy also emphasized IW in its other publications. Its 1993 paper *Sonata* outlined a vision of space and electronic warfare as part of the Copernicus project, whilst Naval Department Publication (NDP) 6, *Naval Command and Control*, discussed 'the information revolution' and the rise of information warfare 'as a major new area of conflict' (Department of the Navy, 1995b:62; 64). These publications led up to the Navy's most detailed formulation of IW in its 1998 publication *Information Warfare Strategic Plan*, which built on its current understanding and chimed with the ideas and arguments of the other services. As well as the services, the Joint Chiefs of Staff were also developing their concept of contemporary war. Following MOP 30 (1993), *C4I For the Warrior* (1993), *C4I For the Warrior – a 1995 Progress Report* (1995) and the 1995 *Doctrine For Command, Control, Communications and Computer (C4) Systems Support to Joint Operations*, the JCS established an Information Warfare Special Technical Operations Branch (J-39) within the Operations (J3) Directorate and an Information Assurance Branch (J6K) in the Command, Control, Communications and Computer Systems Directorate (J-6).

In February 1996, the JCS published its *Joint Doctrine for Command and Control Warfare* which explained C2W as 'an application of IW in military operations' and examined the different elements of C2W (Joint Chiefs of Staff, 1996a:v). This was followed by the summer 1996 publication of *Information Warfare: A Strategy for Peace ... The Decisive Edge in War*, which offered another vision of the information age, the central role of information today, the potential of adversaries to exploit information systems and the need to engage in both defensive (IW-D) and offensive (IW-O) information warfare. 'IW', the paper concludes, 'is a reality today and in the future; it impacts societies, governments, and the range of military operations, and all levels of war' (Joint

Chiefs of Staff, 1996b:19). The defensive significance of IW was reinforced by the May 1996 issuance of the Chairman of the Joint Chiefs of Staff Instruction 6505.1A, 'Defensive IW Implementation', which argued for a broader infrastructure protection, but the JCS's most significant publication of this era was the July 1996 release of *Joint Vision 2010 – America's Military: Preparing for Tomorrow.*

'Full-spectrum dominance'

Joint Vision 2010 again begins by considering the 'dynamic' post-Cold War world and the new range of threats it includes, emphasizing how modern systems can empower smaller numbers, making them more dangerous and providing 'asymmetrical counters to US military strengths' (Joint Chiefs of Staff, 1996c:10–11). The US response to this includes the establishment of 'information superiority' and the application of both offensive and defensive information warfare (Joint Chiefs of Staff, 1996c:16). The paper argues this will underlie and enable a fundamental change in warfare. It argues that 'Instead of relying on massed forces and sequential operations, we will achieve massed effects in other ways', such as concurrent, 'higher lethality weapons', 'precision targeting and longer-range systems' and 'improved command and control, based on fused, all-source, real-time intelligence' sending targeting information direct to the most effective weapon system (Joint Chiefs of Staff, 1996c:17–18).

This is a vision of a de-massified military, using agile, responsive units enhanced through new technology and employing four new operational concepts. 'Dominant maneuver' is 'the multidimensional application of information, engagement, and mobility capabilities to position and employ widely dispersed joint air, land, sea, and space forces to accomplish the assigned, operational tasks', using the asymmetrical leverage of positional advantage and decisive speed and tempo to apply decisive force; 'precision engagement' is 'a system of systems' enabling forces to locate their target, provide responsive command and control, generate the desired effect, assess success and retain the ability to re-engage if necessary; 'full dimensional protection' is the ability 'to control the battlespace' to ensure freedom of action for forces whilst providing multi-layered defences for them; and 'focused logistics' is 'the fusion of information, logistics, and transportation technologies' to provide effective responses and the delivery of 'tailored logistics packages and sustainment directly at the strategic, operational, and tactical level of operations' (Joint Chiefs of Staff, 1996c:19–24).

The result of these concepts acting together will be the achievement of massed effects from more dispersed forces. As the paper argues, the result will be 'full spectrum dominance', a concept that implies both informational/electromagnetic and real-world dominance: 'Taken together these four new concepts will enable us to dominate the full range of military operations from humanitarian assistance, through peace operations, up to and into the highest intensity conflict' (Joint Chiefs of Staff, 1996c:25). Importantly, however, this dominance isn't dependent purely on technology. *Joint Vision 2010* also emphasizes the importance of the USA's 'functional expertise, core values and ethical standards' and the 'quality' of its servicemen: 'Even during a time of unparalleled technological advances we will always rely on the courage, determination and strength of America's men and women to ensure we are persuasive in peace, decisive in war, and preeminent in any form of conflict' (Joint Chiefs of Staff, 1996c:34). American technological superiority, therefore, merges with moral and ontological superiority.

The concept of 'full spectrum dominance' would come to replace AirLand Battle as the primary military doctrine. The US Army would follow *Joint Vision 2010* with their own *Army Vision 2010* in 1997, though their 1996 FM 100-6 *Information Operations* paper would itself transform thinking on IW, moving debate towards their broader concept of IO. The Air Force adopted this concept for their 1998 AFDD 2–5 *Information Operations*, as did the Department of Defense in their December 1996 DOD directive 3600.1 'Information Operations (IO)' (which replaced TS3600.1 *Information Warfare*), and the Joint Staff with their Joint Publication 3–13 *Joint Doctrine for Information Operations* on 9 October 1998.The latter accepted a definition of IO now as 'actions taken to affect adversary information and information systems while defending one's own information and information systems. They apply across all phases of an operation, the range of military operations, and at every level of war' (Joint Chiefs of Staff, 1998:vii). The JCS paper *Joint Vision 2020*, published on 30 May 2000, foregrounded this concept of information operations.

Others were also discussing informational conflict. Martin Libicki's *What Is Information War?*, published in August 1995, aimed to evaluate the reality and value of the concept. He claims here that 'information warfare, as a separate technique of waging war, does not exist', breaking it down instead to seven component parts: command and control warfare (C2W), intelligence-based warfare (IBW), electronic warfare (EW), psychological warfare, hacker warfare, economic information warfare and cyberwarfare (Libicki, 1995:x). His conclusion is that only some of these forms are real and effective, such as C2W, EW, IBW and psychological warfare, whilst others such as hacker warfare or cyberwar are either currently limited or are fantastical and unlikely. His conclusion is deliberately deflationary: 'almost certainly there is *less* to information warfare than meets the eye' (Libicki, 1995:96). Though information systems are becoming important, attacks on them are not necessarily 'more worthwhile' and information itself is not 'a medium of warfare' except in very narrow circumstances, such as electronic jamming (Libicki, 1995:xi). IW, he argues, 'has no business being considered as a single category of operations', but rather needs analyzing in its component parts. Even then, he says, 'most of what U.S. forces can usefully do in information warfare will be *defensive*, rather than *offensive*' (Libicki, 1995:97).

Military-related thinkers were more optimistic. RAND theorists Arquilla and Ronfeldt followed their 1993 article with the 1996 book *The Advent of Netwar* and two edited collections, *In Athena's Camp* (1997) and *Networks and Netwars* (2001), whilst Campen, Dearth and Goodden published the edited collection *Cyberwar* (1996). Many of the contributors to these texts discuss military information warfare but, importantly, Arquilla and Ronfeldt's own focus shifts now from 'cyberwar' towards their concept of 'netwar', or 'lower-intensity conflict at the societal end of the spectrum'. Netwar, therefore, 'refers to conflicts in which a combatant is organized along network lines or employs networks for operational control and other communications'. Their view that 'netwar is likely to be the more prevalent and challenging form of conflict in the emerging information age' (Arquilla and Ronfeldt, 1996:vii) and their analysis of the forms and modes of non-state networked warfare would prove important over the following years.

'Network-centric warfare'

By the late 1990s, therefore, the US Military and its theorists had created, defined and explored a range of related concepts to explain contemporary and emerging modes of conflict, including C2W, C3I, C4I, cyberwar, netwar, information warfare, information operations and full spectrum dominance. At the heart of all of these was the role of computing technology and computer-processed information. One obvious reason for this new focus was a recognition of the role of computing in the Gulf War and the ongoing IT-based military RMA, but as the repeated policy paper references to the 'information revolution' and new 'information age' demonstrate, the most important influence was the transformations happening *outside* the military, across the span of global daily life.

Home computing had taken off from the late 1970s but by the early 1990s computers remained expensive luxuries, as stand-alone devices primarily used for office work, word-processing, desktop publishing and gaming. Only a small proportion of the developed world even knew about or had used online services or connections. Tim Berners Lee released information about the 'World Wide Web' to newsgroups on 6 August 1991 but it wasn't until the development of graphical browsers (with Mosaic in 1994 and Internet Explorer in 1995), increased public access and the emergence of cheap, home dial-up provision from the mid-1990s that the internet and the WWW took-off. The number of websites grew from 10 in December 1991 to 10,022 in December 1994, then exploded from 1995, reaching 342,081 by August 1996 and almost 20m by August 2000. Prices of computers fell, as did the cost of home connections, and networked computing became part of everyday life.

But these changes were only part of a broader digital revolution. In *The Language of New Media* (2001) Lev Manovich identifies the 'two trajectories' of mass media and computing, tracing their emergence and development in the nineteeth century and their meeting and merger to create digital media by the end of the twentieth century. The transition was rapid: by the mid-1990s every major analogue medium was moving towards the digital form in its production, distribution and consumption. Cinema effects, editing and postproduction were joined by digital distribution and projection, with *The Last Broadcast* becoming the first feature released digitally, via satellite, to US theatres in October 1998. Newspaper and print publishing had been transformed by IT in the 1980s and by the end of the next decade production, editing and printing were all digital. The rise of the MP3 in the late 1990s and the launch of Napster in 1999 would transform the music industry again, whilst DVDs were launched in 1996 and would replace video cassettes within a few years. Digital cameras and camcorders came down in price and began overtaking analogue sales from the mid-1990s. Standards were agreed for digital radio and television in the early 1990s, with digital radio transmissions beginning from 1995 and digital television beginning in the USA and UK in November 1998.

Changes were noticeable in every person's life. Access to information was easier, many activities and behaviours moved online, communities and friendships formed and one's relationship to the world was transformed. The monopoly over information and experience held by major companies in the broadcast era was broken and new technologies unleashed the productive and distributive capacities of the individual, empowering them as commentators, and critics and producers and sharers of their own material or messages. What the military theorists and strategists were responding to, therefore, wasn't simply a military transformation but more fundamentally *a global, societal and personal revolution* in information technology and information. The influence of this social transformation upon military thinking can be clearly seen in the final re-theorization of conflict in the 1990s: Cebrowski and Garstka's 'network-centric warfare' (NCW).

Arthur K. Cebrowski and John J. Garstka's article 'Network-Centric Warfare: Its Origin and Future' appeared in the US Naval Institute *Proceedings* in January 1998. Their opening claims emphasize the role of social transformations in information in driving military change: 'Here at end of a millennium we are driven to a new era in warfare. Society has changed. The underlying economics and technologies have changed. American business has changed. We should be surprised and shocked if America's military did not' (Cebrowski and Garstka, 1998 (no pagination in online version)). 'We are in the midst of a revolution in military affairs (RMA) unlike any seen since the Napoleonic Age', they argue, quoting Chief of Naval Operations Admiral Jay Johnson's claim of 'a fundamental shift from what we call platform-centric warfare to something we call network-centric warfare'. For Cebrowski and Garstka, this shift is due to 'changes in American society' and in particular changes in business. Echoing the Tofflers they argue that 'nations make war the way they make wealth', hence changes in the economy necessitate changes in military strategy.

Cebrowski and Garstka offer an overview of contemporary changes in economic practices. They note the shift by businesses to 'network-centric operations', which are characterized by a sophisticated backplane architecture that is used to generate informational superiority that can be rapidly acted upon and translated into competitive advantage, 'locking-out' competitors and securing success. Their examples include the US multinational retail corporation Walmart and the (then) Deutsche Morgan Grenfell investment bank, both of which use high-powered networks for timely information and 'to generate a higher-level awareness' within their ecosystems. This allows a real-time coordination of resources, responding immediately to changes in the market.

These same digital-era business strategies would pay off militarily, Cebrowski and Garstka argue: 'Network-centric operations deliver to the US military the same powerful dynamics as they produced in American business'. 'Tactically, speed is critical', they argue. Hence NCW would be built upon 'three critical elements: sensor grids and transaction (or engagement) grids hosted by a high-quality information backplane' that together would give real-time information about the battlespace to units and commanders. Thus, 'operationally, the close linkage among actors in business ecosystems is mirrored in the military by the linkages and interactions among units and the operating environment'. NCW, therefore, 'enables a shift from attrition-style warfare to a much faster and more effective warfighting style characterized by the new concepts of speed of command and self-synchronization'.

'Speed of command' is defined by Cebrowski and Garstka as 'the process by which a superior information position is turned into a competitive advantage'. Essentially, real-time information brings information superiority, enabling faster decision making which, together with the speed and precision of forces, allows you to dominate the battlespace as businesses dominate the market, '"locking-out alternative enemy strategies and "locking-in" success'. With these tactics, they argue, smaller, faster, connected forces can defeat a larger enemy, offsetting 'a disadvantage in numbers, technology, or position'. 'Self-synchronization' is key to this. An efficient information network allows a decentralization of well-informed forces with a high degree of operational independence, able to self-organize from the bottom up and work together for success.

Thus, the authors argue, the model for network-centric warfare has emerged:

> The entry fee is a high-performance information grid that provides a backplane for computing and communications. The information grid enables the operational architectures of sensor grids and engagement grids. Sensor grids rapidly generate high levels of battlespace awareness and synchronize awareness with military operations. Engagement grids exploit this awareness and translate it into increased combat power.
>
> Cebrowski and Garstka (1998)

Many elements of this system are available now, the authors argue, only requiring 'the co-evolution of that technology with operational concepts, doctrine and organization'. NCW, therefore, is an extension of earlier doctrines, including AirLand Battle, cyberwar, IO and full spectrum dominance. Once again it promotes a networked military form, a decentralized structure with connected agents moving rapidly and decisively to disrupt the plans and capacities of larger, more unwieldy opposing forces, and once again it emphasizes an information system allowing coordination and oversight.

The theory of NCW was developed over subsequent years in Alberts, Garstka and Stein's *Network-Centric Warfare* (1999), Alberts, Garstka, Hayes and Signori's *Understanding Information Age Warfare* (2001), and Alberts and Hayes' *Power to the Edge* (2003), but it was the political take-up of the idea that was most significant. The idea of network-centric warfare was adopted by George W. Bush as a presidential candidate in 1999, for whom it represented a two-fold gain of increasing military effectiveness through high-technology and a reduced spending on and the possibility of cuts to the armed forces. It was especially promoted by Donald Rumsfeld who, as Secretary of Defense from 2001, began the project of transforming the military to a lighter, more mobile, informational force. On 27 July 2001, the Department of Defense presented a report on NCW to Congress and on 29 October Rumsfeld instituted a new, Pentagon 'Office of Force Transformation' under the directorship of retired Vice Admiral Cebrowski to promote the transformation of the US military according to NCW principles.

Conclusion

Thus, by the end of the decade the US military had not only rethought warfare but had also begun the process of remaking it. The Gulf War, therefore, represented only an early stage on the path of development. Information warfare and information operations would begin to be systematically used by the USA in the 1999 Kosovo campaign (see chapters 1 and 9), although the absence of a ground offensive limited the force that could be used. It would be in the new millennium, in the War on Terror in Afghanistan and Iraq, therefore, that the full flowering of this military transformation – of 'full spectrum dominance' and network-centric warfare – would be unveiled. NCW would be applied in Afghanistan and would form the basis for Rumsfeld's military plan for the 2003 invasion of Iraq where the USA's light, mobile forces rapidly defeated the Iraqi Army. Securing and holding the entire country against a networked insurgency, however, would prove to be a very different challenge.

Key reading

The references above give the complete list of the military essays, reports and books I've discussed in the chapter. If you wanted to just read the most important I would suggest focusing on Arquilla and Ronfeldt (1993), Cebrowski and Garstka (1998), Department of the Air Force (1995, 1996a, 1996b, 1998), Department of the Army (1995, 1996), and the Joint Chiefs of Staff (1996b, 1996c, 2000). Rattray (2001) and Fredericks (1997) provide overviews of information warfare that review this military literature, but as most of the reports are listed above and easily available online then I would recommend going to the original sources. If you want to follow how Arquilla and Ronfeldt's ideas developed, then look at their later book (1996) and edited collections (1997, 2001). Libicki (1995) and Campen, Dearth and Goodden (1996) also provide important contemporary discussions of information warfare, whilst the debates surrounding Cebrowski and Garstka's idea of 'Network-Centric Warfare' can be followed through Alberts, Garstka, Hayes, and Signori (2001), Alberts, Garstka, and Stein (2000), the Department of Defense (2001) and Alberts and Hayes (2003).

Beyond this, background information about the birth of computing can be found in Agar (2001), Barrett (2006), Campbell-Kelly and Aspray (2004), Ceruzzi (2000), Frauenfelder (2005) and Hally (2005), whilst Hafner and Lyon (1996) and Naughton (2000) cover the invention of networked computing and the internet. Merrin (2014) provides an overview of the 1990s-digital revolution and its broader impact. Rattray (2001) contains information about the military take-up of computing and Dickson (2012), Gibson (2000), Uziel (2015) and Cockburn (2015) provide the best account of Vietnam-era technical developments. Issues around computer security and intrusion are covered in Rattray (2001), Healey (2013) and Schwartau (1996). Important developments in 1970s–80s military strategy can be found in Boyd (1986) and the Department of the Army (1981). Chapman (2003) and Krepinevich and Watts (2015) provide an introduction to the RMA whilst Campen (1992) discusses the use of IT in the 1991 Gulf War.

References

Agar, J. (2001) *Turing and the Universal Machine*, Cambridge: Icon Books.

Alberts, D. S., Garstka, J. J., Hayes, R. E., and Signori, D. A. (2001) *Understanding Information Age Warfare*, CCRP Publication Series, http://www.dodccrp.org/files/Alberts_UIAW.pdf.

Alberts, D. S., Garstka, J. J., and Stein, F. P. (2000) *Network Centric Warfare*, CCRP Publication Series, http://www.dodccrp.org/files/Alberts_NCW.pdf.

Alberts, D. S. and Hayes, R. E. (2003) *Power to the Edge*, CCRP Publication Series, http://www.dodccrp.org/files/Alberts_Power.pdf.

Arquilla, J. and Ronfeldt, D. (1993) 'Cyberwar is Coming!' *Comparative Strategy*, 12(2), pp. 141–55, https://www.rand.org/content/dam/rand/pubs/reprints/2007/RAND_RP223.pdf.

Arquilla, J. and Ronfeldt, D. (1996) *The Advent of Netwar*, Santa Monica, CA: RAND.

Arquilla, J. and Ronfeldt, D. (1997) *In Athena's Camp. Preparing For Conflict in the Information Age*, Santa Monica, CA: RAND.

Arquilla, J. and Ronfeldt, D. (eds.) (2001) *Networks and Netwars*, Santa Monica, CA: RAND.

Barrett, N. (2006) *The Binary Revolution*, London: Weidenfeld and Nicolson.

Boyd, J. (1986) 'Patterns of Conflict', Military Presentation, December, http://www.ausairpower.net/JRB/poc.pdf.

Campbell-Kelly, M. and Aspray, W. (2004) *Computer: A History of the Information Machine*, Oxford: Westview Press.

Campen, A. D. (1992) *The First Information War: The Story of Communications, Computers and Intelligence Systems in the Persian Gulf War*, New York: AFCEA International Press.

Campen, A. D., Dearth, D. H., and Goodden, R. T. (eds.) (1996) *Cyberwar: Security, Strategy and Conflict in the Information Age*, New York: AFCEA International Press.

Cebrowski, A. K. and Garstka, J. J. (1998) 'Network-Centric Warfare: Its Origin and Future', *Proceedings*, USS Naval Institute, *January*, pp. 28–35, http://www.kinection.com/ncoic/ncw_origin_future.pdf.

Ceruzzi, P. E. (2000) *A History of Modern Computing*, London: MIT Press.

Chapman, G. (2003) 'An Introduction to the Revolution in Military Affairs', XV Amaldi Conference on Problems in Global Security, Helsinki, Finland, http://www.lincei.it/rapporti/amaldi/papers/XV-Chapman.pdf.

Cockburn, A. (2015) *Kill Chain: Drones and the Rise of High-Tech Assassins*, London: Verso.

Department of Defense. (1996) 'Global Command and Control System Adopted', News Release, No. 552–96, 26 September, http://www.defense.gov/Releases/Release.aspx?ReleaseID=1049.

Department of Defense. (2001) *Network Centric Warfare Report to Congress*, 27 July, Washington, DC, http://www.dodccrp.org/files/ncw_report/report/ncw_main.pdf.

Department of the Air Force. (1995) *Cornerstones of Information Warfare*, August, Washington DC: Headquarters, Department of the Air Force, http://www.c4i.org/cornerstones.html.

Department of the Air Force. (1996a) *Information Warfare*, October, Washington, DC: Headquarters.

Department of the Air Force. (1996b) *Global Engagement: A Vision for the 21 Century*, November, Washington, DC: Headquarters, Department of the Air Force, http://www.au.af.mil/au/awc/awc gate/global/global.pdf.

Department of the Air Force. (1997) *Air Force Basic Doctrine*, Air Force Doctrine Document 1, September, Washington, DC: Headquarters, Department of the Air Force, http://www.globalsecur ity.org/military/library/policy/usaf/afdd/1/afdd1.pdf.

Department of the Air Force. (1998) *Information Operations*, Air Force Doctrine Document 2–5, 5 August, Washington, DC: Headquarters, Department of the Air Force, http://www.globalsecurity. org/military/library/policy/usaf/afdd/2-5/afdd2-5.pdf.

Department of the Army. (1981) *US Army Operational Concepts. The AirLand Battle and Corps 86. TRADOC Pamphlet 525-5*, 25 March, Fort Monroe, VA: Headquarters, United State Army Training and Doctrine Command, http://cdm16040.contentdm.oclc.org/cdm/ref/collection/ p4013coll9/id/656.

Department of the Army. (1994) *Force XXI Operations. TRADOC Pamphlet 525-5*, 1 August, Fort Monroe, VA: Headquarters, United States Army Training and Doctrine Command. PDF URL: 314276-4.

Department of the Army. (1995) *Force XXI, America's Army of the 21 Century*, 15 January, Fort Monroe, VA: Office of the Chief of Staff of the Army, Louisiana Maneuvers Task Force, http:// www.dtic.mil/dtic/tr/fulltext/u2/a402570.pdf.

Department of the Army. (1996) *FM 100-6, Information Operations*, 27 August, Washington, DC: Headquarters, Department of the Army, http://www.bits.de/NRANEU/others/amd-us-archive/ fm100-6(96).pdf.

Department of the Navy. (1995a) *Copernicus... Forward C4I For the 21st Century*, 9 January, Washington, DC: Headquarters, Department of the Navy, http://www.dtic.mil/dtic/tr/fulltext/u2/ a390355.pdf.

Department of the Navy. (1995b) *Naval Command and Control*, Naval Doctrine Publication 6, 19 May, Washington, DC: Headquarters, Department of the Navy, http://www.dtic.mil/dtic/tr/fulltext/ u2/a304321.pdf.

Dickson, P. (2012) *The Electronic Battlefield*, Takoma Park, Maryland: Fox Acre Press.

Frauenfelder, M. (2005) *The Computer: An Illustrated History*, London: Sevenoaks.

Fredericks, B. E. (1997) 'Information Warfare at the Crossroads', *Joint Force Quarterly (JFQ)*, Spring, pp. 97–103, http://www.dtic.mil/dtic/tr/fulltext/u2/a529170.pdf.

Gelb, L. H. (1992) 'Foreign Affairs: Who Won the Cold War?' *New York Times*, 20 August, http:// www.nytimes.com/1992/08/20/opinion/foreign-affairs-who-won-the-cold-war.html.

Gibson, J. W. (2000) *The Perfect War: Technowar in Vietnam*, New York: The Atlantic Monthly Press.

Hafner, K. and Lyon, M. (1996) *Where Wizards Stay Up Late: The Origins of the Internet*, New York: Touchstone.

Hally, M. (2005) *Electronic Brains. Stories From the Dawn of the Computer Age*, London: Granta Books.

Healey, J. (ed.) (2013) *A Fierce Domain: Conflict in Cyberspace 1986–2012*, Vienna: Atlantic Council, Cyber Conflict Studies Association.

Joint Chiefs of Staff. (1995) *Doctrine For Command, Control, Communications and Computer (C4) Systems Support to Joint Operations*, Joint Pub 6-0, Washington, DC: Joint Staff, http://www.bits. de/NRANEU/others/jp-doctrine/jp6_0(95).pdf.

Joint Chiefs of Staff. (1996a) *Joint Doctrine for Command and Control Warfare (C2W)*, Joint Pub 3-13.1, 7 February, Washington, DC: Joint Staff, http://www.iwar.org.uk/rma/resources/ c4i/jp3_13_1.pdf.

Joint Chiefs of Staff. (1996b) *Information Warfare. A Strategy For Peace ... The Decisive Edge in War*, Washington, DC: Joint Staff. PDF URL: ADA318379.

Joint Chiefs of Staff. (1996c) *Joint Vision 2010 – America's Military: Preparing For Tomorrow*, Washington, DC: Joint Staff, http://www.dtic.mil/jv2010/jv2010.pdf

Joint Chiefs of Staff. (1998) '*Joint Doctrine for Information Operations*,' Joint Pub 3-13, 9 October, Washington, DC: Joint Staff, http://www.c4i.org/jp3_13.pdf.

Joint Chiefs of Staff. (2000) *Joint Vision 2020. America's Military: Preparing For Tomorrow*, Washington, DC: Joint Staff, http://www.pipr.co.uk/wp-content/uploads/2014/07/jv2020-2.pdf

Krepinevich, A. and Watts, B. (2015) *The Last Warrior: Andrew Marshall and the Shaping of Modern American Defense Strategy*, New York: Basic Books.

Libicki, M. (1995) *What is Information Warfare?* Washington, DC: National Defense University Press.

Merrin, W. (2014) *Media Studies 2.0*, Abingdon, Oxon: Routledge.

Naughton, J. (2000) *A Brief History of the Future. The Origins of the Internet*, London: Phoenix Books.

Rattray, G. J. (2001) *Strategic Warfare in Cyberspace*, London: MIT Press.

Schwartau, W. (1996) *Information Warfare*, New York: Thunder's Mouth Press.

Uziel, D. (2015) 'Igloo White: The Automated battlefield', *The Future of Things*, http://thefutureof things.com/3902-igloo-white-the-automated-battlefield/.

4 The War on Terror
Reporting 9/11 and Afghanistan

Breaking news ....

CNN's coverage is typical, breaking abruptly into the tail-end, early-morning financial news to show images of the burning tower. The smooth, predictable flow of breakfast news, designed to wake you up slowly and ease you into the working day, is shattered. The soft-focus, still-sleepy ads for mortgage products and loans cut without warning to those ash-grey images of the smoking tower against the heat-haze of the morning sky: '… This just in, you are looking at, er obviously a very disturbing live shot there. That is the World Trade Centre and we have unconfirmed reports this morning that a plane has crashed into one of the towers of the World Trade Centre' (CNN, 2001). The shift in tonality is like a shift in reality itself. There is the immediate confusion of the live event and that lag behind the real as the anchors try to comprehend what they're seeing. Their voices are calm but unbelieving as they search for the facts and words to describe what's happening. Eyewitnesses are quickly found and questioned about the plane and its size, trajectory and appearance, all against the background of the silent, smoking images: 'It is a remarkable scene, as we're seeing right now. Flames still coming out of the windows, black smoke billowing from what appears to be all sides, er, obviously windows shattered and steel jutting out from the structure …'.

Those watching at home are thrown into the heart of the event: into the unpredictability, uncertainty, horror and excitement of an unfolding, real-time emergency. CNN already has the headline 'World Trade Centre Disaster', but the seriousness of the event isn't under-stood yet. Live television can't even follow what's happening. Fifteen minutes into CNN's coverage, during a telephone interview, a plane swoops in from the right to hit the second tower. CNN's anchor misses it completely, attributing the on-screen fireball to the original plane. It takes a few seconds before he's corrected –

> Umm, I … we're getting word that perhaps … er, let me just put Winston on hold for just a moment … Now one of our producers said perhaps a second plane was involved and let's not, let's not speculate to that point but at least put it out there that perhaps that may have happened.

He's unconvinced, but notices the second tower is on fire – 'now this was not the case, am I correct? a couple of moments ago …' – and says now they'll check on the second plane theory, 'if perhaps that had happened'.

Even with the second plane confirmed they still don't understand what's happening. Replaying the video, the anchor says,

> … there is the second plane; *another* passenger plane hitting the World Trade Centre. These pictures are frightening indeed. These are just minutes between each other, so naturally you will guess and you will speculate and perhaps ask the question if some type of navigating equipment is awry that two commuter planes would run into the World Trade Centres at the same time …

Others were faster to grasp the significance. BBC1 waited until the end of *Neighbours*, going over to BBC News 24 at 2.10pm (9.10 EDT) soon after the second plane struck. Within minutes a commentator suggests, 'certainly, looking at that second plane, I think we are now talking about a terrorist attack, unless that was some bizarre coincidence' (BBC, 2001a). By then the images were being carried internationally and over the following hours millions around the world would watch this global event unfold. Few at the time had heard of the name al-Qaeda, but that would soon change.

Box 4.1 al-Qaeda 1988–2001

Al-Qaeda is a Sunni Muslim, radical Salafist-Jihadist, Islamist terrorist organization founded by Osama bin Laden in 1988, though not operationalized until 1998. It was responsible for a number of significant anti-western attacks in the late 1990s and organized and executed the 9/11 attacks. Commentators disagree, however, as to al-Qaeda's size, its real level of organization and its threat. For some it constitutes a dangerous, ongoing, highly organized force at the heart of a global web of terror, whilst others see it as a smaller, more amorphous organization, offering support to individuals drawn to its 'brand' whose status as al-Qaeda operatives is more complex. The roots of al-Qaeda lie in the Afghan mujahideen resistance against the December 1979 Soviet invasion and the ideas of its founder and leader, Bin Laden.

Osama bin Laden was born in Saudi Arabia in 1957. His father was a Yemeni who had moved to Saudi Arabia and built a family fortune through a successful construction company, establishing in the process close connections with the Saudi royal family. Influenced by Palestine and the global situation of Muslims, Osama bin Laden became more religious and more radical. He joined the Muslim Brotherhood, read the works of the anti-western founder of radical Islam, Sayyid Qutb (1906–1966) and was taught by Qutb's brother at university. Following the 1979 Soviet invasion of Afghanistan, Bin Laden visited Peshawar, Pakistan, for the first time in 1980 thereafter splitting his time between Saudi Arabia and Peshawar as he devoted himself to fundraising for the Afghan mujahideen resistance. It was in Peshawar that he met the radical Islamist Palestinian Abdullah Azzam (1941–1989), whose influential 1979 fatwa *Defence of the Muslim Lands, the First Obligation After Faith* had famously promoted offensive jihad, declaring the fight to free Afghanistan from the Soviets the personal duty of every Muslim. In 1984 Bin Laden settled in Peshawar, establishing, with Azzam, Maktab al-Khidamat (MAK), or the 'Services Bureau', an organisation for jihadi recruitment, aid and fundraising. Its primary aim was to build an Arab jihadi force, serving as a hostelry for Arab fighters and a base for Azzam's publishing. With his family fortune, Bin Laden became the chief financier of the Arab Afghan jihad, also helping construct training camps in Pakistan, and later Afghanistan.

From Azzam, Bin Laden developed an international perspective. Unlike the Afghan mujahideen, whose main concern was liberating their own country, Azzam saw Afghanistan as a starting point for the recapture of all lost Muslim lands and even for a war against all unbelievers. Azzam also promoted the idea of martyrdom and its rewards among the Arab fighters, extolling both the life and death of the jihadi in his writings. Although Arab fighters played only a minor role in the Afghan conflict and were disliked by the mujahideen, by 1986 Bin Laden's Arab brigade was involved in fighting around their 'Lion's Den' camp in Afghanistan and their defeat of Soviet troops convinced Bin Laden that he was building a vanguard for a global jihad. Bin Laden was increasingly drawn, however, towards two Egyptian militants, Ayman al-Zawahiri and Dr Fadl (Sayyed Imam al-Sharif), the leaders of 'Islamic Jihad'. Their more radical interpretations of faith, especially concerning the expelling of Muslims for apostasy ('takfir') and use of this concept to justify killing other Muslims, led them into conflict with Azzam. Azzam wanted to take jihad back to Israel next and opposed takfir as a regressive idea, creating 'fitna', or strife between Muslims.

The Soviet Union began to withdraw from Afghanistan in 1988, leading to a debate in the jihadi movement as to its future. Azzam had written in 1987 of the need for a 'vanguard' to carry jihad forward, though his claim that 'This vanguard constitutes the strong foundation (*al qaeda al-sulbah*) for the expected society' referred only to a mode of activism rather than a group. His idea, however, inspired a meeting in Peshawar on 11 August 1988 where Bin Laden and other radicals, including al-Zawahiri, Dr Fadl and Mohammed Atef, decided to form an organization that would carry on global jihad after the Soviet withdrawal. On the 20 August, the group met again to establish al-Qaeda al-Askiriya ('the military base') with Bin Laden as its 'Emir', due to his wealth and influence. The fifteen people who attended these two meetings formed the core of what was soon shortened to al-Qaeda ('the base'). Azzam himself was increasingly marginalized and his death by a car-bomb on 24 November 1989 has often been blamed on Islamic rivals. After the Soviets left Afghanistan in 1989, Bin Laden returned to Saudi Arabia, although his anger at the stationing of US troops there during the 1991 Gulf War and his support for radical Islamist reformers led him into conflict with the authorities. He returned briefly to Pakistan in 1992 then accepted an invitation that year to join Hassan al-Turabi's new Islamic society in Sudan.

During this time, Bin Laden's hatred of 'the far enemy' increased. Whereas most radicals focused on 'the near enemy' of corrupt or hostile Muslim governments, Bin Laden's al-Qaeda blamed the west and Israel (the 'Crusaders' and the Jews) for the global position of Islam, seeing the USA especially as the leading contemporary anti-Islamic force. In interviews and speeches in the 1990s Bin Laden listed his major grievances with the USA as their military presence in Saudi Arabia, their support for Israel and their policy towards Iraq. Bin Laden's role in terrorism at the time was primarily as a source of funds and although a number of terrorist acts of the era were subsequently attributed to al-Qaeda its involvement has been disputed. Al-Qaeda is claimed to have carried out its first bombings on 29 December 1992 in response to US intervention in Somalia, at two Yemeni hotels where US troops had been staying, but others suggest Bin Laden merely provided funds for an attack planned and executed by Tariq al-Fahdi. Bin Laden also provided funds for Sheikh Omar Abdul Rahman's 'Islamic Group' that carried out the 1993 World Trade Centre bombing although he wasn't involved in that plot. Al-Qaeda fighters have been claimed to have taken part in the 'Battle of Mogadishu' in Somalia in October 1993 that left 18 American troops dead, although little evidence has been provided. Similarly, two bomb blasts in Saudi Arabia in 1995 and 1996, that killed five and 19 US servicemen respectively, are sometimes, erroneously, attributed to al-Qaeda.

Bin Laden's stay in Sudan was unsuccessful and costly and international pressure on Turabi forced him to leave so he returned to Afghanistan on 18 May 1996. He had no relationship with the Taliban, however, who had emerged in his absence. They were suspicious of his activities, especially after his declaration of war against Americans for occupying Saudi Arabia on 23 August 1996. His global vision of jihad clashed with the parochial Taliban, whose only interest was the imposition of Sharia law in Afghanistan. By 1997 the failure of the Islamic Jihad group in Egypt pushed al-Zawahiri towards closer relations with al-Qaeda and towards Bin Laden's internationalist position. Al-Qaeda would eventually absorb the Islamic Jihad group in June 2001 with al-Zawahiri as the second in command. The new centrality of al-Zawahiri to al-Qaeda was demonstrated by the publication of his 'World Islamic Front For Jihad Against Jews and Crusaders' as fatwa on 23 February 1998. Signed by Bin Laden, it offered a religious justification for the killing of Americans everywhere.

One of the weapons Bin laden would use for this strategy would be the Afghan training camps. There were over 120 camps operating in Afghanistan and Pakistan by 2001, run by a variety of militant groups and funded by Gulf States, by Pakistan and by private donors. Most were training militants to fight for the Taliban against the Northern Alliance, or to return home to fight local battles. Between 10 and 20,000 recruits passed through Afghan camps between 1996 and 2001. Militant training involved both physical and ideological elements. Physical training included a boot camp and basic military training, with the possibility of extra classes (such as guerrilla war and assassination), whilst ideological training involved inculcation into hard-line Deobandi Islam or radical-utopian, Salafist jihadism. Bin Laden was associated with two camps in particular, al-Farooq and Abu Jindal, which operated as specialist training camps for those desiring or selected for terrorist attacks on the 'Zionist-Crusader alliance'. The question of what al-Qaeda was, therefore, is complex. It primarily served as a 'base' for training, funding and organizational help for independent militants who were expected to initiate and execute their own plans. Although it comprised a permanent core of militants, those it trained were not necessarily 'al-Qaeda fighters'. Al-Qaeda's level of involvement varied from providing training and help to a small number of acts which were more directly-orchestrated as part of its war on 'the far enemy'.

Although al-Qaeda had existed as a nominal organization since 1988, the creation of the 'World Islamic Front' marked the moment of its real launch and operationalization as a terrorist group. Its first acts were the twin bombings of the East African US embassies in Nairobi, Kenya and Dar es Salaam, Tanzania on 7 August 1998, which killed 213 and 11 people respectively, injuring over 4,500 in Nairobi. These attacks drew the ire of the USA. On 20 August President Clinton responded with 'Operation Infinite Reach': cruise-missile attacks on a Sudanese chemical plant and Afghan training camps. Bin Laden wasn't present, however, and the strikes had little effect other than to boost his global reputation among radicals. The Taliban regime came under pressure from the USA to hand over Bin Laden. UN Resolution 1267, passed on 15 October 1999, designated Bin Laden and his associates as terrorists and instituted a sanctions regime against the Taliban government. Bin Laden's funds, the ancient Pashtun honour code 'Pashtunwali' and notions of 'nanawatai' (sanctuary) left them obliged to protect him, but this became harder as al-Qaeda operations continued. Although the bombing of the USS *The Sullivans* in Aden harbor, Yemen, failed, on 12 October 2000 a suicide skiff-bomb attack on the USS *Cole* in the same harbour killed 17 US sailors and wounded 39.

Bin Laden's anti-western world-view was not widely shared by radical Islamic groups in the era, most of whom were focused on national issues and domestic rather than transnational jihad. Few Muslims genuinely believed in his claim of a western war on Islam and there is evidence Bin Laden planned 9/11 to provoke a US over-reaction and create precisely that war that he claimed existed. Emboldened by the Jihadi-movement's defeat of one superpower, the USSR, in Afghanistan, he believed a weak USA, unable to stomach casualties (as Beirut in 1983 and Mogadishu in 1993 had shown) would soon capitulate, leading to the success of Islamism across the Muslim world. By the Spring of 1999 Bin Laden had agreed to Khaled Sheikh Mohammed's idea of using hijacked planes as weapons. Two cells, one in Hamburg and one comprised of two Saudis and two Yemenis, began to be organized and trained to carry out the plot. On 9 September 2001, al-Qaeda suicide bombers killed Ahmed Shah Massoud, the leader of the anti-Taliban 'Northern Alliance', establishing the Taliban's debt to Bin Laden two days before 9/11. Bin Laden and al-Zawahiri then fled into the mountains to watch the effect of their attacks on September 11 using a TV and a satellite dish.

9/11 and after

On September 11, 2001 the USA suffered four coordinated suicide attacks by Islamic terrorists trained, funded and organized by al-Qaeda. Nineteen hijackers took control of four passenger planes and aimed them at key economic and political landmarks. American Airlines Flight 11 hit the World Trade Centre North Tower (1 WTC) at 8.46am EDT, United Airlines Flight 175 hit the South Tower (2 WTC) at 9.03am and at 9.37am American Airlines Flight 77 hit the west side of the Pentagon in Arlington County. High in the hills of Logar province, Afghanistan, the leader of al-Qaeda, Osama bin Laden, and his entourage were watching US news via a satellite receiver. After the first plane hit, Bin Laden held up two fingers to indicate more was coming; after the second plane, he held up three fingers; after the Pentagon was struck he held up four fingers. The final attack never came. At 10.03am United Airlines Flight 93 aiming for Washington's Capitol Hill, crashed into a field near Shanksville, Pennsylvania after passengers stormed the cockpit.

The attacks had stopped, but the images continued to come: pictures of the smoking Pentagon, of ground-level panic, of the emergency services at the scene, of falling people, of the collapsing South Tower at 9.58am, of the lone tower still burning, of its collapse at 10.28am, of the collapse of 7 WTC at 5.20pm and the post-apocalyptic television images of the dust and ash-covered public streaming, walking and running out of the debris.

Through the whole day the world followed the confusion and panic, the false reports of new attacks and incidents, the lockdown of the USA, the passage of President Bush back to the White House, the repeated television images, the ongoing rescue attempt, the street interviews with Manhattanites, the commentary and speculation, and the disbelief at the changed skyline. In total 2,996 people were killed in the attacks, including the 19 hijackers, 343 firefighters and 72 law enforcement officers. Over 6,000 people were injured. The remains of the World Trade Centre burned for a hundred days.

The success and scale of the attacks unnerved Bin Laden who had expected, at most, the destruction of the planes and the floors of the tower they hit. No-one had foreseen the towers' fall, the number of casualties (believed at the time to be in the tens of thousands) or the extent of the US anger. Hence, perhaps, Bin Laden's attempt to deny involvement in his first public statement on the events on 16 September 2001. However, whilst the attacks on the World Trade Centre represent the deadliest terrorist attack to date, the physical death and damage was only one part of the event. The power and significance of the attacks owed as much to its spectacular media dissemination and to the impact of those images around the world. 9/11 was more than a physical attack on Manhattan and the Pentagon: the coverage and its images were experienced as a shock and assault and a symbolic humiliation.

The link between terrorism and the media has been widely-noted. In her speech to the American Bar Association on 15 July 1985 British PM Margaret Thatcher complained that terrorists thrive on media coverage, famously arguing 'we must try to find ways to starve the terrorist and the hijacker of the oxygen of publicity on which they depend' (Thatcher, 1985). Almost a century earlier, Gustave Le Bon had recognized the power of spectacle on the public mind, especially in regard to public symbols. Writing in *The Crowd* (1895) Le Bon argues that a recent Parisian outbreak of influenza that killed 5,000 people made little impression as it 'was not embodied in any visible image' but was only apparent statistically. However:

> An accident which should have caused the death of only five hundred instead of five thousand persons, but on the same day and in public, as the outcome of an accident appealing strongly to the eye, by the fall, for example, of the Eiffel Tower, would have produced, on the contrary, an immense impression on the imagination of the crowd.

His conclusion is that the facts about the world are less important than 'the way in which they take place and are brought under notice', with the most memorable event being that which possesses a 'startling image which fills and besets the mind' (Le Bon, 2005:63–4).

There had been a previous attempt to bring down the World Trade Centre when, on 26 February 1993, a group of radical Islamists led by Ramzi Yousef (Abdul Basit Mahmoud Abdul Karim) detonated a truck bomb in a public garage beneath the towers, killing six and injuring 1,042. Bin Laden had no direct connection with this attack although Yousef had received terrorist training in Afghanistan and financing from his uncle Khalid Sheikh Mohammed who would later become 'the principal architect of 9/11' (National Commission on Terrorist Attacks Upon the United States, 2004:145). Writing in 1997 Paul Virilio described this attempt as inaugurating a new 'age of large-scale terrorism', for had it succeeded it could have caused 20,000 deaths – 'the equivalent of a strategic missile strike' (Virilio, 1997:170). But destruction alone isn't enough, he says: destroying the World Trade Centre without anyone knowing about it would be 'pointless'. Hence, Virilio says, the terrorists 'anticipated the information war described by the Pentagon', recognizing the importance of that new, fourth, military front of information and understanding that the physical explosion must be 'simultaneously coupled to a multimedia explosion' (Virilio, 1997:174). September 11, 2001 was the realization of that terroristic infowar and multimedia explosion.

As in infowar, the 9/11 terrorists seized control of both the airspace and airwave space: they not only hijacked the planes, they hijacked the media too. Just as they used the airspace to deliver their explosive weapons – the impact of the planes – so they used the airwaves to deliver their implosive weapons – the impact of the images across the globe. Hence Baudrillard's description of 9/11 as 'the absolute event' (Baudrillard, 2002:3). 'What stays with us, above all else, is the sight of the images' (Baudrillard, 2002:26), he says. Indeed, the power of the event was precisely that of the image and its global transmission: 'the image consumes the event, in the sense that it absorbs it and offers it for consumption' – it gives it 'unprecedented impact', but impact as 'an image-event' (Baudrillard, 2002:27). Thus 'the fascination with the attack is primarily a fascination with the image' (Baudrillard, 2002:28–9), Baudrillard argues, albeit it an image with the real 'superadded' to it, 'like a bonus of terror, like an additional *frisson*: not only is it terrifying, but, what is more, it is real (Baudrillard, 2002:29)'.

Television contributed to this *frisson*. In its live broadcast of the attacks, in its real-time images, iconic shots of the smoking skyline, circling helicopter close-ups of the towers, commentary, speculation, repeated shots of the plane strike and the collapsing towers, and in the mixture of horror and awe with which people watched, television offered a spectatorial, scopophilic pleasure. As Slavoj Žižek argues in *Welcome to the Desert of the Real* (2002), 'The same shots were repeated *ad nauseam*, and the uncanny satisfaction we got from it was *joissance* at its purest' (Žižek, 2002:12). Sky News, for example, continually replayed the video of the second plane, slowing it down to relish the detail and commenting how 'slow-motion pictures reveal the *full-force* and *horror*' of the crash...' (Merrin, 2005:102).

The German composer Karlheinz Stockhausen was widely-criticized for his comments days later that 9/11 was 'the greatest work of art that has ever existed' and his comparison of the attacks to an artistic 'performance' (BBC, 2001b), but there was some truth in this. Just as the twentieth-century avant-garde used cabarets, events and 'happenings' in the world to break through the structures of everyday life and reveal reality anew, so terrorism now uses the media to the same effect. By virtue of its spectacular enactment, its break into reality and the global transmission that overturned our world, 9/11 was a spectacular performance of violence. Baudrillard makes this point, describing terrorism as our contemporary 'theatre of cruelty': one confronting and challenging us with the violence of its spectacle. As he says about the images of 9/11, here, 'the spectacle of terrorism forces the terrorism of spectacle upon us' (Baudrillard, 2002:30).

Thus, the spectacle was central to the event, forming and framing our experience. Many compared the events to cinema and especially to those Hollywood disaster movies that had anticipated the destruction of the city. Interviewed on NBC moments after the second lane struck, Jennifer Oberstein, for example, could only stutter, 'I've never seen anything ... It looks like a movie....' (NBC, 2001). As Žižek says: 'for us, corrupted by Hollywood, the landscape and shots of the collapsing towers could not but be reminiscent of the most breath-taking scenes in big catastrophe productions' (Žižek, 2002:15). For Žižek, this was not an intrusion of the real shattering our illusory bubble but 'quite the reverse':

it was before the WTC collapse that we lived in our reality, perceiving Third World horrors as something which was not actually part of our social reality, as something which existed (for us) as a spectral apparition on the (TV) screen – and what happened on September 11 was that this fantasmatic screen apparition entered our reality. It is not that reality entered our image: the image entered and shattered our reality.

Žižek (2002:16)

What the attacks did was bring into the safe, ordered, everyday life of the west the images of aerial destruction and death it was used to seeing on television happening in other countries.

In many ways, therefore, 9/11 represented the reversal of the west's own model of war. As in the Gulf, this was a unilateral, technologically-realized, devastating aerial strike that admitted no possibility of response, instantly crippling and humiliating the symbolic heart of American economic and political power, bringing an instant, live, televisual victory. It was an attack that not only used but was also amplified by the media coverage, whose endless replays 'created a montage effect not of a single cruise-missile strike but of dozens of strikes on New York City: of an *urban storm* reversing that 'desert storm' unleashed on the Arab world in 1991' (Merrin, 2005:106).

Except here it was American not Arab lives on the ground and the pilots were committed not to their own safety but their certain death. For Baudrillard the efficacy of the action lies in this combination of modern technology with 'the absolute weapon of death'. In a 'zero-death' culture, the gift of one's own life is 'a definitive act', creating 'an irreducible singularity' in a system that cannot cope with it (Baudrillard, 2002:8–9; 16). Thus, this wasn't a purely destructive act but rather a symbolic one, Baudrillard says. Nor was it 'cowardly'. In contrast to the west who seek 'the impersonal elimination of the other' in their wars and conflicts, this involved a 'dual pact' with the enemy (Baudrillard, 2002:25–6). It was a personal 'sacrificial' (Baudrillard, 2002:17) act aimed at reversing humiliation and humbling and destroying a hegemonic power: 'It is that power which has humiliated you, so it too must be humiliated' (Baudrillard, 2002:26).

9/11 represented, therefore, a dual attack on the USA, encompassing a physical attack upon its people and the imagic, globally-mediated humiliation of its power. Although the former was the primary spur to subsequent US action, the latter would play an important part in US foreign policy over the following years.

Once the attacks were over the USA faced three tasks: (1) of ascertaining who was responsible for them and what broader allegiance or support they had, (2) of interpreting the attacks and deciding upon an appropriate response, and (3) of enacting that response, in particular the legal, diplomatic or military actions necessary to hold those responsible to account and to guarantee the USA's future security.

Although no one claimed responsibility for the attacks, the identification of the terrorists and their affiliation proved easy. Intelligence and security experts on the day suggested al-Qaeda as the most likely group behind the attacks whilst the NSA and German intelligence agencies intercepted communications that pointed to Bin Laden. Within a matter of days, the USA's 'Operation PENTTBOM' had identified the hijackers (from their reported seat numbers, from the undisguised flight and credit card records, from evidence from a rented car, from evidence in Mohamed Atta's luggage and from recovered passports etc.) and the intelligence services quickly linked them to al-Qaeda, mapping their connections to radical Islamic activities and groups.

The background to the attacks was eventually pieced together, with its roots lying in the failed, January 1995 'Bojinka plot'. Following the 1993 bomb attempt on the WTC the escaped bomber, Ramzi Yousef, and his uncle Khalid Sheikh Mohammed plotted a three-phase terrorist attack, funded in part by Bin Laden, that would include the assassination of Pope John Paul II, the bombing of 11 airliners en route to the USA and the hijacking and crashing of a plane into the CIA headquarters. 'Oplan Bojinka' was discovered, but in 1996 Khalid Sheikh Mohammed took the idea of using hijacked planes as weapons to Bin Laden, who would later authorize it in a 1999 meeting as an al-Qaeda operation. The attacks were funded by Bin Laden, planned by Khalid Sheikh Mohammed and organized by al-Qaeda's military chief Mohammed Atef. In a video-tape released on 29 October 2004, Bin Laden publicly claimed responsibility for the attacks for the first time.

For America, the meaning of the attacks was also simple. After the second plane hit, it became obvious that, as Chief of Staff Andrew Card told President Bush, 'America is under attack', but what was more significant was the conceptualization of these attacks as an act of 'war'. Tom Brokaw on NBC responded to the second plane with the words, 'This is war'; Senator John McCain told CNN this was 'an act of war', and the same phrase appeared the next day on the covers of the *New York Post* and *USA Today*. In a statement on 12 September, President Bush agreed: 'The deliberate and deadly attacks, which were carried out yesterday against our country, were more than acts of terror. They were acts of war' (Bush, 2001a). References to Pearl Harbor were also common. The last words Bush wrote in his journal on September 11 before sleeping were 'The Pearl Harbor of the twenty-first century took place today', and the metaphor, with its historical resonance as the moment when the USA joined World War II, helped frame the events as an implicit declaration of war.

The implications of this framing were important. Whereas a terrorist act could be prosecuted legally as a crime, an act of war demanded an equivalent national and military response. Bush's claim on 12 September that 'We will make no distinction between the terrorists who committed these acts and those who harbor them' already suggested a broader target, and this was confirmed on 20 September in his address to Congress with his declaration of 'a war on terror' (Bush, 2001b, 2001d). This war 'begins with al-Qaeda', he says, 'but it will not end until every terrorist group of global reach has been found, stopped, and defeated'. Bush had already presented the war in near-theological terms, claiming on 12 September, 'This will be a monumental struggle of good versus evil, but good will prevail' (Bush, 2001a), and he now expanded upon the geo-political implications of this bipolarity: 'Either you are with us, or you are with the terrorists. From this day forward, any nation that continues to harbor or support terrorism will be regarded by the United States as a hostile regime.'

Although Pearl Harbor could provide a patriotic model for public mobilization, it couldn't provide the military model for this new challenge. In contrast to World War II's mass-deployment of troops against conventional forces and physical space, Bush recognized that this would be 'a different kind of conflict against a different kind of enemy': a conflict 'without battlefields or beachheads' against an enemy 'who believe they are invisible' (Bush, 2001c). This was an accurate vision of the long-term challenges facing the USA and the complexity of the task, but fortunately the question of the immediate response was simple: Osama bin Laden was responsible for 9/11 and he was being sheltered by the Afghan Taliban regime. Either they cooperated with the USA's campaign against terror or they would become the first targets of its military reprisal.

This simple understanding of the event, of who did it, and of what action had to be taken was uncritically accepted by the mainstream media, whose coverage was as important for what it *didn't* report as for what it did. Firstly, there was no significant discussion in the media of why 9/11 happened. Such discussions are often avoided for fear of being seen to justify terrorist action, but the resulting focus on the facts of the event, the tragedy and the human stories de-contextualizes and de-historicizes what are always political acts with political causes. A deeper understanding of legitimate Muslim grievances and the appeal of radical Islam might have led to a more informed and sensitive post-9/11 foreign policy, instead of one that did so much to reinforce and extend those grievances and radicalize so many more.

Instead, the open-wound of 9/11 shut down all critical discussion and historical or political analysis. As Philip Knightley points out, the political debate played out like a traditional American western 'revenge' narrative (Keeble and Mair, 2010:107), and this idea of an originary 'wrong' that had to be paid for was accepted by the mainstream media. Their reverent treatment of the event ensured that hardline positions dominated, contributing towards a situation in which military action was accepted as inevitable, justified and even moral. BBC journalist Kevin Marsh attributes journalism's failure at this time to 'the political consensus in the face of the 9/11 attacks ... and the speed at which events careered towards war' (Keeble and Mair, 2010:81), whilst the BBC broadcaster Huw Edwards refers to 'the stifling context of 9/11 and the global response to that event' and the resulting 'rock-solid political consensus which propelled us towards war' (Keeble and Mair, 2010:x). Both highlight how difficult it was for any dissenting opinion to find either space or an audience in those months.

There was also insufficient debate as to what a 'war on terror' was, and its basis in international law. There were important questions here as to how you fought such a 'war': how you defined or identified terrorists, what forces or tactics you used against them, how civilian casualties would be avoided, and how success would be evaluated or victory achieved. What were the rules of such warfare? How did it relate to international sovereignty? What was the status and rights of the combatants? Why also were states that terrorized their own populations or subject minorities not included here? And how do you justify attacking selected states when so many states have long histories of funding and supporting violent resistance groups, including the United States itself with its sponsorship of militant groups in Central America? Noam Chomsky was one of the few critics of US hypocrisy, pointing out that cities such as Boston had provided funds for IRA campaigns and asking does that mean the UK has 'the right to bomb the United States?' (Chomsky, 2002).

Such nuances were lost in a public debate dominated by President Bush's simple division of 'Good' and 'Evil'. This was especially seen in the reporting of the Taliban. Media explanations of who they were focused on their violence, their oppression of women, their harsh Islamic punishments and the fear they instilled in the public, thereby establishing them as co-partners in evil with al-Qaeda itself. However, as stories of ordinary Afghanis in books such as Anand Gopal's *No Good Men Among the Living* (2014) and James Fergusson's *Taliban* (2011) reveal, many initially welcomed the Taliban in the mid-1990s for ending the arbitrary violence of the warlord-era and for bringing peace and order to the country. For a time in the mid-late 1990s the USA had implicitly supported the Taliban, alongside their allies Saudi Arabia and Pakistan, as part of a regional competition for influence against both Iran and Russia, and the US government had even courted the Taliban for a time as part of a plan by the US oil company Unocal for a pipeline through Afghanistan.

In fact, in a country ravaged by decades of war, every faction in Afghanistan was compromised. After the fall of Kabul in 1996 the Taliban imposed 'the strictest Islamic system in place anywhere in the world' (Rashid, 2001:50), and in their subsequent battle for control of the north they carried out numerous massacres, but the repression of civilians and massacres of both defeated troops and civilians preceded their rule and were a feature too of the anti-Taliban opposition. Indeed, many of the 'Northern Alliance' warlords who found themselves in favour with the Americans after 9/11 and who were returned to power and privilege after the Afghanistan War had themselves been responsible for massacres of civilians and prisoners of war both in the civil-war era and in the fight with the Taliban, making simple moral distinctions between the two sides difficult to justify.

Even the contemporary media image of al-Qaeda was a distorted one. Commentators such as Jason Burke in *Al-Qaeda* (2003) and *The 9/11 Wars* (2011) have criticized the political and media exaggeration of its threat. The idea of al-Qaeda as a large, highly-organized network with agents throughout the west, ready to follow the orders of their leaders was a false one, he argues. Burke questions whether 'al-Qaeda' existed in any meaningful way until the late 1990s and even then, he says, it was a much more 'complex' organization than we took it to be. Writing in 2015, he argues that following 9/11:

> a series of misconceptions about Osama bin Laden and al-Qaida became widely accepted. Some focused on the person of Bin Laden himself – his wealth, health and history. The group that he led, until then relatively marginal with no real support base and only a few hundred members, was portrayed as a sprawling global terrorist organisation, with obedient 'operatives' and 'sleeper cells' on every continent, and an ability to mobilize, radicalize and attack far beyond its real capacities. Historic incidents with no connection to the group or its leader were suddenly recast as 'al-Qaida operations'. Any incident, anywhere in the world, could become an al-Qaida attack
>
> Burke (2015:8).

Thus, Burke says, 'The threat posed by al-Qaida was described in apocalyptic terms, and a response of an equally massive scale was seen as necessary'. The result of this caricature was 'the global war on terror', which Burke describes as 'a monumentally misconceived strategy that is in part to blame for the spread of radical Islamic militancy over the past decade'. That 'War on Terror' began against Afghanistan on 7 October 2001.

Box 4.2 The origins of the Afghan War

The Afghan War is an armed conflict that began with the US attacks on Afghanistan on 7 October 2001 and formally ended with the official withdrawal of US troops on 31 December 2014 (although both the US military presence and the war have effectively continued to the present). It was launched as a response to 9/11 and the governing Taliban's refusal to hand over Osama bin Laden, although its roots lie in the longer history of Afghanistan that brought the Taliban to power.

The last King of Afghanistan, Mohammed Zahir Shah, ruled from 1933 until his overthrow by his cousin Mohammed Daoud Khan on 17 July 1973. Daoud established 'The Republic of Afghanistan' with himself as the first President, and became known for his progressive politics and his attempts to modernize the country with help from both the USA and USSR. Daoud was overthrown and killed in the Saur revolution on 28 April 1978 that brought the pro-Soviet, communist People's Democratic Party of Afghanistan (PDPA) to power and established 'The Democratic Republic of Afghanistan'. Under the leadership of Nur Muhammad Taraki the new government instituted a series of modernizing reforms and an oppression of opponents that led to widespread revolt among the Muslim population. Taraki was ousted and killed by Hafizullah Amin in September 1979 but Amin's rule saw armed opposition to the government increase. By December 1979 the government had lost control of all territory outside the cities to the Muslim 'mujahideen'. The USSR invaded Afghanistan on 24 December 1979, to shore up the communist government, killing Amin and replacing him with Babrak Karmal. Mujahideen guerrilla opposition to the Soviet-backed government continued, however, and by the mid-1980s the USSR was deploying 115,000 troops to fight the opposition.

The mujahideen were supported by the Arab Gulf States and by the USA, who saw the opportunity to turn Afghanistan into the Soviet Union's 'Vietnam'. Radical Islamic, foreign Arab fighters were also drawn to the mujahideen cause, inspired in part by the inspirational jihadi writings of Abdullah Azzam, who, together with Osama bin Laden, was in Pakistan helping with the recruitment and funding of foreign fighters. Azzam promoted jihad against the atheist Communist USSR as a duty of all Muslims and Afghanistan soon eclipsed Palestine as the leading site of Muslim struggle and, Azzam believed, the precursor to a global fight against all unbelievers. The Arab fighters played only a small part in the mujahideen resistance, however, and their radical ideology, desire for martyrdom and cultural differences made them unpopular with the Afghan rebels.

Like the nineteenth-century colonial superpower, Great Britain, the Soviet Union found itself unable to defeat the Afghans. Using asymmetric warfare, and with the aid of Saudi Arabia and the CIA (who, from September 1986, provided portable FIM-92 Stinger missiles for use against Soviet and government aircraft), the mujahideen inflicted considerable damage on Soviet forces. By mid-1987 the Soviet leader Mikhail Gorbachev announced a planned withdrawal of troops and the last Soviet soldiers left on 15 February 1989. Afghanistan had cost them 14,453 troops, with 53,753 wounded, whilst up to 1.5m Afghanis had been killed during the same period. Karmal had been replaced by the Soviets with Mohammad Najibullah in November 1986. Against expectations, his communist government survived for several years after the Soviet withdrawal until Kabul finally fell to the mujahideen on 16 April 1992.

Following the fall of Kabul, the Afghan political parties signed a peace and power-sharing deal, the Peshawar Accord, on 24 April 1992. Afghanistan became 'The Islamic State of Afghanistan' with Burhanuddin Rabbani appointed its president in June. The Peshawar Accord was opposed, however, by the powerful warlord Gulbuddin Hekmatyar and his Hezb-e Islami party and a civil war developed between rival mujahideen warlord factions, with Iran, Saudi Arabia and Pakistan all offering support to different groups who controlled different ethnic territories. The period between 1992 and 1996 was marked by continual fighting, numerous massacres, widespread violence, looting, rape and systematic corruption. What ended this chaos was the rise of the Taliban.

In 1994 Mullah Mohammed Omar founded a Pashtun madrassa (a religious school) in Singesar, near Kandahar, with around 50 armed students (Arabic, 'taleb') who formed the basis of a new fundamentalist, radical Islamic militia, the 'Taliban'. They were initially popular for opposing government corruption and imposing peace under Sharia law. Seeking a regime favourable to their interests, the Pakistan Inter-Services Agency (ISI) helped its development with recruits from the Saudi-funded, Afghan refugee madrassas in Pakistan, which taught a Deobandi-based radical Sunni Islam, echoing Saudi Wahhabism. With ISI support the Taliban became a national movement, capturing Kandahar by November 1994 and Herat province by 5 September 1995. On the night of 26 September 1996, they took Kabul and established 'The Islamic Emirate of Afghanistan'. Although they brought peace and security to a war-torn Afghanistan, the period of their rule from 1996 to 2001 became known for its religious fundamentalism and its human rights and cultural abuses (such as the destruction of the sixth-century Bamiyan Buddhas in March 2001). The remaining mujahideen, who only controlled a small amount of territory in the north, banded together to form the 'United Islamic Front for the Salvation of Afghanistan' (or the 'Northern Alliance'), under Ahmed Shah Massoud – 'the Lion of Panjshir' – a famous warlord and anti-Soviet militia leader who had served as Minister of Defence in the post-1992 government.

Bin Laden's return to Afghanistan in May 1996 wasn't welcomed by the Taliban. Their main concern was securing control of the country and imposing Sharia law and they had little interest in radical Islam's global jihad. Although his support and wealth helped their military campaign, over the following years they grew increasingly annoyed by Bin Laden's activities, especially his media pronouncements and his promotion of anti-western terrorism. The August 1998 US Embassy attacks angered the Taliban by bringing retaliatory cruise-missile strikes, international isolation and, from October 1999, UN sanctions against the regime. Mullah Omar was increasingly infuriated by Bin Laden but tribal laws of hospitality, the importance of his wealth and fund-raising power, and his oath of fealty to Omar in April 2001 bound the Taliban closer to their guests. Ultimately a government that defined itself as having realized true Islamic law couldn't be seen to be submitting to secular western values and justice. The October 2000 attack on the USS *Cole* and 9/11 focused US attention on Afghanistan. On 20 September 2001 Bush announced the USA's 'War on Terror' and the US government demanded the Taliban extradite Bin Laden and expel al-Qaeda. The Taliban refused, asking for proof of his guilt for 9/11. The USA rejected this and launched 'Operation Enduring Freedom' on 7 October 2001. A Taliban offer to try Bin Laden in an Afghani Islamic court or to hand him over to a neutral country if the USA ceased its campaign was rejected by the USA.

The Afghan War and its media coverage

In 1997 President Clinton had issued an executive order authorizing the CIA to capture Bin Laden, using lethal force if necessary. He was spotted at a hunting camp in 1999 but, fearful of an error, Clinton decided against a missile strike. The CIA drew up new plans to kill Bin Laden based upon helping the Northern Alliance to overthrow the Taliban, allowing a CIA manhunt, and these were accepted by President Bush with a $125m budget being approved on 3 September 2001. Although too late to stop 9/11, the plans were immediately adopted as the blueprint for the USA's war in Afghanistan. By early October, with public diplomacy and backdoor talks with moderates stalled, US patience wore out and it launched 'Operation Enduring Freedom' against Afghanistan.

Whilst the Weinberger and Powell doctrines had emphasized military preparedness and overwhelming military force, the CIA plan was far lighter, based around an air-campaign against the Taliban and a Northern Alliance ground campaign supported by CIA commanders and, later, Pentagon-controlled special operations forces. The air campaign targeted Taliban command, control, communications and intelligence (C3I) targets in Kabul, Kandahar and Jalalabad (especially military command posts, bunkers, air-defences and airfields and the training camps). After two weeks, it moved on to the Taliban frontlines, weakening them enough to allow the Northern Alliance to advance. This tactic proved successful, with Mazar-i-Sharif falling on 9 November and the Taliban fleeing Kabul on the 12th, leaving it to be occupied by the Northern Alliance the following day. By the end of 26 November Kunduz had also fallen.

On 26 September Bush had declared the USA was 'not into nation-building', being focused on 'justice' (Fairweather, 2014:31), but the State Department soon recognized that a pro-western regime was needed in Afghanistan to prevent al-Qaeda's return and that some reconstruction was necessary. On 5 December 2001, prominent Afghans at the UN-organized Bonn Conference agreed to establish the Afghanistan Interim Administration (AIA) with the USA's preferred leader, the moderate Pashtun Hamid Karzai, as President. The refusal to invite the Taliban to these talks and the US rejection of a Karzai-brokered peace deal with Mullah Omar would be cited by many years later as a fundamental mistake, missing the opportunity to end the conflict and disarm the Taliban. The USA, however, was committed to its 'War on Terror' and to the elimination of the Taliban, not their inclusion in government. The Bonn agreement was, as Fairweather says, 'a victor's peace' (Fairweather, 2014:51), with the eventual interim government dominated by the non-Pashtun Northern Alliance.

The final Taliban stronghold, Kandahar, fell on 7 December when Mullah Omar and his circle fled, although the city was taken not by Karzai, as the Americans wanted, but by the rival warlord Gul Agha Sherzai. The last battle in the Afghan War centred around the Tora Bora cave complexes where al-Qaeda-trained foreign fighters and Bin Laden himself had taken refuge. The battle was conducted using massive aerial bombing and, from 3 December, special operations forces and Afghan ground troops. The US refusal to put in its own troops and its faith in Afghan fighters proved to be a major mistake. Although initially successful in pushing al-Qaeda back to higher positions, the warlords were uninterested in the US manhunt, unwilling to commit to a serious fight, and were easily bribed. In all likelihood, Bin Laden 'walked straight through US lines ... courtesy of America's Afghan allies', through 'some kind of underhand deal with either Hazrat Ali and Haji Mohammed Zaman Ghamsharik or some of their minions' (Fairweather, 2014:64). Unaware of his escape, the clear-out of the Tora Bora complex continued afterwards, with fighting in the region continuing until 17 December when the war was declared over. On 22 December, Hamid Karzai was sworn in as Afghanistan's interim President.

From one perspective, the Afghan War was extremely successful. As General Tommy Franks wrote in his memoirs: 'In Operation Enduring Freedom, Central Command and its Afghan allies had defeated the Taliban regime and destroyed Osama bin Laden's al-Qaeda terrorist sanctuary in seventy-six days' (Bolger, 2014:68).

As in Kosovo, where the USA had achieved a war with zero combat fatalities, the costs of this war were low, with the US military suffering only 12 casualties from 10 October to 5 December 2001. Eight of these were listed as 'non-hostile' (including accidents, injuries and weapons discharges) and three were killed by friendly fire. Only one American was killed by 'hostile fire'.

The situation, however, was more complex than this. The Taliban had not been defeated and much of their force and leadership, including Mullah Omar, had escaped. The al-Qaeda leadership, including Bin Laden and al-Zawahiri, had also escaped capture, and they too weren't alone. According to Fergusson, Pakistan's President Musharraf, the War on Terror's new ally, persuaded the USA to help it evacuate its intelligence officers from Kunduz, where they had been helping the Taliban. The USA agreed a pause in the fighting for a plane to land but,

> what was supposed to be a minor extraction turned into a major airlift. As many as a thousand-people boarded the Pakistani planes, including many Taliban and Al-Qaeda fighters ... some analysts believe that more foreign terrorists escaped from Kunduz than they did from Tora Bora.
>
> Fergusson (2011:157)

Afghanistan was nominally free of the Taliban and al-Qaeda but it was back under the control of mujahideen warlords, few of whom represented the ethnic Pashtun majority and against whose arbitrary violence and systematic corruption the Afghan population had turned in 1994 in support of the Taliban. On 20 December 2001, the UN authorized an International Security Assistance Force (ISAF), initially to help security in Kabul and the surrounding region. By February 2002 18 countries were contributing towards the peace-keeping mission in what became a major nation-building exercise. In 2002 the AIA became the Afghan Transitional Administration of the new 'Islamic Republic of Afghanistan'. In August 2003 NATO took control of the ISAF operation and on 9 October 2004 Karzai won the first democratic Presidential election.

The media coverage of the Afghan War replicated the top-down model of government and military control deployed in the Gulf and Kosovo wars, except here the control was greater than ever. The military campaign was conducted with almost total secrecy and media organizations were reliant on government and military briefings that offered little information. There were no western journalists in Afghanistan and when the bombing began international networks had to rely on pictures from Al-Jazeera's Kabul office. In the following weeks journalists scrambled to join the Northern Alliance and as the Taliban were pushed back others crossed the border from Pakistan, but most of the conflict went unreported. As Matheson and Allan argue, western authorities proved uninterested in helping the media: 'The US and allied forces, in turn, gave few journalists access to their operations, which largely comprised long-range bombing sorties and secretive special operations campaigns, and sought to control other information flows as well' (Matheson and Allan, 2009:60). This included the Pentagon buying-up commercial satellite images of the country, 'to reduce the visual evidence of the effects of bombing being seen in the west' (Matheson and Allan, 2009:60–1). Much of the on-the-ground reporting came from Pakistan and relied heavily on official accounts augmented by refugee testimony.

Philip Knightley points out that the fact that the war began at night was appropriate, as the US and UK governments 'seemed determined to keep their citizens in the dark about what was happening in Afghanistan'. Journalists in London interviewed those in Pakistan or experts in the studio but 'the truth was that nobody outside government knew anything': war correspondents, he says, 'had no real access to the war' (Keeble and Mair, 2010:108–09). US networks had no more luck. Interviewed by Sherry Ricchiardi, editors complained about the lack of information from the Pentagon, with one commenting, 'This administration has clamped the most severe information freeze I've seen in 35 years of reporting' (Ricchiardi, 2001). The military control of the battlefield now included the control of any media organization operating within it, hence the destruction of Al-Jazeera's Kabul office by a US missile on 13 November.

The US networks had initially struck a deal with Al-Jazeera to carry its footage, although they were selective about what they used. As Samuel-Azran says, US networks 'self-censored all counter-hegemonic news material' from Al-Jazeera, especially images of civilian casualties (Samuel-Azran:2010:42). Hence on 23 October whilst Al-Jazeera broadcast graphic images of casualties from air strikes on two villages near Kandahar and Quetta that had killed 122 civilians, the US networks reported the story with less emotive images of people being treated in hospital. The subject was a highly-sensitive one for the US government as they knew images of civilian casualties from one of the poorest countries on earth could undermine the morality of their campaign. Al-Jazeera's footage rattled the administration leading, on 28 October, to Secretary of Defense Donald Rumsfeld condemning their coverage as 'propagandistic and inflammatory' and accusing them of *manufacturing* images: 'What they do is when a bomb goes down they grab some children and some women and pretend that the bomb hit the women and the children' (Samuel-Azran, 2010:49).

When Al-Jazeera was subsequently bombed, none of the US networks, including their content-partners CNN and ABC, provided any significant coverage or criticized the attack. Fox did cover it, but described it as 'tough luck' for 'Osama bin Laden's network of choice', saying it was hit, military officials said, because 'the building was being used by the Al-Qaeda terrorist network' (Samuel-Azran, 2010:46–7). As Philip Knightley says, Rear Admiral Craig Quigley was open about the bombing. For him:

> Nothing had gone wrong. Quigley said that the Pentagon was indifferent to media activity in territory controlled by the enemy, and that the Al-Jazeera compound in Kabul was considered a legitimate target because it had 'repeatedly been the location of significant al-Qaeda activity'. Al-Jazeera said that this activity consisted of inter-views with Taliban officials, something that it had hitherto thought to be normal journalism, and that it believed that its office was bombed in revenge for acting as a broadcast conduit for tapes from Osama bin Laden
>
> Keeble and Mair (2010:114).

The Al-Jazeera attack can be understood as part of the USA's 'information warfare' strikes on the Taliban's C3I infrastructure. The USA had spent the 1990s refining their concepts of C3I, IW and IO and this was an application of these concepts on the battlefield. When the USA spoke of 'full spectrum dominance', this included the electromagnetic broadcasting spectrum and *any* attempt to produce and distribute information not sanctioned by or supportive of the US action. Philip Knightley's conclusion – that 'In the history of war reporting, Afghanistan will, I believe, be regarded as a turning point, the moment marking the military's final triumph over the media' (Keeble and Mair, 2010:114) – was accurate.

As Matheson and Allen argue, the war was also notable for the use of new digital tools. In the absence of footage from the ground, 'Information graphics and other technologies

of simulation took on new importance to represent the otherwise unseen conflict'
(Matheson and Allan, 2009:61). Online newspapers employed 'interactive maps' with
satellite imagery that audiences could zoom in on, but graphics such as these, which
seemed to give a closer experience of the conflict only offered a grossly-simplified vision
of the conflict, ignoring both the realities of war and the complexity of Afghan society and
culture.

The other major development was the use of satellite phones and video-phones, first
used journalistically on 31 December 1999 when CNN broadcast live video of a breaking
story using a video camera connected to a satellite uplink unit that digitized the signal and
sent it through an Inmarsat satellite phone. Now journalists were freed from a crew of
technicians and bulky equipment, being given a new mobility as individual investigators
reporting 'with much greater ease from any location on the globe' (Matheson and Allan,
2009:63), even somewhere as inhospitable as Afghanistan. Matheson and Allan note this
also brought increased independence from political and military control as 'portable
communications such as these breathed new life into the tradition of the independent
foreign correspondent', allowing reporters to speak directly to audiences over the heads of
editors and officials (Matheson and Allan, 2009:63).

Satellite phone reporting was also much more frequently live, 'privileging the dramatic
image of the journalist in the midst of conflict as the 'authentic' testimony'. The journalist
themselves increasingly became the story 'as actors' in the war whose experience could be
followed by the audience (Matheson and Allan, 2009:64). The phone's low-fi aesthetic
was even part of their appeal, coming to represent their liveness and hurriedly-grasped
reality: as Matheson and Allan conclude, 'The live, mobile report from the war zone, with
its pixelated images and distorted sound sustained an epistemological claim to speak
directly from the reality of war' (Matheson and Allan, 2009:64).

But the 'reality' of this war was sometimes highly questionable. The battle to depose
the Taliban and find Bin Laden were real objectives, but the destruction of 'al-Qaeda
fighters' and 'bases' was always a problematic concept. Bin Laden's group in Afghanistan
was small and many of those labelled 'al-Qaeda fighters' were simply foreign fighters
trained to help the Taliban. The best example of the gross-overestimation of al-Qaeda,
however, is the coverage of their Tora Bora 'fortress'. On 26 November, *The New York
Times* carried an account by an ex-Russian soldier of an elaborate cave-complex in
Zhawar in the 1980s, with the paper claiming that Bin Laden had upgraded it in the
1990s. By the following day *The Independent* had moved the base to Tora Bora in the
present, under the headline 'al-Qaeda almost "immune to attack" inside its hi-tech underground lair'.

Its vast size and 'hotel'-like amenities, home to 2,000 suicidal militants, caught the
imagination of the American press, who embellished the story at will, before *The Times*
outdid everybody with a specially-commissioned, carefully-labeled, cut-away illustration
of the mountain range portraying a secret underground-base commensurate with the most
grandiose Bond villain. The high-point of the absurdity came when NBC's 'Meet the
Press' showed the graphic to Donald Rumsfeld who commented, 'Oh, you bet. This is
serious business. And there's not one of those. There are many of those' (Epstein, 2015).
When the US troops cleared out Tora Bora two weeks later they found nothing but the
most basic caves and dug-outs: Bin Laden's 'secret lair' was a media-created myth.

The media coverage was also lacking when it came to human abuses and civilian
casualties. Firstly, there was little coverage of the Northern Alliance's human abuses. Now
the Taliban were in retreat, the warlords were free to follow old grudges, with the Tajik

and Uzbek militias exacting revenge upon all Pashtuns, about a million of whom lived in the north. Fergusson says, 'more than half of these now fled south, as the looting, raping, kidnapping and revenge killings of the early 1990s made an all-too-predictable comeback' (Fergusson, 2011:150). The quid pro quo to Northern Alliance help against al-Qaeda and with finding Bin Laden, he says, was 'to turn a blind eye to their excesses' (Fergusson, 2011:150). Another issue was their treatment of POWs. The prisoner uprising at Qala-i-Jangi fort was put down by the warlord Abdul Rashid Dostum and by allied airpower with overwhelming force, leaving only 86 of the 300–500 prisoners alive, though the allied authorities justified the necessity of this action. Less justifiable was the mass-asphyxiation and execution of prisoners in the Dasht-i-Leili massacre.

Here, up to 7,500 POWs from the siege of Kunduz were loaded onto sealed containers by Dostum's forces for transport to Shergerghan prison. In December 2001 allegations surfaced in the *New York Times* of prisoner abuse, being explored more fully in an August 2002 *Newsweek* article and Jamie Doran's 2002 documentary *Afghan Massacre: The Convoy of Death*, the latter of which included eyewitness testimony of hundreds, if not thousands, dying in the containers; of guns being fired into the containers when prisoners screamed for air-holes, and of executions in the desert, some claimed to have been carried out in the presence of US soldiers. Although those claims have been questioned, a 2008 freedom of information request revealed the Defense and State Departments believe about 1,500–2,000 prisoners were killed. Not only was this massacre unknown and unreported at the time, but there are allegations that the Bush administration refused to investigate the CIA-backed warlord and hampered later journalistic investigations into the massacre.

Afghanistan's civilians also attracted limited media attention. Afghanistan was already suffering a humanitarian crisis as three years of drought had led to the collapse of the rural economy, widespread famine and a refugee crisis, with up to four million people leaving their homes by Spring 2001. This story was dropped by western media after 9/11, however, as sympathy for Afghanistan didn't fit into the dominant political mood. When bombing began in October the USA highlighted its air drops of food to demonstrate its campaign wasn't aimed at the Afghan people, but this was a small, symbolic gesture and overall the war exacerbated both the suffering of ordinary Afghanis and the refugee crisis. The air drop also produced some of the most surreal images of warfare to date with news footage of Afghanistan being bombed being followed by footage of food and medicine being dropped. Although the drops were aimed at different groups, the incongruity and absurdity of the juxtaposition escaped the media.

The ongoing refugee crisis was largely ignored. On 3 January 2002, *The Guardian* wrote that up to 100 people a day were dying in the Maslakh refugee camp, west of Herat. As Media Lens pointed out, ITN and BBC News led instead with extensive coverage of Australian bushfires in which no-one had died: 'Like reports of hundreds of civilian deaths from US bombing over the Christmas period, both ITN and BBC TV news have deemed this mass death of Afghan refugees unworthy of attention' (Media Lens, 2002). They contrast this lack of coverage with the space accorded the Kosovo refugee crisis in 1999, concluding that whereas in the Kosovo War 'the plight of refugees was used as powerful propaganda supporting NATO's assault' in Afghanistan the west's war 'bears considerable responsibility for the refugee crisis, for the disruption of food supplies, and for the mass-suffering and death both inside and outside the Afghan camps'. The crisis was ignored, therefore, because here it undermined rather than supported western military action. The west's narrative of the defeat of evil in the name of humanitarian values would have been undermined by exposure of the effects of the war.

The civilian casualties of the bombing were rarely discussed in the media reports of the war. For some media organizations, they simply didn't matter. On 31 October CNN Chairman Walter Isaacson sent a memo ordering his staff to remind viewers of 9/11. 'It seems perverse to focus too much on the casualties or hardship in Afghanistan', he wrote. Instead of promoting the Taliban's viewpoint 'we must talk about how the Taliban are using civilian shields and how the Taliban have harbored the terrorists responsible for killing close to 5,000 innocent people' (Kurtz, 2001). Isaacson defended this as journalistic 'balance', but in effect a major US media organization was justifying civilian deaths.

UK media coverage of these deaths was also limited. As David Edwards and David Cromwell of Media Lens point out, ITN did at least cover the human-interest story of 'Marjan, the One-Eyed Lion in Kabul zoo', on the 9, 13, 22 and 26 January 2002. His 'battered image touched people around the world', ITN said, 'his plight a symbol of his maltreatment by the Taliban', in a report that ended with the news that vets had flown out to help and with footage of Marjan chewing happily on a large piece of meat (Cromwell and Edwards, 2006:80). Elsewhere in famine-stricken Afghanistan, however, people were reduced to eating grass to survive. After Marjan's state was highlighted by western tabloids more than £270,000 was donated to a fund set up by the World Society for the Protection of Animals. His death of old age by the end of January was reported around the world, in stark contrast to the unnamed Afghanis, who enjoyed far less western concern.

Just as CBS 'liberated' Kuwait City in the 1991 Gulf War, so in Afghanistan the BBC reporter John Simpson arrived in Kabul on 13 November, telling the *Today* programme: 'It was only BBC people who liberated this city. We got in ahead of Northern Alliance troops ... I can't tell you what a joy it was. I felt very proud indeed to be part of an organization that could push forward ahead of the rest'. Although Simpson's triumphalism was ridiculed, the report highlighted how western media privileged *being-there* – reporting live from key events – as the essence of journalism, and as giving an insight into the reality of war. The reality of ordinary Afghans, however, was not present in these excited broadcasts.

By coincidence, on the same day that the number of people killed on 9/11 was revised down to 3,234, a report was published estimating the number of civilians killed in the Afghanistan War as 3,767 (Milne, 2001). On 2 July 2002 John Pilger wrote that 'this is now estimated to have passed 5,000 civilians deaths: almost double the number killed on September 11' (Keeble and Mair, 2010:111). These numbers should have led to considerable soul-searching in the west. If 3,234 lives lost in New York City were an atrocity, then why weren't the deaths of 3,767 Afghanis? Le Bon's point that for deaths to matter they need to crystallize in a striking image is relevant here, as the unrecorded bombing of some of the poorest people in the world would never arouse as much sympathy as 9/11, but the lack of interest in the Afghan casualties compared with those in the west makes it hard to escape the conclusion that Third World lives mattered less than those in the First World. A war begun in the name of Liberal humanist values ended by exposing their racist hypocrisy.

Conclusion

The Afghan War was launched as the first phase of a 'War on Terror' that would punish those responsible for 9/11 and the state supporting them. Although the easy victory meant the political mood was triumphal, the Taliban were undefeated and their leaders had escaped, whilst Bin Laden and Zawahiri and their circle had also disappeared. In his absence Bin Laden would take on an even more spectral and iconic presence, permeating western popular culture: Bin Laden would appear in *South Park, Family Guy* and *American Dad*; Eminem could be seen dressed as Bin Laden, getting down in a cave with his homies at the end of his 2002 *Without Me* video; Aaron Barschak gate-crashed Prince William's birthday party in June 2003 in full Bin Laden costume; and the 2006 book *Where's Bin Laden?*, based upon the popular *Where's Wally?* cartoons, spoofed the ongoing manhunt.

If the political and military success of the War on Terror could be questioned, so too could its imagic success. The Afghan War did little to reverse the symbolic humiliation of 9/11. Against the global spectacle of those attacks, this war could only offer the sight of the richest nation on the planet systematically bombing one of the world's poorest. There wasn't even enough for the USA to bomb: as the Captain of the USS *Enterprise* admitted, Afghanistan was 'not a target-rich environment', whilst another General described the military campaign as merely 'turning big bits of rubble into small bits of rubble' (Merrin, 2005:107). Bombed, ramshackle training camps, cities already destroyed by decades of civil war, mountain-warfare and a ground campaign conducted with Afghani fighters visually indistinguishable from the Taliban came a poor second to the real-time, epoch-defining images of the imploding Twin Towers. Despite Bush's promise of 'a different kind of war', Afghanistan was a traditional campaign that failed even to match the spectacle of earlier media-wars. The USA still needed a spectacular, global spectacle of their power to combat the memory of 9/11 and deter future attacks. This explains in part their return to Iraq in 2003.

Key reading

To understand the political and military consequences of 9/11 I'd recommend beginning with the event itself. The facts are easy to find but I'd suggest watching the footage of the day for a better idea of the impact upon the American psyche. Watch the live broadcast from the start and follow the hours of coverage so you can understand the shock, confusion and horror of those on the streets and the watching public. I've focused here on CNN's coverage (2001), but YouTube has hours of footage from NBC, CBS, Fox, ABC, BBC News 24 and BBC Worldwide. For background information on the rise of al-Qaeda and the planning of 9/11 see Bergen (2002), Burke (2003), Gerges (2011) and Wright (2011). These are essential reading, providing an excellent framework for understanding much of what has followed in the decade after 9/11. The aftermath of 9/11 is well-covered in Burke (2011) and Bergen (2011), whilst the response of cultural theorists to the events can be followed in the work of Baudrillard (2002) and Žižek (2002). President Bush's speeches are available online and are important (2001a, 2001b, 2001c, 2001d), although his rhetoric of good versus evil should be balanced by a more nuanced understanding of Afghanistan's problems, of its political history, of the rise of Taliban and of their brief regime (see Rashid, 2001; Fergusson, 2011; Gopal, 2014). Information on the invasion phase of the Afghan War can be found in Fairweather (2014) and Bolger (2014), whilst the media coverage of the war is discussed in Cromwell and Edwards (2006), Matheson and Allan (2009), Keeble and Mair (2010), and Samuel-Azran (2010).

References

Baudrillard, J. (2002) *The Spirit of Terrorism*, London: Verso.

BBC. (2001a) 'How BBC1 Interrupted Programmes on 9/11', *YouTube*, 2 January 2013, https://www.youtube.com/watch?v=toPXdfK9Eck.

BBC. (2001b) 'Barbican Stands by Stockhausen', *BBC News*, 21 September, Friday, http://news.bbc.co.uk/1/hi/entertainment/1556137.stm.

Bergen, P. L. (2002) *Holy War*, London: Phoenix.

Bergen, P. L. (2011) *The Longest War*, New York: Free Press.

Bolger, D. P. (2014) *Why We Lost*, New York: Harcourt Mifflin Harcourt.

Burke, J. (2003) *Al-Qaeda*, London: I.B. Tauris.

Burke, J. (2011) *The 9/11 Wars*, London: Penguin.

Burke, J. (2015) *The New Threat From Islamic Militancy*, London: The Bodley Head.

Bush, G. W. (2001a) 'Text of Bush's Act of War Statement', *BBC News*, 12 September, http://news.bbc.co.uk/1/hi/world/americas/1540544.stm.

Bush, G. W. (2001b) 'Bush Addresses Nation: Full Text', *BBC News*, 12 September, http://news.bbc.co.uk/1/hi/world/americas/1539328.stm.

Bush, G. W. (2001c) 'The President's Radio Address', *The American Presidency Project*, 15 September, http://www.presidency.ucsb.edu/ws/?pid=25001.

Bush, G. W. (2001d) 'Text: Bush Address to Congress', *BBC News*, 21 September, http://news.bbc.co.uk/1/hi/world/americas/1555641.stm.

Chomsky, N. (2002) 'On Afghanistan. Noam Chomsky Interviewed by Tom Sebastian', *Hard Talk*, 2 February, http://www.chomsky.info/interviews/20020227.htm.

CNN. (2001) 'September 11 2001 As It Happened – CNN Live 8.40am-10.11am', *YouTube*, 10 August 2012, at: https://www.youtube.com/watch?v=rsIWPPw-JzU.

Cromwell, D. and Edwards, D. (2006) *Guardians of Power. The Myth of the Liberal Media*, London: Pluto Press.

Epstein, E. J. (2015) 'Netherworld. Fictoid 3: The Lair of Bin Laden', http://www.edwardjayepstein.com/nether_fictoid3.htm.

Fairweather, J. (2014) *The Good War*, London: Penguin Random House.

Fergusson, J. (2011) *Taliban*, London: Corgi Books.

Gerges, F.A. (2011) *The Rise and Fall of Al-Qaeda*, Oxford: Oxford University Press.

Gopal, A. (2014) *No Good Men Among the Living*, New York: Metropolitan Books.

Keeble, R. and Mair, J. (eds.) (2010) *Afghanistan, War and the Media: Deadlines and Frontlines*, Bury St Edmunds, Suffolk: Arima Publishing, at: http://eprints.lincoln.ac.uk/3357/1/Afghanistan_final.pdf.

Kurtz, H. (2001) 'CNN Chief Orders 'Balance' in War News. Reporters are told to Remind Viewers why U.S. is Bombing', *Washington Post*, 31 October, http://www.globalresearch.ca/articles/KUR111A.html.

Le Bon, G. (2005) *The Crowd. A Study of the Popular Mind*, Milton Keynes: Filiquarian Publishing.

Matheson, D. and Allan, S. (2009) *Digital War Reporting*, Cambridge: Polity.

Media Lens (2002) 'Media Ignores the Mass Death of Civilians in Afghanistan', 3 January, http://www.medialens.org/index.php/alerts/alert-archive/2002/166-media-ignores-the-mass-death-of-civilians-in-afghanistan.html.

Merrin, W. (2005) *Baudrillard and the Media: A Critical Introduction*, Cambridge: Polity.

Milne, S. 2001, 'The Innocent Dead in a Coward's War', *The Guardian*, 20 December, p. 16.

National Commission on Terrorist Attacks Upon the United States. (2004) *The 9/11 Commission Report*, 22 July, Washington, DC, http://www.9-11commission.gov/report/911Report.pdf.

NBC. (2001) 'NBC News Coverage of the September 11, 2001, Terrorist Attacks (Part 1 of 2)', *YouTube*, https://www.youtube.com/watch?v=OtZKEjr-Sfg.

Rashid, A. (2001) *Taliban*, London: Yale University Press.

Ricchiardi, S. (2001) 'Sherry Ricchiardi: Media Coverage of the War in Afghanistan', *CNN.com/Community*, http://edition.cnn.com/2001/COMMUNITY/10/24/ricchiardi/.

Samuel-Azran, T. (2010) *Al-Jazeera and US War Coverage*, New York: Peter Lang.

Thatcher, M. (1985) 'Speech to the American Bar Association', *Margaret Thatcher Foundation*, 15 July, http://www.margaretthatcher.org/document/106096.

Virilio, P. (1997) *Pure War*, New York: Semiotext(e).

Wright, L. (2011) *The Looming Tower. Al-Qaeda's Road to 9/11*, London: Penguin.

Žižek, S. (2002) *Welcome to the Desert of the Real*, London: Verso.

5   Shock and awe

Reporting the Iraq War

Targeting Iraq

Following the success of the Afghanistan campaign, Iraq was chosen as the next target in the War on Terror. Its links with al-Qaeda terrorists and its advanced weapons of mass destruction (WMD) programme meant, the USA argued, that it posed an immediate risk to the safety of the region and to world peace. The replacement of Saddam Hussein with a democratic regime was seen as an urgent necessity.

Except this chronology is deceptive. It was later revealed that Iraq was perhaps the first target in the War on Terror. Notes from one of Defense Secretary Donald Rumsfeld's aides revealed in 2002 that within hours of the attack on the Pentagon on 9/11, with all evidence pointing to Bin Laden, Rumsfeld ordered the military to begin plans to strike Iraq, asking his staff to find ways to link it to Saddam. Other notes revealed in 2007 quote Rumsfeld as saying, 'my interest is to hit Saddam at the same time as we go after al-Qaeda'. President Bush felt the same way, asking counter-terrorism expert Richard A. Clarke to find 'any shred' of evidence Saddam was involved (see Roberts, 2002; Borger, 2006; Raw Story, 2007; Packer, 2007:40).

In fact, the plans to invade Iraq predated 9/11. Within days of taking power on 20 January 2001, the Bush administration had discussed deposing Saddam. Their first National Security Council meeting on 30 January focused on Iraq and ended with Bush asking the military to draw up contingency plans for a ground invasion in support of a native insurgency, whilst the second meeting on 1 February included a secret memo entitled 'Plans for a Post-Saddam Iraq', which discussed peace-keeping troops, war crimes tribunals and contracts for Iraq's oil wealth. Former Treasury Secretary Paul O'Neil later commented, 'From the very first instance it was about Iraq. It was about what we could do to change this regime. Day one, these things were laid and sealed' (Leung, 2004). This desire for regime-change wasn't new. It was the culmination of a 'neoconservative' campaign that had run throughout the 1990s.

Neoconservativism is a US right-wing, idealistic, hawkish political movement that believes in US 'exceptionalism', seeing the USA as superior in its political system and values and as having a moral duty to promote these globally and to oppose tyranny. It believes the USA should be militarily dominant with the right to take preemptive action to defend itself and assert its hegemony and democratic values. Its origins are traceable to the 1950s and 60s, to disillusioned Democrats who moved to the right in opposition first to the Soviet Union and then to the New Left and the anti-war movement. The pejorative term 'neoconservative' was coined by Michael Harrington in 1973 and was taken up as a positive identity by 'the godfather of neoconservativism', the US writer and editor Irving Kristol. The two main strands of neoconservative thought revolve around a critique of liberalism's selfish individualism in favour of strong social values (ideas derived from the work of Daniel Bell and Leo Strauss) and a belief in the need to globally defend these values through a strong foreign policy backed with military might (ideas influenced by the political analysts Fritz G. A. Kraemer and Albert Wohlstetter).

Neoconservatives were active in the Nixon and Ford administrations but opposed the turn to détente, emphasizing instead the continued Soviet threat and need for a strong military. They served under Reagan too, being impressed by his rebuilding of post-Vietnam American morale and renewal of the arms race, but they were disappointed by his support for non-democratic regimes and turn to détente in alliance with the reformist Soviet leader Mikhail Gorbachev. The Berlin wall's collapse in 1989 appeared to vindicate the neoconservative vision of liberal-democracy's supremacy, though Iraq's invasion of Kuwait in 1990 highlighted the new threat of rogue states destabilizing regions through military aggression or their desire for chemical or nuclear arms. Hence the neoconservatives supported President Bush Sr's decision to use force to repel Iraq in 1991, though they were immediately infuriated by his refusal to finish the job and depose Saddam. Bush's decision was a pragmatic one, resisting a long-term troop commitment and retaining Iraq as a stable counterweight to the USA and Saudi Arabia's regional enemy, Iran. For the neoconservatives, however, Saddam's brutal crushing of internal revolts after the war and the continued need to police his regime in the following years demonstrated the evil of his regime and convinced them this was an unfinished war. The pro-Israeli and idealistic neoconservatives came to believe that replacing Saddam with a pro-western government would increase US hegemony, bolster Israel's security and become an example of democratic change that would foment a broader Arabic transformation.

Out of office through the 1990s following President Clinton's victories in 1993 and 1996, the neoconservatives used the time to hone their ideological world-view and campaign for their foreign policy aims. They became a significant force in public debates through the decade through their articles in right-wing magazine and newspapers and the activities of right-wing think tanks such as the Hudson Institute, The American Enterprise Institute (AEI) and their primary mouthpiece The Project for a New American Century (PNAC), founded by William Kristol (son of Irving) and Robert Kagan. Many of those who signed PNAC's statement of principles on 3 June 1997 would serve in the Bush administration.

The 'neocons' attacked both Clinton's ineffective humanitarian foreign policy and containment of Saddam and the Republican party's isolationist stance. Kristol and Kagan's influential 1996 *Foreign Affairs* article, 'Toward a Neo-Reaganite Foreign Policy' summed up their unhappiness. It asserted the USA's 'strategic and ideological predominance' and the need for its 'benevolent global hegemony', and argued for military expansion not withdrawal. 'American hegemony is the only reliable defence against a breakdown of peace and international order', they claimed, and it should be promoted through 'military supremacy and moral confidence' (Kristol and Kagan, 1996). By 1996 Iraqi regime change had become the leading neoconservative cause. Regular calls for war appeared in articles and publications and letters to Clinton from PNAC on the 26 January 1998 and the neoconservative-linked Committee for Peace and Security in the Gulf on 19 February both demanded the overthrow of Saddam. The accumulated pressure led to the October 1998 Iraq Liberation Act, which declared it was US policy to support Iraqi regime change and offered assistance to opposition groups.

George Bush Jr wasn't an obvious choice for neoconservative support for the November 2000 election but he proved open to education, with Richard Perle, Paul Wolfowitz and Condoleezza Rice all schooling him in the conservative world-view. Taking power in January 2001, Bush's administration had a strong conservative base, with Dick Cheney as Vice-President, Donald Rumsfeld as Secretary of Defense and neoconservatives in important positions, including Paul Wolfowitz as Deputy Secretary of Defense, Richard Armitage as Deputy Secretary of State and Richard Perle as Chairman of the Defense Policy Board Advisory Committee. The neoconservatives weren't dominant but their influence upon foreign policy would prove crucial. As former Defense Intelligence Agency analyst W. Patrick Lang argues,

> The sincerely held beliefs of a small group of people who think they are the "bearers" of a uniquely correct view of the world, sought to dominate the foreign policy of the United States in the Bush 43 administration, and succeeded in doing do through a practice of excluding all who disagreed with them.
>
> Lang (2004:39)

War with Iraq wasn't, however, inevitable as the State Department, Joint Chiefs of Staff and CIA were unconvinced of Saddam's threat, but 9/11 changed that. With it, the public caught up with the neoconservatives' hawkish patriotism and the desire for military action moved from being an ideological niche to a mainstream, popular credo.

On 15 September 2001 Bush rejected Wolfowitz's argument for invading Iraq in favour of striking Afghanistan, but on the 17 he signed a 'Top-Secret' directive that 'ordered the Pentagon to begin preparing military options for an invasion of Iraq' (Lang, 2004:46). The decision wouldn't be made public but the grounds for it would be prepared in Bush's public speeches. Having declared an initial 'War on Terror', Bush began to broaden his vision of this in his subsequent speeches. His State of the Union Address on 29 January 2002 subtly shifted the War on Terror towards 'terrorists and regimes who seek chemical, biological or nuclear weapons', naming Iraq, Iran and North Korea as 'an axis of evil' posing 'a grave and growing danger' and hinting at preemptive action to protect the USA from their threats (Bush, 2002a). Soon after, in his Commencement Address at West Point on 1 June, Bush offered an important reformulation of both terrorism and the response it required. He argued the USA faced 'a threat with no precedent' as modern technology gave weak states and terrorists 'a catastrophic power to strike great nations' at any time. Deterrence, therefore, no longer works and a more offensive response is required that takes the battle to the enemy. This requires both a 'military strength beyond challenge' and a military 'ready for preemptive action when necessary to defend out liberty and defend our lives' (Bush, 2002b).

This 'Bush doctrine' echoed the so-called 'Wolfowitz doctrine' presented in the Bush Sr administration's 1992 *Defense Planning Guidance* document. Its claim then of the need to retain US military superiority through the use of preemptive, unilateral force against emerging threats proved so controversial that it had had to be rewritten. Now it became policy as Bush's September 2002 *National Security Review* argued for the right of 'anticipatory self-defense' and the promotion of democratic regime-change. With these elements in place, the neoconservative dream of ousting Saddam came closer. All that was needed was to convince the public.

Selling Iraq

The USA's public case to invade Iraq rested upon two claims: (1) Iraq's continuing programme to develop weapons of mass destruction in defiance of UN resolutions and (2) its support for Islamist terrorists. Saddam's relations with the world were governed by UN resolutions, especially resolutions 687 and 1284. Resolution 687 on 3 April 1991 had ordered the destruction of Iraq's chemical and biological weapons and ballistic missiles with a range over 150km, had demanded Iraq agree not to develop nuclear weapons, and had established a UN Special Commission (UNSCOM) inspections regime to oversee this disarmament. This operated in Iraq until Saddam forced their withdrawal in December 1998, which prompted resolution 1284 on 17 December 1999, which replaced UNSCOM with UNMOVIC (the United Nations Monitoring, Verification and Inspection Committee). Iraq's continuing refusal to allow UNMOVIC access was now sufficient pretext for the USA for military action.

Bush found a supporter in the UK's Prime Minister Tony Blair, whose left-wing, Christian-humanitarian belief in confronting global evil meshed perfectly with Bush's neoconservative, religious belief in military intervention for democratic regime-change. Bush won Blair over, most famously in a summit in Crawford, Texas on 6–7 April 2002 where Blair agreed to support military action for regime change on the condition that the USA built a coalition and demonstrated that options to remove Saddam's WMDs through the weapons inspectors had been exhausted. Blair also needed parliamentary support, which depended upon UN confirmation of the right to proceed.

On 3 November 2002 Iraq submitted a report to the UN Security Council and International Atomic Energy Agency (IAEA) stating it had no WMDs, but the UN dismissed this and stated on 8 November that it was in 'material breach' of its disarmament obligations, passing the US-UK-penned resolution 1441 which offered Iraq 'a final opportunity' to comply, with a 30-day timeframe for UNMOVIC verification of its WMDs. UNMOVIC inspectors arrived in Iraq in late November but the USA had little interest in their conclusions. On 16 October Bush had signed into law the 'Authorization for the Use of Military Force Against Iraq Resolution of 2002', passed by Congress days earlier; on 19 December the USA declared Iraq was in 'material breach' of resolution 1441, and in January 2003 the USA and UK began moving military forces into the Gulf.

On 5 February 2003 US Secretary of State Colin Powell gave a presentation of the USA's case to the UN, stating that 'Iraq's behaviour demonstrates that Saddam Hussein and his regime have made no effort – no effort – to disarm as required by the international community'. Instead he was following 'a policy of evasion and deception', moving documents and WMDs 'to keep them from being found by inspectors'. There was 'no doubt', Powell said, that Saddam retained a huge stockpile of biological and chemical weapons and was trying to acquire nuclear weapons or that he had long-standing relations with al-Qaeda, supporting the recent operations in northern Iraq of the militant and Bin Laden associate Abu Musab al-Zarqawi (Powell, 2003).

There was significant opposition to the war. France and Germany opposed the USA's plans whilst many UK Labour MPs wanted a second resolution from the UN before they would give support. A popular anti-war movement also emerged whose protests culminated in a global day of action on 15 February 2003 that was the largest protest in history, encompassing 6–10m people in 60 countries and over 600 cities, with Rome attracting three million protestors and London about one million. This movement was largely ignored by the Labour government. Ultimately, the UK Attorney General's opinion on 7 March that military action was legal under resolutions 1441 and 678, Jacques Chirac's public claim on 10 March that France would oppose any new resolution (rendering another UN vote moot), and Blair's victory in a Commons war vote on 18 March sealed the UK's support. The USA dismissed Hans Blix's UN testimony that Iraq was cooperating and testimony from IAEA-head Mohammed Elbaradei that there was 'no evidence or plausible indication of the revival of a nuclear weapons programme in Iraq' and on 17 March Bush appeared on television telling Saddam and his sons (in his best Sherriff impression), 'They have 48 hours to get out of town' (Lang, 2004:60). On 19 March 2003 'Operation Iraqi Freedom' began.

Prior to the invasion the American press and public were notably pro-war. A January 2003 US survey found 68% of respondents believed Iraq gave support to al-Qaeda or was involved in 9/11, with 13% believing the USA had found 'conclusive evidence' of this link (Kull, 2003:2, 13). Tumber and Palmer argue that the UK's pre-invasion reporting was more balanced, but concede specific issues such as the al-Qaeda links, the anti-war movement and political alternatives to war received less attention. Des Freedman, however, is more critical, arguing,

> The use by the British and US governments of lies, distortion and misrepresentation to justify their war on Iraq was largely echoed and amplified by the media of those countries. Despite a clear anti-war majority in Britain and a significant anti-war minority in the USA, most media outlets supported the war and failed systematically to challenge the arguments for an invasion or to expose the brutality and consequences of the war.
>
> Miller (2004:63)

The extent of the mainstream media's failure became apparent later as evidence emerged of the extent to which the US and UK governments had manufactured and manipulated evidence to promote a war they'd already agreed upon. The Bush administration cut out advisors who disagreed with its world-view, opposed the assessments of the intelligence services and developed its own intelligence sources, 'stove-piping' their claims straight to the President's office. It also developed the PCEG (the Policy Counterterrorism Evaluation Group) and the OSP (the Office of Special Plans) specifically to deliver raw intelligence to the White House and build a case for war, and set up the White House Iraq Group to promote the case for war with the public.

Much of the intelligence these groups used came from the 'Information Collection Programme' run by the émigré group the Iraqi National Council (INC) whose leader, Ahmed Chilabi, had strong neoconservative ties. Many of the defectors the INC used were discovered to be unreliable or lying. Claims by Rafid Ahmed Alwan al-Janabi (codename 'Curveball') were widely repeated by the Bush administration, although in 2011 al-Janabi admitted he'd lied, saying, 'They gave me this chance. I had the chance to fabricate something to topple the regime' (Chulov and Pidd, 2011). There are claims too that the PCEG and OSP cherry-picked evidence, highlighted 'the most inflammatory' claims, changed the context and dates of information and made spurious connections to advance the White House's agenda (Lang, 2004:51–2).

Meanwhile, the administration leaked hundreds of stories to the press to push their case and the Pentagon groomed military analysts 'to shape terrorism coverage from inside the major TV and radio networks' (Barstow, 2008). Dick Cheney visited members of Congress, making exaggerated claims of Iraq's threat, regularly repeated these on television and pressured the CIA over the intelligence his sources were producing. Fearful of mistakes and 'demoralized' and 'beaten down' by the administration, the CIA was forced into dropping its caution (Hersh, 2005:224).

The administration also used dubious intelligence to bolster its case. Erroneous claims that Iraq possessed drones capable of delivering chemical or biological weapons, that it had purchased 'yellow-cake' uranium from Niger, and that aluminum tubes it had purchased were intended for a nuclear programme were all repeatedly used by key administration figures in congressional briefings, television appearances and speeches. Media-friendly soundbites such as Condoleezza Rice's 'We don't want the smoking gun to be a mushroom cloud' (Rich, 2006:243) trumped careful intelligence analysis, helping to construct a sense of imminent threat that had little basis in reality.

The UK government also manipulated evidence, publishing a series of documents arguing for the urgent danger of Saddam's regime. A dossier on 24 September 2002 claimed Iraq had WMDs it could deploy within 45 minutes, that it had reconstituted its nuclear programme and that it had sought uranium from Africa, all of which were false, whilst a second dossier on 3 February 2003 was discovered to be plagiarized from an online PhD. It was later revealed that from the late 1990s to the 2003 invasion MI6 had run 'Operation Mass Appeal' designed to plant stories about Saddam's WMDs in the media and 'gain support for sanctions and the use of military force in Iraq' (Rufford, 2003), whilst 'data of dubious quality' provided to MI6 by UN chief weapons inspector Scott Ritter was passed on to the media, appearing on front pages internationally. The government used these reports, Ritter says, to promote 'the perception that Iraq was a nation ruled by a leader with an addiction to WMDs' (BBC, 2003).

Later leaks also exposed the pre-existing arrangements between Bush and Blair for war. A White House memo from 5–7 April 2002 confirmed Blair's agreement to follow the USA into Iraq long before the parliamentary vote; the 'Downing Street memo' from 23 July included the head of MI6's assessment that 'military action was now seen as inevitable' and that in the USA 'the intelligence and the facts were being fixed around the policy'; and a note from Blair to Bush on 28 July began with the words, 'I will be with you whatever'. A White House memo from 31 January 2003 described Blair as 'solidly with the president and ready to do whatever it took to disarm Saddam' and included discussions about both a deal to invade Iraq regardless of WMDs and the possibility of provoking Saddam into a confrontation (Rycroft, 2002; Norton-Taylor, 2006; Owen and Lowther, 2015; Booth, 2016).

On 6 July 2016 the Chilcot Report, the UK government inquiry into Britain's involvement in Iraq from 2001–09, was published. It concluded that the UK had chosen to join the invasion before peaceful options had been exhausted, that Blair deliberately exaggerated the threat posed by Saddam Hussein, that Blair personally promised Bush Britain's support, that the decision to invade was made in unsatisfactory circumstances, that there were insufficient plans for the postwar period, that the UK intelligence agencies produced 'flawed information' and that there was no imminent threat from Saddam (Chilcot, 2016).

In time, the claims of Iraqi WMDs would prove wrong. The UNSCOM team proved to have been highly-effective at destroying Saddam's weapons after 1991. Saddam's dissimulation over the following years was the result of his need to maintain his regional standing and deter Iran, though the side-effect of this was that few later believed him when he claimed to have fulfilled the UN's demands. Claims of Saddam's links to terrorism were equally dubious. The neoconservatives had seized on Laurie Mylroie's 2000 American Enterprise Institute publication, *Study of Revenge: Saddam Hussein's Unfinished War Against America*, which claimed Iraqi involvement in the 1993 World Trade Centre bomb and which, as Bergen says, saw Saddam as 'the mastermind of a vast terrorist conspiracy against the United States against virtually all evidence and all expert opinion' (Bergen, 2011:135). In July 2003 Mylroie would tell the 9/11 commission that Saddam's intelligence agents were also behind those attacks, a belief shared by senior White House officials who, Burke says, repeatedly suggested 9/11 'had depended upon Iraqi assistance or even been an Iraqi plan from the beginning' (Burke, 2011:102).

Crucially, the neoconservatives still viewed the world of dangers in terms of heavily militarized enemy states and so believed major terrorist attacks required state support. Hence Paul Wolfowitz's response to Richard Clarke's April 2001 administration briefing about al-Qaeda: 'You give Bin Laden too much credit. He could not do all these things like the 1993 attack on New York not without a state sponsor' (Clarke and Knake, 2004:231–2). Out of office through the 1990s and obsessed with toppling Saddam Hussein, the neocon-servatives failed to recognize changes in both terrorism and Islamist militancy.

Even after 9/11 they failed to understand Islamism, remaining convinced of its links with Saddam. In fact, contacts between al-Qaeda and Iraq in the 1990s had floundered over their fundamental differences and although Powell was correct that Islamists were operating in northern Iraq in 2003 they were opposed by Saddam and were actually being *protected* by the coalition's restrictions on Iraqi military actions in the area. Saddam's regime had embraced Islam since the Gulf War, but this was a populist ploy to increase support for a fundamentally secular government whose Baathism brooked no alternative source of authority. Regionally and internally, Saddam functioned as a bulwark against Islamist militancy. Ironically, in removing him, the USA would produce in Iraq the very thing they claimed to oppose.

Neither of these issues really mattered, however, because as soon as the war began the issue of WMDs and al-Qaeda links was forgotten.

Box 5.1 The Iraq War 2003

The Iraq War is an armed conflict that began with the US invasion of Iraq on 19 March 2003 and ended with the US declaration of victory on 15 April and the occupation of the country. Its roots lay in the aftermath of the 1991 war that had left Saddam's regime in power. UN resolution 687 (3 April 1991) had authorized a United Nations Special Commission (UNSCOM) to ensure the destruction of Iraq's chemical and biological weapons, any nuclear programme and ballistic missile with a range over 150km. Iraq was also subject to harsh sanctions, as well as southern and northern 'no-fly-zones', imposed by the USA after Saddam's repression of post-war Shia and Kurdish uprisings and regularly enforced with USAF strikes on Iraqi air-defences. Over the following years Iraq suffered ongoing punitive military attacks with major US cruise missile strikes in June 1993 (in response to an Iraqi plot to assassinate George Bush Sr), September 1996 (in response to an Iraqi offensive against the Kurds), December 1998 (in response to Saddam's failure to comply with UN disarmament demands), and in February 2001 (in response to an increased threat from Iraqi defence systems). Iraq's resistance to the UN inspections forced the UNSCOM team's withdrawal in 1998 and Iraq refused to recognize UN resolution 1284 (17 December 1999) or allow the new UNMOVIC inspection team it authorized into the country.

Saddam's survival and flaunting of UN resolutions angered the USA. The neoconservatives especially saw Iraq as an unfinished war and campaigned for Saddam's removal through the 1990s. 9/11 provided the opportunity, with the Bush administration using the pretext of WMDs and claimed Islamist links to extend the War on Terror. Preparations for the invasion began in June 2002 with the launch of 'Operation Southern Focus', designed to soften up air defences, and the penetration of northern Iraq in July by a CIA team to persuade Iraqi commanders not to resist an invasion and to organize a Kurdish northern offensive. The USA and UK began a military build-up in the Gulf in January 2003. Rumsfeld, a fan of cost-efficiency and 'network-centric warfare' (see chapter 3) rejected the army's plan for 500,000 troops and insisted a smaller, more mobile force of 130,000 was sufficient to defeat Iraq. Backed by a 46-nation (by April, 48) coalition and headed by US General Tommy Franks, the final invasion force comprised 148,000 US troops, with 45,000 British troops and around 3,500 from Australia, Poland and Spain. At 5.34am Baghdad time on 20 March (9.34pm EST, 19 March) 'Operation Iraqi Freedom' began. After an initial attempt to kill the Iraqi leadership, the main air campaign began on the 21, apparently employing 'shock and awe' to sap the enemy's will to resist. The main ground offensive came from Kuwait, with the US 1st Marine Expeditionary force and the armored 3rd Infantry Division moving in a double-thrust north towards Baghdad. In the west, an air and amphibious assault was launched to secure the oil fields and ports of the al-Faw peninsula and UK troops moved to take Umm Qasr and Basra. The CIA/JSOC forces and 70,000 Kurdish militia constituted the northern front against the Islamists and the Iraqi army.

Unlike in 1991, there was strong resistance. The UK fought battles at Umm Qasr and Basra, and the USA defeated Iraqi forces at Nasiriyah, Najaf and Karbala. After a delay to secure supply lines and for a sandstorm, the USA reached the outskirts of Baghdad in early April. After fierce fighting to secure Baghdad International Airport on 3–4 April the USA launched armoured reconnaissance raids into the city on the 5th—6th. Fighting took place around three key road interchanges on the 7th and Iraq counterattacked over the Jumhariya bridge on the 8th, but by 9 April the USA had taken control of the city. The most effective resistance through the war was by the 'Fedayeen Saddam' paramilitary force, Syrian foreign fighters and remnants of the Republican Guard. The main force of the Iraqi Army mostly refused to fight. The USA declared victory on 15 April after Tikrit fell and on 1 May 2003 President Bush arrived on the aircraft carrier USS *Abraham Lincoln* to declare 'the end of major combat operations' beneath a banner reading 'Mission Accomplished'.

On the surface the invasion appeared successful, removing Saddam's regime in 21 days and avoiding a bloodbath, with only 139 US and 33 UK troops killed. In October 2003 invasion-phase Iraqi casualties were estimated at 10,800–15,100, with 3,200–4,300 of those being civilians. In 2005 an Iraq Body Count report estimated 6,882 civilian deaths from US-led forces up to 1 May. Problems soon became obvious, however. Widespread looting brought chaos, Saddam had escaped and the last stand of 'former regime elements' turned out to be the beginning of a planned Baathist guerilla war that would soon coalesce with a Sunni insurgency and an emerging jihadi movement. The USA's failure to plan for the war's aftermath, mishandling of key decisions, inability to provide basic security or services and continued 'occupation' soon turned Iraqis against their 'liberators'. The Iraq War finished on 15 April 2003 but the violence was just beginning.

Mediating Iraq

The 2003 Iraq War represented the high-water mark of the system of military media management introduced in the 1991 Gulf War, again employing a top-down model aiming at the control of battlefield information and victory in the media sphere. In two important ways, however, this control went further: first, in the extent to which they integrated the mainstream media into their operation, and second in the way in which the military became media producers. This informational control and media production was visible in the opening bombing campaign.

As in 1991, the outbreak of war was carried live on television, but this time CNN viewers could watch live images of the assault on Baghdad. The bombing campaign was promoted on television as 'Shock and Awe' – the application of the idea of 'rapid dominance (Ullman and Wade, 1996)' developed in 1996 by Harlan K. Ullman and James Wade Jr that argued for spectacular, overwhelming displays of force to shock and incapacitate the enemy – but the reality was more complex. What we were actually watching was a failed attempt to assassinate the Baathist leadership. After an initial strike on a presidential palace failed to kill Saddam the Iraqi leadership went into hiding. The USA decided to hit their original 55 targets anyway and it was these images that were broadcast around the world under the headline 'Shock and Awe', though, as Sanders explains, 'it was a meaningless pyrotechnic display, the Americans knowing full well the buildings were empty' (Sanders, 2013). The real function of the attack, following the failed spectacle of the Afghan War, was to reestablish the USA's global reputation by repeating their greatest media triumph. The real-space 'shock' of the munitions was less important than the real-time 'awe' of the live images. As Rich argues, this was a spectacle for mediated consumption: 'On-screen the pyrotechnics of Shock and Awe looked like a distant fireworks display, or perhaps the cool computer graphics of a *Matrix*-inspired video game, rather than the bombing of a large city. None of Baghdad's six million people were visible' (Rich, 2006:75).

The US authorities meticulously planned and controlled their informational activities. Early morning conference calls between US departments and the UK established the day's message that would be disseminated in daily press conferences Washington, London and from Central Command's $1.5m media Centre at Camp As Sayliyah, Near Doha, Qatar. The latter, with a backdrop created by a Hollywood set-designer and fully equipped for media reporting, was criticized by journalists for its complex access and security procedures, its windowless, brightly-lit, air-conditioned, warehouse-like building and its 'unreal detachment' from the outside world and from the war (Tumber and Palmer, 2004:67). Sealed inside, journalists spent their time talking to each other or watching television. Brigadier General Vincent K. Brooks' media briefings were criticized for his restrictive approach and for the hierarchy of seating and questioning that favoured the US networks, the BBC and the international news agencies. Journalists soon realized that Brooks' briefings were largely 'spin', being tightly-controlled PR presentations that gave little away and avoided difficult questions, and many stopped attending them

Alongside these briefings, the second element of the military-informational system was the military 'embedding' of journalists. Whilst this appeared to represent a remarkable openness, reversing the policy of Kosovo and Afghanistan by giving journalists unprecedented access to the military and the battlefield, in reality it proved an extension of the system of media control, ensuring journalistic compliance. Close wartime links between journalists and the military weren't new but embedding went even further than the 1991 pool system, threatening to erase the distinction between the two. By March 2003 up to 775 journalists were embedded with the military; a process controlled by the US Department of Defense. Those selected were made to sign a contract with the military, with their reporting covered by a 50-point Pentagon plan. Although they were free to leave, they could not move or report independently or meet with other units.

Embedded journalists travelled and lived with their units. Sending copy back was easier than in 1991, with no censors or transmission units and advances in technology allowing more mobile, flexible and even solo reporting. The US military presented embedding as an example of coalition transparency, versus the disinformation of the enemy. General Tommy Franks lauded precisely this, saying, 'I believe that the greatest truth that's available to the world about what's going on is found in the pictures that come from the front lines where the war is being fought'. For him embedding was an example of First Amendment freedom of speech, letting the country 'know the truth' about the war (Tumber and Palmer, 2004:51). The major benefit for reporters was access to the battle-field. As Tumber and Palmer explain, 'It gave them a prestige – "we were there" – with their audience, something which was rare in previous conflicts' (Tumber and Palmer, 2004:18). This live reporting under fire was remarkable to watch and attracted significant ratings. Overall the embed project proved to be a huge success for the military: as David Miller says, 'Embedded journalists were the greatest PR coup of the war' (Miller, 2004:89).

Embedding did attract criticism from politicians and the military. The live coverage was blamed for reducing the complexity of war, for jeopardizing operational security by giving too much information about operations, for creating unrealistic expectations about the speed and success of the war and for putting pressure on the military. In a speech on 2 April 2003 the UK Foreign Secretary Jack Straw said, 'Had the public been able to see live reports from the trenches, I wonder how long the governments of Asquith and Lloyd George could have maintained the war effort. Imagine the carnage of the Somme on Sky and BBC News 24'. Could the spirit of Dunkirk 'have withstood the scrutiny of 24-hour live news?', he asked (Straw, 2003). There were also occasional tensions between journalists and their units, with some being asked to leave for compromising army operations whilst others left due to safety issues or for lack of stories.

Embedding was also criticized for its limitations. As in 1991, real-time images eclipsed information, giving only tiny slices of action, with the excitement of the moment replacing, facts, context or any understanding of the broader sweep of the conflict. 24-hour news now included a range of gimmickry, including split-screen and multi-window effects, audience voting, email feedback and interactive buttons such as Sky News' red 'war' button in the top corner that allowed a personally-tailored experience of the conflict and the reporting. All of this seemed to realize the dream of *live war* but, as in 1991, the reality was still absent from our screens: viewers saw live firefights but no casualties, no injuries and no dead bodies. It remained a sanitized, technologically-hyperrealized simulacrum of warfare.

Despite Franks' claims, embedding was no guarantor of truth. The western media made fun of the Iraqi Information Minister Mohammed Saeed al-Sahaf, dubbing him 'Comical Ali' for his on-air Panglossian stories about the war's progress, but their own reports proved as fictive. For at least four days from 20 March the port of Umm Qasr was reported to have been taken by the coalition, despite continued fighting there, whilst the death or capture of Ali Hassan al-Majid ('Chemical Ali') was repeatedly reported. For all the benefits of embedding, the military themselves often despaired of the pressures and 'view-through-a-smartie-tube' focus of the news, which elevated minor incidents to major global events and treated any engagement as a 'setback'.

The biggest criticism of embedding, however, was that it marked the loss of journalistic independence and objectivity. Journalists had an obvious situational bias in only seeing things from the same perspective of the military and having no contact with the Iraqi population, a situation exacerbated by the fact that almost none spoke the language of the country they were in or had any specialist regional knowledge. Moreover, the military assumed the journalists' role was to promote their view of the war: as CNN's Bob Franken commented, 'there was frequently a belief on the part of the military that we were there to represent the home-team' (Tumber and Palmer, 2004:49). A bigger problem than all this, however, was in journalists themselves 'going native'.

Embedded journalists became part of the military, adopting its uniform, acronyms and slang, making friends with the soldiers and ultimately identifying with their lives and mission. As war correspondent Chris Hedges says, depending on the troops for their food and quarters and protection, journalists identify with and want to protect them in turn: 'They become part of the team. It is a natural reaction. I have felt it' (Tumber and Palmer, 2004:61). Keith Harrison confirmed this identification and integration, arguing, 'We look and sound like soldiers ... We answer to the Commanding Officer, we follow orders, we share the rations, we eat where the soldiers eat and we sleep where they sleep ... We're becoming indoctrinated and recognize the sights and sounds of military life instantly' (Tumber and Palmer, 2004:54).

Effectively, embedded journalists traded personal and professional distance for access to stories and the event. Once integrated into the military, however, they were unable or unwilling to assert any independent or critical opinion. Personal attachment overrode professional detachment, creating a powerful self-censorship, turning journalists into cheerleaders for the war and preventing any interrogation of the event. Sometimes the embeds went further. The BBC's Gavin Hewitt, for example, describes acting as a spotter for his unit, picking out targets for them to attack, thereby completing the transformation of the journalist into an active military combatant.

One of the most insightful critiques of the UK coverage came from the satirist and broadcaster Armando Iannucci, whose article 'Shoot Now, Think Later' appeared in *The Guardian* on 28 April. Whereas Baudrillard had suggested the Gulf War hadn't happened, Iannucci says, 'it's perhaps more true to say of the recent war in Iraq that an explanation of it never happened' (Iannucci, 2003). Why, despite all the access and live reporting was the coverage so inaccurate and confused? The reason, he suggests, is that the it was in thrall to the latest media format, reality TV:

> Programmed over the past few years to watch endless numbing footage of ordinary people trapped by television producers in an artificial situation, we, the viewers, felt we were watching Reality-TV on a massive scale. As Rageh Omar, John Simpson and Bill Neely swapped smart suits for combat gear, we realized we were watching a People Show. Raggi and John and Bill were Jo Public, ordinary bods whom we'd had in our living rooms once or twice a week over the past few years. Now we were watching them placed in a hostile environment, and have cameras pointed at them 24 hours a day while a whole nation gawped.
>
> Iannucci (2003)

Embedding led 'to great footage, but lousy reportage': no journalist living alongside the military was going to report 'the troops I'm living with are disgruntled. Their equipment doesn't work, they're probably blowing up children, and one or two of them are going to die'. Instead, 'objectivity melted faster than a division of the Republican Guard'. The emphasis upon immediacy and action ensured questions about the war's legality were left behind 'like yesterday's old news'. The need to fill the screen and the 'automatic reflex' of the broadcasters that once it had begun that this was 'our war' led, Iannucci concludes, to 'a total breakdown of intelligent and dispassionate analysis'.

The links with reality-television were widely-noted. Embeds such as David Bloom, who filmed himself riding on the back of his 'Bloom mobile', turned the war into a *Survivor*-style challenge show in which *their* story – their journey through the war – was the central motif. As Stahl says, 'Embedding itself became the plot and embedded reporters the main characters' (Stahl, 2010:86). These links weren't a coincidence. Stahl notes that from 2001–02 reality TV had already been transformed 'into a robust arm of the military-entertainment complex' (Stahl, 2010:73) through a range of military-themed and co-produced reality TV challenge shows, whilst the idea of embedding emerged from one such military-produced show.

The 2002–03 ABC-Pentagon joint venture *Profiles From the Front Lines*, was the brainchild of communications consultant Victoria Clarke, the Assistant Secretary of Defense for Public Affairs and previously director of the Washington office of the PR firm Hill & Knowlton (the firm who had devised the baby-incubator deaths story prior to the 1991 war). Designed to be pro-American and pro-military and employing talents such as Jerry Bruckheimer, producer of *Top Gun* (1986) and *Black Hawk Down* (2001), and Bertram van Munster, creator of the reality show *COPS*, the show followed US troops in Afghanistan, telling their stories. For Clarke, the idea was to increase public understanding and support: 'we knew the more people saw the US military, the more they would understand the mission and how they were going about their jobs' (Tumber and Palmer, 2004:13). As Stahl explains, 'for the Pentagon the news was a liability. An entertainment show like *Profiles*, on the other hand, existed in the twilight dimension of representation, having some claim on the truth but not bogged down with the potentially "disloyal" ethic of journalistic objectivity' (Stahl, 2010:84–85). In avoiding policy issues and focusing on characters and their stories *Profiles* depoliticized the conflict whilst serving as perfect PR in creating a close audience identification with the troops.

Profiles came to an end eight days before the Iraq War began, with its central premise laying the basis for embedding. From the beginning, therefore, embedding was a military-designed, reality TV production designed for propaganda and public relations purposes. It served to assimilate journalism, to control its reporting, to limit public information, criticism and debate, and to challenge the traditional role of the press as an independent force and a check on power. What we saw instead, Stahl argues, was 'militainment': the experience of watching a 'virtual soldier' and 'civilian contestant' survive in a warzone under the gaze of real-time television; a mode of media infusing conflict 'with the interactive excitations and pleasures of reality television' (Stahl, 2010:90).

Following the principles of 'information warfare' and 'full spectrum dominance', the US army treated any alternative source of battlefield information as hostile. This included 'unilaterals' – western journalists operating outside the embedding programme. Left to fend for themselves they were viewed by the military as unpredictable, dangerous and off-message, and were fired on by both coalition and Iraqi forces. The Iraq War proved one of the deadliest for journalists with 17 killed or dying in the six-week invasion. The casualties included ITN's Terry Lloyd who, an October 2006 inquest ruled, was 'unlawfully killed' by US troops on 22 March 2003, being shot in the head in a makeshift ambulance after being hit by crossfire in a firefight near the Shatt al Basra bridge. The National Union of Journalists (NUJ) described his death as a 'war crime' and 'a despicable, deliberate, vengeful act', saying 'US forces appear to have allowed their soldiers to behave like trigger-happy cowboys in an area where civilians were moving about' (BBC, 2006).

Al-Jazeera was again a target for US anger. From the beginning, it offered an alternative vision of what it called 'The War on Iraq', referring to 'invasion forces' rather than 'coalition forces', employing the term 'resistance' and emphasizing colonialism above the western rhetoric of 'liberation'. It showed images of the historic city of Baghdad burning, covered middle-eastern anti-war protests and offered a broader coverage across Iraq, including from places such as Basra western reporters couldn't reach. Separate from both the military and from patriotic cheer-leading Al-Jazeera was able to offer better, more independent and even-handed journalism. It treated the coalition and Iraqi governments the same, expressing journalistic skepticism at claims by each side, discrediting fake reports when it discovered them and showing Iraqi as well as coalition statements and press conferences. Its fact-checking was more thorough than the western media's and it included on air a broader range of commentators. As Miles concludes, in Iraq Al-Jazeera 'had broken the hegemony of the Western networks and, for the first time in hundreds of years, reversed the flow of information, historically from West to East' (Miles, 2005:278).

Al-Jazeera's balance attracted criticism and threats from each side, with the west treating it as 'an enemy news station' and as 'Saddam's voice' (Miles, 2005:256, 259), and Iraq blaming it for subverting Arab support. Al-Jazeera's primary commitment, however, was to depicting the war accurately, which included emphasizing humanitarian issues such as the civilian casualties of the western bombing as well as showing images of captured US troops and dead US and UK soldiers. The Pentagon and MOD were outraged and accused Al-Jazeera of disregarding the Geneva Convention, despite the USA removing all convention rights for 'enemy combatants' in the War on Terror and the fact that Iraqi prisoners had been widely shown on western media.

As in Afghanistan, this hostility translated into military action. On 8 April, an Al-Jazeera reporter was killed and a cameraman injured when the USA bombed their Baghdad television station as well as Abu Dhabi television. On the same day two journalists were killed and four wounded when a US tank opened fire on the Palestine Hotel, the base for western journalists. The USA's initial explanation was that the tank was receiving small arms fire from the hotel whilst 'the Al-Jazeera office that was hit was a mistake' (Tumber and Palmer, 2004:44), although it later asserted that an Iraqi mortar-spotter had been operating from the hotel. Brigadier General Brooks responded to Al-Jazeera's protests with the claim 'This coalition does not target journalists. We do not know every place journalists are operating on the battlefield. It's a dangerous place indeed' (Pew Research Centre, 2006), but the USA had a history of targeting communications and knew the precise locations of the television stations, whilst the Palestine Hotel was world famous as a base for journalists. Many concluded that non-embedded war correspondents were being deliberately targeted to discourage independent reporting.

Two other events showed the military becoming active media producers themselves. The first revolved around the rescue of the injured Private Jessica Lynch who had been captured by Iraqi forces on 23 March and who was rescued by US special forces in a raid on Nasiriyah hospital on 1 April. The next day the USA held a press conference, releasing a military-filmed and edited five-minute video-movie of the raid: 'an action-packed montage of the guns-blazing Special Operations raid to rescue Lynch, bathed in the iridescent green glow of night-vision photography' (Rich, 2006:81). The 'Saving Private Lynch' video was a powerful propaganda tactic, designed to distract from criticism of the war's progress, explicitly following *Profiles From the Front Lines* by focusing upon an individual's story and heroism. Influenced by reality TV and the 2001 film *Black Hawk Down*, and referencing Spielberg's 1998 movie *Saving Private Ryan*, the video was 'one of the most stunning pieces of news management yet conceived' (Kampfner, 2003), demonstrating the military's mastery of public relations and sentiment. As Stahl notes, 'The drama was simple, unquestionably noble, successful, and freed from the cumbersome need to explain the rightness of the Iraq invasion' (Stahl, 2010:81).

The story soon unravelled, however, as the circumstances of Lynch's capture, her resistance, her injuries and her treatment in hospital were all disputed. It was also discovered that hospital staff had apparently tried to return Private Lynch once but their ambulance was fired on by US troops, that Iraqi forces had fled the hospital the night before the raid, and that the special forces knew they would be unopposed. What was filmed, therefore, was a branch of reality TV 'militainment': an operation staged for the cameras, designed for global consumption and with a human-interest focus perfect too for the ABC-Pentagon joint-production, made-for-TV movie, *Saving Jessica Lynch* (2003), it would later become.

Just as the Lynch story was being questioned, a new feel-good image arose to replace it: the toppling of Saddam's statue. On 9 April Iraqis began attacking the statue of Saddam opposite the Palestine Hotel in Firdos Square, Baghdad, before the USA eventually brought it down with an M88. There was a brief faux pas when a marine placed the US flag over Saddam's head before it was hastily replaced by the more PR-friendly Iraqi flag and the celebrations could commence. The event lasted about two hours with every major network and news company staying with it live. The images were repeated all day (between 11am and 8pm Fox replayed the toppling every 4.4 minutes and CNN every 7.5 minutes) and they dominated the next day's front-pages. Although a weak symbolic counterpoint to the toppling of the Twin Towers, the media seized upon the images, inflating a militarily irrelevant sideshow into a 'historic' global event marking the fall of Saddam himself. 'You think of seminal moments in a nation's history … indelible moments like the fall of the Berlin Wall', Bill Hemmer said on CNN, 'and that's what we're seeing right now' (Rich, 2006:83).

Later research exposed this event too as, in part, a military production for the world's cameras. Bryan McCoy, the marine commander deployed to secure the Palestine hotel, 'was fully cognizant that he was about to move into an area where there were a lot of journalists and there were going to be opportunities' (Maass, 2011). Coalition forces had been destroying images of Saddam so when Iraqis asked to borrow a sledgehammer he recognized the opportunity for psychological propaganda: 'I realized this was a big deal. You've got all the press out there and everybody is liquored up on the moment. You have this Paris, 1944, feel. I remember thinking: The media is watching the Iraqis trying to topple this icon of Saddam Hussein. Let's give them a hand' (Maass, 2011). Realizing the event would fail and the Iraqis would drift away without US help, McCoy got authorization for the marines to finish the job.

The media coverage was misleading. Wide-angle shots showing the square ringed with US vehicles and revealing a crowd only a few hundred strong were rarely repeated. Around half the crowd were journalists and marines but these were avoided in the close-up shots, which focused upon the Iraqis whose own enthusiasm and level of participation in the event seemed to follow the presence of the cameras. As well as obscuring the scale and nature of the event, the media's non-stop coverage, celebratory tone and emphasis on the toppling of the statue as the end of Saddam were profoundly disconnected from the reality of the war and of a conflict whose violence had barely begun. This was a deliberate choice by the networks. As Peter Maas says, with Saddam absent, no leadership to surrender and no conclusive battles to report, the media were looking for 'some sort of culminating event':

> Primed for triumph, they were ready to latch onto a symbol of what they believed would be a joyous finale to the war. It was an unfortunate fusion: a preconception of what would happen, of what victory would look like, connected at Firdos Square with an aesthetically perfect representation of that preconception.
>
> Maass (2011)

Hence a globally-photogenic but ultimately irrelevant media event became, by self-fulfilling prophecy, a 'historic', 'where-were-you-when' event. The saturation coverage and celebrations fuelled the idea the war was over and had been successful, and thus diverted attention and interest from Iraq at a time when it was needed. It prepared the ground for Bush's 'mission accomplished' speech on 1 May and for the fundamental misreading of the emerging Baathist guerrilla war in Iraq as merely the last-gasp of defeated 'former regime elements'.

Bush's May Day speech was perhaps the greatest media performance of the invasion phase. Channelling *Top Gun*, Bush arrived on the flight deck of the USS *Abraham Lincoln* in a S-3B Viking fighter-jet, emerging in a full naval aviator flight uniform to announce the end of major hostilities beneath a 'mission accomplished' banner. Again, not everything was as it seemed. The television shots concealed that the carrier was only forty miles off the coast of San Diego, with a skyline that would have been visible from another angle, whilst Bush needn't have arrived by plane and wasn't required to wear a flight suit that was a far cry from the National Guard uniform of his only military experience.

The Iraq War remained primarily a television and newspaper war, dominated by live 24-hour coverage and the main television news summaries, with newspapers providing next-day overviews. Online it was the same established news providers who dominated. The top US news site during the war was CNN.com with 26 million unique users in March 2003, whilst in the UK the top sites were the BBC with three million users and *The Guardian Unlimited* with 1.3 million users. Much of the reporting replicated that available on television or in the papers but the rise of affordable broadband meant that a range of multimedia experiences could be offered, including clips, raw videos, archived reports, journals and interactive maps and slide-shows. One important effect of the net was in allowing users to find a broader range of information including accessing international media coverage and alternative news sources such as Al-Jazeera's English-language website, or progressive sites such as Alternet, Counterpunch and CommonDreams.

The internet also empowered users as producers of content. Anyone could comment on the war in forums and chat-rooms, or on their own web-pages or 'weblogs'. The first blogs appeared in the mid-1990s, taking the form of online diaries that were manually updated on websites, and by 1999 specialist blogging tools such as LiveJournal and Blogger began to appear. 9/11 was a spur to political blogging and the first 'warblogs' – blogs dealing specifically with war – appeared during the Afghanistan campaign. Iraq saw the take-off of the 'milblog': blogs written by serving soldiers in the field, such as 'Live From the Sandbox', written by 'LT Smash' (Scott Koenig) about his experiences as an army reservist in the Gulf. It was receiving 6,000 hits a day by mid-2003, with feedback and commentary on his posts producing an ongoing conversation with his readership. As Melissa Wall argues, blogs now emerged as 'a new form of journalism', establishing their credibility through their personal experience and voice, their distance from power and their direct relationship with their audience (Wall, 2005).

Some used the format to develop independent journalism. Freelancer Chris Allbritton used his blog, 'Back to Iraq', to ask for funding for travel and expenses, offering his readers exclusive reports, the chance to request stories and a journalism that 'owes its sole allegiance to the readers' rather than editors or proprietors (Matheson and Allan, 2009:80). By the end of March, he was receiving over 14,000 unique users a day and over $10,000 had been pledged. Some mainstream organizations such as the BBC also picked up on the interest in blogging and began publishing journalist's diaries, journals and personal reflections in addition to their filed reports. Although obviously subjective, their situated-ness and candour gave them a credibility ordinary embedded reporting lacked.

The most famous invasion-phase blog was by the Iraqi 'Salam Pax' (Salam Al-Junabi). He had created *Where is Raed?* in June 2002 to communicate with a friend and discuss disappearances under Saddam, but traffic to his site picked up in October when he began posting about US invasion plans. When the war started his blog attracted global attention, becoming the most linked-to diary on the net, attracting 20,000 readers a day until the regime cut off internet access ten days in. Pax was important as almost the only voice describing the experience of the conflict and impact of the bombing on the lives and experiences of ordinary Iraqis. The internet had allowed the bottom-up voices of civilians to be heard before but Pax's blog was a milestone in the movement away from military informational control and the dominance of mainstream media, anticipating the participa-tive battlefields to come.

Digital technologies, therefore, were transforming the global media ecology, although it should be remembered that the audience for these websites and blogs were small. The Iraq War occurred on the cusp of the 'Web 2.0' revolution that, from 2004–05, would transform the experience of the occupation and postwar Iraq. The invasion remained a war dominated by television and print journalism and was marked, above all else, by their failure. It was the mainstream news organizations that failed to challenge the ideological agendas of their governments, that failed to question claims of Iraq's imminent threat, that failed to reflect anti-war sentiment, that failed to challenge the claim that war was the only remaining option, and that failed to query the absence of plans for the war's aftermath. It was the mainstream media's journalists that eagerly donned combat fatigues and climbed onto military vehicles, filing breathless live reports about the war's progress without any mention of the missing WMDs or the Iraqis whose country it was.

Many journalists later recognized this. In his 2010 documentary *The War You Don't See* John Pilger got CBS News Anchor Dan Rather and BBC embed Rageh Omaar to admit to their mistakes in Iraq. 'I didn't really do my job properly', Omaar accepts. Pilger's conclusion in an accompanying 2010 article is simple and clear: 'Had journalists questioned the deceptions that led to the Iraq War, instead of amplifying them, the invasion would not have happened' (Pilger, 2010a, 2010b).

Key reading

As with chapter 4, Wright (2011) and Burke (2011), read together, provide the best political background to the War on Terror and thus also to the Iraq War, although I'd also recommend Keegan (2004, chapter 5) for an easy-to-read overview of the specific issues around the Iraq invasion. The political machinations behind the US and UK decision to go to war are well covered in Lang (2004), Hersh (2005, chapter 5) and Rich (2006). Keegan (2004) provides a useful broader history of Iraq and a detailed overview of how the Iraq War unfolded militarily, although his concluding assessment written soon after the war's end that 'objectively the world was undoubtedly a safer place …' has proven naïve. Tumber and Palmer (2004), Lewis et al. (2006) and Hoskins (2004) offer the best overview of issues around the media during the war including the military's informational arrangements and control, embedding, journalist safety, the unilaterals, and the newspaper and television coverage. Stahl (2010, chapter 3) is excellent on the reality TV origins of embedding whilst Miller (2004) offers many important critical essays on the propaganda role of the media in the war. Pilger (2010a, 2010b) offers a polemical critique of the journalist's role in the Iraq War.

References

Barstow, D. (2008) 'Behind TV Analysts, Pentagon's Hidden Hand', *New York Times*, 20 April, http://www.nytimes.com/2008/04/20/us/20generals.html.

BBC. (2003) 'MI6 ran "Dubious" Iraq Campaign', *BBC News*, 21 November, Friday, http://news.bbc.co.uk/1/hi/uk/3227506.stm.

BBC. (2006) 'Iraq Reporter Unlawfully Killed', *BBC News*,13 October, Friday, http://news.bbc.co.uk/1/hi/uk/6046950.stm.

Bergen, P. L. (2011) *The Longest War*, New York: Free Press.

Booth, R. (2016) '"With you, Whatever': Tony Blair's Letters to George Bush', *The Guardian*, 6 July, https://www.theguardian.com/uk-news/2016/jul/06/with-you-whatever-tony-blair-letters-george-w-bush-chilcot.

Borger, J. (2006) 'Blogger bares Rumsfeld's Post-9/11 Orders', *The Guardian*, 24 February, http://www.theguardian.com/world/2006/feb/24/freedomofinformation.september11.

Burke, J. (2011) *The 9/11 Wars*, London: Penguin.

Bush, G. W. (2002a) 'George W. Bush, State of the Union, Washington, DC, January 29, 2002', http://www.presidentialrhetoric.com/speeches/01.29.02.html.

Bush, G. W. (2002b) 'George W. Bush. Commencement Address at the United States Military Academy at West Point. West Point, New York. June 1 2002', http://www.presidentialrhetoric.com/speeches/06.01.02.html.

Chilcot, J. (2016) *The Report of the Iraq Inquiry*, 6 July, Ordered by the House of Commons, London: Her Majesty's Stationery Office, http://www.iraqinquiry.org.uk/the-report/.

Chulov, M. and Pidd, H. (2011) 'Defector Admits to WMD Lies That Triggered Iraq War', *The Guardian*, 15 February, http://www.theguardian.com/world/2011/feb/15/defector-admits-wmd-lies-iraq-war.

Clarke, R. A. and Knake, R. A. (2004) *Against All Enemies*, London: Simon & Schuster Ltd.

Hersh, S. M. (2005) *Chain of Command*, London: Penguin.

Hoskins, A. (2004) *Televising War*, London: Continuum.

Iannucci, A. (2003) 'Shoot Now, Think Later', *The Guardian*, 28 April, https://www.theguardian.com/media/2003/apr/28/broadcasting.iraqandthemedia.

Kampfner, J. (2003) 'The Truth about Jessica', *The Guardian*, 15 May, https://www.theguardian.com/world/2003/may/15/iraq.usa2.

Keegan, J. (2004) *The Iraq War*, London: Random House.

Kristol, W. and Kagan, R. (1996) 'Toward a Neo-Reaganite Foreign Policy', *Foreign Affairs*, July/August, http://carnegieendowment.org/1996/07/01/toward-neo-reaganite-foreign-policy/1ea.

Kull, S. (2003) 'Misperceptions, Media and the Iraq War', PIPA/Knowledge Networks Poll, 2 October, http://www.pipa.org/OnlineReports/Iraq/IraqMedia_Oct03/IraqMedia_Oct03_rpt.pdf.

Lang, W. P. (2004) 'Drinking the Kool-Aid', *Middle East Policy*, 11(2), pp. 39–60, at: https://www.blackwellpublishing.com/content/BPL_Images/Journal_Samples/MEPO1061-1924~11~2~152%5C152.pdf.

Leung, R. (2004) 'Bush Sought "Way" to Invade Iraq?' *CBS News*, 60 Minutes, 9 January, http://www.cbsnews.com/news/bush-sought-way-to-invade-iraq/.

Lewis, J., Brookes, R., Mosdell, N., and Threadgold, T. (2006) *Shoot First and Ask Questions Later*, New York: Peter Lang.

Maass, P. (2011) 'The Toppling: How the Media Inflated the Falloff Saddam's Statue in Firdos Square', *Pro Publica/The New Yorker*, 3 January, https://www.propublica.org/article/the-toppling-saddam-statue-firdos-square-baghdad.

Matheson, D. and Allan, S. (2009) *Digital War Reporting*, Cambridge: Polity.

Miles, H. (2005) *Al-Jazeera*, London: Abacus.

Miller, D. (ed.) (2004) *Tell Me Lies: Propaganda and Media Distortion in the Attack on Iraq*, London: Pluto Press.

Norton-Taylor, R. (2006) 'Blair-Bush Deal before Iraq War Revealed in Secret Memo', *The Guardian*, 3 February, http://www.theguardian.com/world/2006/feb/03/iraq.usa.

Owen, G. and Lowther, W. (2015) 'Smoking Gun Emails Reveal Blair's 'Deal in Blood' with George Bush Over Iraq War was Forged a YEAR before the Invasion had Even Started', *The Mail on Sunday*, 17 October, http://www.dailymail.co.uk/news/article-3277402/Smoking-gun-emails-reveal-Blair-s-deal-blood-George-Bush-Iraq-war-forged-YEAR-invasion-started.html.

Packer, G. (2007) *The Assassin's Gate: America In Iraq*, London: Faber and Faber.

Pew Research Centre. (2006) 'Al-Jazeera Timeline', 22 August, http://www.journalism.org/2006/08/22/al-jazeera-timeline/.

Pilger, J. (Dir.) (2010a) 'Why are Wars not being Reported Honestly?', *The Guardian*, 10 December, https://www.theguardian.com/media/2010/dec/10/war-media-propaganda-iraq-lies.

Pilger, J. (Dir.) (2010b) 'Why are Wars not being Reported Honestly?', *The War You Don't See*, Network [DVD], 10 December, https://www.theguardian.com/media/2010/dec/10/war-media-propaganda-iraq-lies.

Powell, C. (2003) 'Full Text of Colin Powell's Speech', Address to the United Nations Security Council, 5 February, http://www.theguardian.com/world/2003/feb/05/iraq.usa.

Raw Story. (2007) 'Hours after 9/11 Attacks, Rumsfeld allegedly said, "My Interest is to Hit Saddam"', 2 May, http://rawstory.com/news/2007/Hours_after_911_attacks_Rumsfeld_allegedly_0502.html.

Rich, F. (2006) *The Greatest Story Ever Sold*, London: Penguin.

Roberts, J. (2002) 'Plans for Iraq Attack Began on 9/11', *CBS News*, 4 September, http://www.cbsnews.com/news/plans-for-iraq-attack-began-on-9-11/

Rufford, N. (2003) 'Revealed: How MI6 Sold the Iraq War', *The Sunday Times*, 28 December, http://www.informationclearinghouse.info/article5433.htm.

Rycroft, M. (2002) 'The Secret Downing Street Memo', 23 July, http://nsarchive.gwu.edu/NSAEBB/NSAEBB328/II-Doc14.pdf.

Sanders, R. (2013) 'The Myth of "Shock and Awe": Why the Iraqi Invasion was a Disaster', *The Telegraph*, 19 March, http://www.telegraph.co.uk/news/worldnews/middleeast/iraq/9933587/The-myth-of-shock-and-awe-why-the-Iraqi-invasion-was-a-disaster.html.

Stahl, R. (2010) *Militainment Inc*, Abingdon, Oxon: Routledge.

Straw, J. (2003) 'Straw's Iraq Speech: Full Text', *BBC News*, 2 April, http://news.bbc.co.uk/1/hi/uk_politics/2908389.stm.

Tumber, H. and Palmer, J. (2004) *Media at War: The Iraq Crisis*, London: Sage.

Ullman, H. K. and Wade, J. jr. (1996) *Shock & Awe: Achieving Rapid Dominance*, National Defense University/Institute For National Strategic Studies, http://www.dodccrp.org/files/Ullman_Shock.pdf.

Wall, M. (2005) '"Blogs of War": Weblogs as News', *Journalism*, 6(2), pp. 153–72.

Wright, L. (2011) *The Looming Tower. Al-Qaeda's Road to 9/11*, London: Penguin.

6 From Abu Ghraib to Facebook

The end of military informational control

The digital revolution

When President Bush declared the end of major hostilities in Iraq on 1 May 2003 the media response was congratulatory. The aftermath was chaotic and Saddam and the WMDs were still missing, but the regime had been quickly removed and the predicted 'bloodbath' hadn't materialized. The USA had shown it could not only fight a war broadcast live on television but also that its model of media-management could even generate a large level of public support. The military's control of the war's narrative, its integration of mainstream media, its successful propaganda activities and its threats to independent journalism achieved an important informational victory. This model of media management had served the USA well since its introduction in 1991 but within a year of the 2003 Iraq invasion it would find itself overturned by an ongoing digital revolution. The two key elements behind this revolution were the rise of cheap, networked computing and the digitalization of older analogue media forms.

The original ARPAnet had been created in 1969 and in the mid-1970s engineers had created the protocols to link networks together into what they called an 'internet', but by the early 1990s online communication remained the preserve of a skilled minority. It was Tim Berners Lee's 'World Wide Web' (WWW), created whilst working for CERN in Zurich, that changed this. The WWW software, released the same year as the Gulf War (being made available free on CERN's computers on 17 May 1991 and officially made public when Berners Lee posted about it to news-groups on 6 August) was a set of tools allowing easy online publication and hypertext linking. When combined with graphical browsers such as Mosaic in 1993, it made online communication easier, helping democratize internet access and use. With growing interest in this 'new' technology, easily-available browsers (such as Microsoft's Internet Explorer, free with Windows 95), increased public access and dial-up ISPs emerging, 'the web' took off from 1995. Helped by ever cheaper and more powerful computers, by the millennium the internet had become an important domestic medium and a central part of everyday life.

Accompanying these changes in computing and networking was the digitalization of older, analogue forms. Media theorist Lev Manovich explains the rise of 'new media' as the result of the historical development of two trajectories – mass media and computing – and their meeting and merger at the end of the twentieth century (Manovich, 2001). Essentially digital technology became a meta-medium, transforming previously separate analogue forms into types of digital content. With this digitalization, computing became the basis of our informational and media experiences. Although each individual form had its own longer history of digital experimentation, from the mid-1990s we can trace the rapid commercial digitalization of cinema, newspapers and print, music, home-video, photography, video-camcorders, telephony, radio and television. By the millennium, every broadcast medium had been transformed by digital technology in some or all aspects of its production, distribution and consumption.

These developments were slow to impact on conflict and our understanding of it. The internet had empowered users in Kosovo and Iraq, especially, to speak of their experiences, but problems of access and language-barriers limited its global impact. Even in the west the internet was still in an early phase of development. Chat-rooms, forums and email had opened up peer-to-peer communication but few people had a permanent, public online presence as personal web-pages required money and coding skills and the new modes of interactivity hadn't yet had a revolutionary cultural impact. By 2003 cheap broadband improved the web experience and made using images, audio and video easier, but the web of the Iraq War remained a largely read-only medium.

The military authorities, therefore, gave little thought to the internet. For them control over global public perception meant control over the mainstream broadcast media. Even when, during the Iraq War, servicemen in the field began blogging about their experiences the military didn't appreciate the implications of this. As a top-down institution, used to issuing orders to masses of people, it remained tied to and in tune with top-down mass media, which, it assumed, held the key to the public mind. It was only in April 2004, when images of US abuse of Iraqi prisoners at Abu Ghraib went global, that the military understood the threat of the empowered digital user.

Abu Ghraib

The photographs were horrific. There was 'the hooded man', showing a hooded prisoner on a box, arms out, with wires attached to his hands; more hooded men on boxes with their arms handcuffed behind their legs or handcuffed over railings; men with female underwear on their heads; a female military policewoman (MP) holding a leash round the neck of a nude man on the floor; a nude man, arms outstretched, covered in faeces; prisoners in orange recoiling in terror from aggressive dogs; a soldier sitting on a man between two stretchers; nude prisoners piled up in 'human pyramids' next to proud and smiling MPs; hooded, naked prisoners lined up with a female MP pointing and laughing at their genitals; prone groups of prisoners with smiling soldiers over them, fists raised to strike; photos of bloodied floors and walls; more dogs and naked and cowering prisoners; and images of MPs posing with a corpse, giving the camera the thumbs-up. The corpse was that of Manadel al-Jamadi, an Iraqi suspect in a bomb attack who was murdered by CIA operatives during an interrogation on 4 November 2003. Placed into deep-freeze, he was nicknamed 'the iceman', 'Mr Frosty' and 'Bernie' (referencing the 1989 film *Weekend at Bernie's*) and taken out for photos.

The images were taken by US military police at Baghdad Central Prison, better known as 'Abu Ghraib', 20 miles outside of Baghdad. The prison had been a symbol of Saddam's repressive regime, becoming known for the abuse, mutilations, torture and mass executions that took place there. Standing empty following a prisoner amnesty in October 2002, it was reopened by the USA in July 2003 due to the need to rebuild Iraq's criminal and prison system and for a high-security institution to hold the 'terrorist' insurgents resisting the coalition. By March 2004 it held 7,490 prisoners, the majority swept up from the streets in confused US security operations. Only a small proportion were security detainees. Most were innocent civilians (including children) or petty criminals, though their legal processing was nearly useless and few were ever charged.

There had been earlier reports of abuse at the prison as well as in the detention systems at Afghanistan and Guantanamo Bay but little proof, with post-9/11 patriotism and the 'War on Terror' overriding public interest in prisoner welfare. What made Abu Ghraib different was the photographic evidence. The scandal was exposed on 13 January 2004 when Spc. Joseph Darby reported CDs of photographs he had been given to the Army's Criminal Investigation Command (CID). The Army launched an investigation, the *Army Regulation 15–6 Investigation of the 800th Military Police Brigade* under Major General Antonio Taguba, which reported in March that it had found evidence of 'systemic and illegal abuse' at the prison (Taguba, 2004; Hersh, 2007). The investigation, however, was limited to the military police, rather than the command structure, and the prison authorities encouraged staff to destroy additional evidence in order to contain the issue. The images leaked, being broadcast on 28 April in a US-TV *60 Minutes II* news report. The Pentagon had asked for broadcast to be delayed but *60 Minutes* feared being scooped by Seymour Hersh whose own investigation of Abu Ghraib was published online by *The New Yorker* on 30 April, seeing print from 10 May.

The images had an immediate, global impact. As Matheson and Allan argue, they 'shattered the norms of Pentagon-sanctioned imagery at a stroke' (Matheson and Allan, 2009:148). It was in the Arab media that they received the heaviest coverage. The primary meaning most took was of US hypocrisy and racism as a war claimed to have been fought for democracy and freedom, to bring the rule of law and respect for human rights to the Arab world, was exposed by images of US soldiers torturing Iraqis in the same prison Saddam had used. The images of men being sexually humiliated by female MPs and posed in female underwear and homosexual acts especially inflamed Arabic opinion. The public relations disaster of Abu Ghraib would act as a spur to the insurgency, helping solidify anti-American feeling and recruit new fighters as well as directly inspiring the murder of US citizen Nick Berg by Abu Musab al-Zarqawi, who released a decapitation video directly referencing Abu Ghraib onto the internet on 8 May.

The USA's response was to distance itself from the images, claiming the behaviour wasn't acceptable, condoned or common. This was a case of 'bad apples' who didn't represent US policy or values. On 6 May President Bush said he was 'sorry for the humiliations suffered by the Iraqi prisoners and the humiliations suffered by their families', whilst the following day Rumsfeld told Congress, 'to those Iraqis who were mistreated by members of US armed forces, I offer my deepest apology. It was inconsistent with the values of our nation … And it was certainly fundamentally un-American' (CNN, 2004; BBC, 2004). Rumsfeld admitted prisoner 'abuse' but refused, however, 'to address the "torture" word' (Hochschild, 2004), whilst the broader American culture was also in denial. The Liberal press was criticized for its lack of patriotism, whilst right-wing commentators hailed the MPs as heroes, risking their lives to fight terrorists who were killing US soldiers and Iraqis. The shock-jock Rush Limbaugh famously dismissed the acts as an 'emotional release', as 'people having a good time' (Sontag, 2004).

It soon became clear that 'the incidents were not exceptional, but examples of routine violent and degrading treatment of prisoners in the prison and elsewhere' (Matheson and Allan, 2009:151). More importantly, much of it was authorized. The policy traced its origins to the British military in Northern Ireland who, to solve the problem of how to legally torture terrorist suspects during interrogations, had developed five techniques designed to inflict pain and distress but avoid the legal definition of torture: hooding, sleep deprivation, stress positions, white noise, and food and drink deprivation. They were tested on 'the hooded men': 12 Catholic men, among over 340 arrested on 9–10 August 1971 who were chosen for experimental 'deep interrogation' (Gallagher, 2015). Although the European Commission of Human Rights ruled in 1976 that these techniques constituted torture, this was overruled on appeal in 1978 by the European Court of Human Rights.

The techniques were rediscovered by the USA after 9/11. On 13 November 2001 President Bush signed an executive order entitled 'Detention, Treatment and Trial of Certain Non-Citizens in the War Against Terrorism', which removed US civil rights from individuals involved in terrorism, establishing military commissions for their trials. Then on 7 February 2002 Bush signed a memorandum entitled 'Humane Treatment of Taliban and Al-Qaeda Detainees', which asserted that enemy combatants in the War on Terror were not 'prisoners of war' and hence were not entitled to protection under the Geneva Convention. Using a loophole in article 5 of the fourth convention the administration declared they were instead 'security detainees' to be interrogated. On 1 August 2002 the Assistant Attorney General Jay S. Bybee, head of the Office of Legal Counsel of the US Department of Justice, signed into effect a set of legal memoranda drafted by the Deputy Assistant Attorney General John Yoo. These 'Torture memos' authorized and defended the use of a set of 'enhanced interrogation techniques' (EIT) for intelligence purposes as part of the War on Terror.

Based upon the British techniques (and adding others, such as waterboarding), these EIT were employed from 2001 to 2003 at the Bagram Airbase detention facility in Afghanistan (where three prisoners were beaten to death), at CIA 'black site' facilities around the world and at Guantanamo Bay (which had been specifically chosen as it wasn't on the US mainland and was outside of US civil law) before being transferred to the Iraqi prison system. Although the Bush administration's legal advisors stated this wasn't torture, the world saw the Abu Ghraib photos differently and eventually Congress would agree. On 9 December 2014, the Senate Intelligence Committee report on CIA Torture was part-published. Investigating CIA EIT-use from 2001 to 2006 it found the systematic use of 'torture' by the CIA alongside other, harsher techniques including making detainees stand on broken bones, mock executions and violent rectal feeding.

Rather than being 'bad apples', therefore, the Abu Ghraib MPs were following orders. This was 'standard operating procedure': their job was to apply these techniques to prisoners to soften them up ready for interrogation by the CIA and military intelligence. This activity was accepted by the whole command structure at the prison, through the military authorities, to the Bush administration itself. Many of the worst images, such as the hooding and stress positions were legal according to the administration and the MPs were not charged for these activities. What the MPs did wrong was to photograph their behaviour and add elements of abuse that weren't sanctioned, such as the leash, the punching, and sitting on or piling up bodies – although even this is a grey area as the sexual humiliation, the use of dogs and numerous other forms of abuse were part of the authorized psychological preparation of the prisoners.

In the end, little actionable intelligence resulted from the Abu Ghraib interrogations, as few prisoners knew anything about the insurgency, let alone global al-Qaeda operations. Eleven MPs were eventually convicted, mostly for minor offences. Only Charles Graner, Ivan Frederick and Lynndie England received substantial sentences, of 10, eight and three years, respectively, although all served far less time than that. The commanding officer of Abu Ghraib, Janis Karpinski, was demoted to Corporal, but there were no charges for officers or for military intelligence. No-one has ever been held accountable for al-Jamadi's death and in August 2012 the USA announced no criminal charges would be brought regarding his murder.

Claims that this was an isolated incident unravelled in the following years. Michael Keller's 2007 Abu Ghraib memoir *Torture Central* alleged torture continued at the prison for a year after the photographs became public, whilst the October 2010 publication of the 'Iraq War logs' by Wikileaks revealed US torture and abuse of Iraqi prisoners continued for years. In January 2005 testimony emerged of other forms of abuse at Abu Ghraib, including urinating on prisoners, beating wounded limbs with metal batons, pouring phosphoric acid on detainees and dragging detainees around with ropes attached to their legs or genitals. There have also been claims of other deaths from shootings or deliberate snake-bites. In February 2006 an Australian network unearthed more photographs from Abu Ghraib and on 15 March 2006 Salon.com published the 'Abu Ghraib files' based on the Army CID investigation, which had discovered

> a total of 1,325 images of suspected detainee abuse, 93 video files of suspected detainee abuse, 660 images of adult pornography, 546 images of suspected dead Iraqi detainees, 29 images of soldiers in simulated sexual acts, 20 images of a soldier with a Swastika drawn between his eyes, 37 images of Military Working dogs being used in abuse of detainees and 125 images of questionable acts.
>
> Benjamin (2006)

Even more material from Abu Ghraib emerged in 2009 and there are claims that worse exists. Seymour Hersh claimed in July 2004 that the US government possessed video of US servicemen raping children in front of their mothers, saying 'the boys were sodomized with the cameras rolling' (Sealey, 2004) – claims supported by a former detainee and witness, Kasim Mehaddi. In May 2009, it was confirmed that images of the rape of female and male prisoners and of sexual assault, including with objects, existed. President Obama backtracked on earlier promises to release all Abu Ghraib material, with over 2,000 images remaining classified by mid-2016 to prevent inflaming anti-American sentiment.

At the core of the authorities' problem was their failure to appreciate the significance of digital photography. In contrast to analogue photography, which was marked by its expense and practical limitations, cheaply-available digital photography made taking, seeing, copying and sharing images easy. As one of the abusers, Javal Davis said, 'Everyone in theatre had a digital camera. Everyone was taking pictures of everything, from detainees to death' (Gourevitch and Morris, 2008:179). The most famous Abu Ghraib images came from three MPs. The photographs were uploaded to their laptops, burnt onto CDs and widely shared among the battalion, all without the authorities' knowledge. Donald Rumsfeld's surprise was genuine:

> We're functioning in a – with peacetime restraints, with legal requirements in a wartime situation, in the information age, where people are running around with digital cameras and taking these unbelievable photographs and then passing them off, against the law, to the media, to our surprise, when they had not even arrived at the Pentagon.
>
> Matheson and Allan (2009:149)

With a top-down mentality focused on controlling mass-media coverage and naturally assuming the hierarchy would have full knowledge of anything happening, Rumsfeld failed to realize the digital revolution had turned everyone into a media producer. Abu

Ghraib showed how images could now be taken and shared at speed by anyone, moving beyond any official control or narrative to appear in the global media.

Although the MPs later claimed they took the photographs to document what was happening, their enthusiastic participation undermines this explanation. As Susan Sontag notes, 'the horror of what is shown in the photographs cannot be separated from the horror that the photographs were taken – with the perpetrators posing, gloating, over their helpless captives' (Sontag, 2004). Like images of American lynchings that show the crowd around the hanging bodies, these are 'souvenirs of a collective action whose participants felt perfectly justified in what they had done'. The images are part of a tradition of war trophies, wherein souvenirs from the body or person are cut and taken as part of the symbolic defeat and destruction of the enemy, except here, Sontag says, the trophies are 'less objects to be saved than messages to be disseminated, circulated'.

For Sontag, this is because the images were an expression of a digital culture in which people continually document and share their daily life: 'Where once photographing war was the province of photojournalists, now the soldiers themselves are all photographers – recording their war, their fun, their observations of what they find picturesque, their atrocities – and swapping images among themselves and e-mailing them around the globe'. These photographs were simultaneously personal documents, serving as individual mementos of their lives and activities, and a means of group bonding. They were the product of work boredom and camaraderie, uniting the MPs whilst serving as entertainment – both during the abuse (of which the photography was a central part), and afterwards, in being shared. What surprised even the photographers themselves was the ease and the speed at which that private, contained world became globally available. These images spread through the same digital technologies that helped capture them, taking on entirely new meanings and connotations in becoming public and as they crossed geographical and cultural boundaries.

It is this global reception that is important. Sontag notes in May 2004 that these images will become 'the defining association of people everywhere with the war'. Already she suspects the torture was systemic, authorized and condoned. Hence, she says (as an American), 'Considered in this light, the photographs are us. That is, they are representative of the fundamental corruptions of any foreign occupation' as well as the specific product of the Bush administration's foreign policies: the quasi-religious, endless, with-us-or-against-us war against evil, the removal of rights from prisoners, and the demonization and dehumanization of anyone declared to be a possible terrorist. Ultimately, Sontag says, the Bush administration was less concerned with the images or the victims than with limiting the public-relations disaster: it was the damage to the USA's reputation that primarily motivated Bush's anger.

In his 2005 essay 'War Porn', Jean Baudrillard similarly emphasizes the damage of these images. Whereas 9/11 represented a humiliation of the USA inflicted from without, 'with the images of the Baghdad prisons, it is worse, it is the humiliation, symbolic and completely fatal, which the world power inflicts on itself'. If war, he says, aims at the symbolic extermination of the other – 'the goal of the war is not to kill or to win, but abolish the enemy, extinguish (according to Canetti, I believe) the light of his sky' – then this is a fate the USA and its power here imposes on itself. The most famous photograph may be that of the hooded man, wired up by his hands, Baudrillard says, but in these images 'it is really America that had electrocuted itself' (Baudrillard, 2005).

Like Sontag, Baudrillard also notes an important transformation from the media coverage of the invasion: 'There is no longer the need for "embedded" journalists because soldiers themselves are immersed in the image – thanks to digital technology, the images are definitively integrated into the war'. What both Sontag and Baudrillard identify is a new media revolution in which the on-message, embedded journalist is now superseded by embedded servicemen documenting their military life and activities themselves, without oversight or control. As Sontag says, this is now 'unstoppable': 'the pictures aren't going to go away… there will be thousands more snapshots and videos'. The success of Web 2.0 platforms and services over the following years proved her right.

The rise of Web 2.0

The Abu Ghraib scandal took place on the cusp of another digital transformation. As we've seen, the rise of networked computing and the digitalization of analogue media forms had already revolutionized our informational experiences, but from 2004 ongoing technological and cultural developments would do so again. Cheaper and more widespread broadband connections, easier interconnection and interoperability of increasingly-powerful and available multimedia devices such as mobile phones, and new online services taking advantage of these capacities would cohere into what became known as 'Web 2.0' (O'Reilly, 2005).

Box 6.1 'Web 2.0'

Coined by Tim O'Reilly in 2004, the term describes the rise of a range of mostly-free, web-based applications and platforms that left behind the largely read-only world of 'Web 1.0', being built instead around new 'architectures of participation', 'rich user experiences', user-generated content (UGC), information sharing and personal networks. These Web 2.0 sites included social media such as Myspace (2003) and Facebook (2005), UGC hosting and sharing sites such as Flickr (2004) and YouTube (2005), micro-blogging sites such as Twitter (2006), collaborative sites such as Wikipedia (2001) and aggregation sites such as Reddit (2005). Together with other UGC phenomena such as blogging, podcasting and easy-to-create websites, these developments produced a cultural revolution that transformed the broadcast-era of mass-media.

Whereas the broadcast-era was dominated by 'big media' – by large companies or organizations with sizable investment in facilities, technology and staff, mass-producing content for mass distribution and mass consumption – the Web 2.0-era has seen the rise of personal production by empowered individuals using free or cheaply available, easy-to-use digital platforms and technologies that enable them to produce and share instantly and globally. The broadcast-era top-down, one-to-many, vertical cascade of products to separate and separated consumers has been transformed by this bottom-up production and distribution and by the new ability to create peer-to-peer, one-to-one, one-to-many and many-to-many communicational relationships outside of the major mass-media systems.

Thus, where the broadcast era was marked by a sharp division between a mass of non-productive individuals whose primary role was as receivers of professional products and a minority of paid-creators, the Web 2.0 era allows every individual to become a media producer. One no longer needs to be experienced or employed, or have proven credentials to add to the media ecology. As Clay Shirky says, 'We are living in the middle of a remarkable increase in our ability to share, to cooperate with one another, and to take collective action, all outside the framework of traditional organizations and institutions' (Shirky, 2008:20).

Critics may dismiss what individuals produce but within the Web 2.0 world it is what is of value to each individual that becomes important: a 'Like' on a Facebook comment can be more interesting and significant to someone than a $300m Hollywood blockbuster, years in production. Unlike mass-media production, the messages and creations of the digital individual don't need a mass audience to justify their public investment or bring profits and they don't necessarily aspire to any cultural weight or significance. They avoid pre-production filters such as commissioning editors, and don't have to fit into established market categories or appeal to specific niches. Digital products are not standardized to fit consumer desire, and aren't uniform, identical or finished, with informational creations being personal, personalized and open to endless sharing, remixing and remaking. Thus, whereas the 'mass' was the centre of mass-media, the individual is the centre of the post-broadcast ecology. This is a move from 'media' to 'me-dia': by which we mean the reorientation of media around the individual, their organization of their own media experiences and the centrality of their personal interests, messages, content and networks (Merrin, 2014:77–92). In contrast to the broadcast emphasis on advertising and marketing 'push', this leads to a world of 'pull': of individuals selecting, taking and creating their own unique media experiences.

Whilst there was little interest in amateur production in the broadcast era, when it was assumed that major companies or organizations would naturally produce material for us, Web 2.0 exposes a huge demand for shared, personal products, content, opinions, messages and experiences. This has impacted upon the cultural power and authority of mass-media. Broadcast-era productions were highly crafted, professional, polished and designed for maximum impact, having a credibility in their source and their air of expertise. Now the ability to broadcast one's own opinions and ideas, and respond to and challenge others undermines that position. A mass-media claim becomes merely one amongst many and is freely debated and criticized. This is a world not of unilateral messages but of conversations, and within the 'long tail' of the global, networked digital audience, the knowledge, 'wisdom' and 'collective intelligence' of the crowd can easily trump that of professional journalists or the authorities (Anderson, 2006; Surowiecki, 2005).

The rise of Web 2.0, therefore, initiated a cultural revolution that challenges the older media systems. Broadcasting has survived, but Web 2.0 has ended the dominance of a single, simple broadcast model, forcing established media organizations to adapt to digital changes and find new ways to reach their audiences and monetize their content. What has emerged is a much more complex and continually changing media ecology in which many of the arrangements, assumptions, models, structures, and economic and political power relationships of the broadcast era have been overthrown. This is also an ongoing transformation. The take-off of internet-enabled smartphones (with the success of the iPhone, released in June 2007) and of 'post-pc' devices such as tablets, the rise of apps and cloud-computing, the development of new services and platforms (such as live video-streaming), and the ubiquity of personal and public internet access have all advanced the Web 2.0 revolution.

Although there are many problems caused by this digital empowerment – in particular in the range of new threats to the individual caused by their connectivity – the structural, societal revolution in communication it has led to is truly revolutionary. The result is that almost every individual and non-state actor in the conflict zone now has a complete set of peer-to-peer and global publishing tools at their disposal, making it impossible for any organization or government to control the flow of information and the story they want to promote. Given that the military operates as a top-down organization, used to giving orders and controlling its informational activities, the rise of Web 2.0 technologies and platforms has hit it hard. Since 2004 it has discovered that any serving soldier can bypass its hierarchy, producing, sharing and distributing any material they want, with important consequences for the military's public image and international relationships. This is the rise of the social-media military embed.

The new military embed

After Abu Ghraib, the USA became more aware of the potential for embarrassment by their own servicemen. Their initial focus was on 'milblogs' – the military blogs run by serving soldiers in the field. These had attracted media and public attention during the Iraq invasion but they had been largely ignored by the US authorities. In subsequent years, however, the numbers of bloggers grew and sites such as 'A Line in the Sand', 'Mudville Gazette', 'BlackFive', 'Froggy Ruminations', '2Slick' and 'My War' began to attract significant pageviews. With ongoing operations in Afghanistan and Iraq and civil conditions there deteriorating under insurgent pressure the military increasingly saw blogs as a risk both to operational security (OPSEC) and to the conflict's public perception. One response was censorship. Jason Hartley's irreverent blog 'Just Another Soldier' was ordered closed in October 2003 and he was demoted and fined for security violations; Colby Buzzell's 'My War' faced censorship from September 2004 and he was excluded from missions by his commanders; whilst in January 2005 Michael Cohen was ordered to shut down his blog '67cshdocs' after a graphic post he wrote about the medical response to a suicide bomber attack was accused of violating army regulations.

Official policies emerged in 2005. On 6 April, the Multi-National Corps-Iraq issued MNC-I Policy #9 'Unit and Soldier Owned and Maintained Websites', introducing new regulations requiring Iraqi bloggers to register with their units and directing commanders to conduct quarterly reviews of unit blogs. In August, the Army authorized the Army Web Risk Assessment Cell (AWRAC) to monitor compliance of the policy whilst an ALARACT (All Army Activities) message by Army Chief of Staff General Peter J. Schoomaker instructed commanders to be aware of security violations from blogging. In September, a rapid revision to OPSEC Army Regulation 530–1 called on army personnel to properly implement OPSEC in their communications, specifically referring to blogs in its list of public forums.

By then there were an estimated 1,200 milblogs, although under close-scrutiny from AWRAC many went quiet through 2006 as the military need for informational control clashed with a younger, educated and vocal generation at home with an open digital culture and wanting to connect with families and friends at home. A DOD 'Information security/website alert' issued on 6 August 2006 revealed the military's new approach to the problem, in stating that 'no information may be placed on websites readily accessible to the public unless it has been reviewed for security concerns and approved in accordance with Deputy Secretary of Defense Memorandum web site policies and procedures' (Department of Defense, 2006). On 19 April 2007, the army issued an updated OPSEC Army Regulation 530–1 that followed the same path, prohibiting soldiers (as well as civilians working with the military, army contractors and even soldier's families) from posting to blogs, commenting online or sending personal emails without clearing the material first with a superior officer.

On 13 May 2007 the US military extended its clampdown to social media and sharing sites, introducing a global ban on its soldiers accessing 13 popular websites, including

Myspace and YouTube, through military networks. The main reasons given were the drain on computing resources and the question of operational security. Ironically, the day after, Colby Buzzell won the £5,000 Lulu Blooker prize for the best book produced from a blog, beating 110 entrants from 15 countries and demonstrating the broader value of milblogs. Other milbloggers defended them too, arguing that sites such as Matthew Burden's 'BlackFive', with its 3 million unique users a year, and 'acutepolitics@blogspot.com', which was written by a roadside bomb disposal specialist, provided some of the best military PR at a time when its overseas operations were proving increasingly unpopular. The British followed the USA. On 4 February 2009, in an order titled 'Contact with the Media and Communicating in Public', the British Army banned the use of social networks in combat zones due to the security risks. They also banned mobile phones in Afghanistan, started providing social media training in barracks and developed a dedicated social media team to advise on guidelines for troops.

The US ban on social networking lasted until 18 May 2009 when an Army operations order made YouTube, Twitter, Facebook and Flickr available to some troops again. The climb-down was partly due to the ubiquity of the sites, but also owed much to the military's own employment of Web 2.0 platforms. As Shachtman says:

> Army public affairs managers have worked hard to share the service's stories through social sites like Flickr, Delicious and Vimeo. Links to those sites featured prominently on the Army.mil homepage. The Army carefully nurtured a Facebook group tens of thousands strong, and posted more than 4,100 photos to a Flickr account. Yet the people presumably most interested in these sites – the troops – were prevented from seeing the material.
>
> Shachtman (2009a)

Despite this, in June 2009 there were reports that the US military was considering 'a near-total ban on Twitter, Facebook, and all other social networking sites throughout the Department of Defense' due to security concerns (Shachtman, 2009b). In February 2010, however, the DoD issued a new policy allowing social networking access from their non-classified network NIPRNET and moving away from the notion of blacklisted sites. Security issues, they said, were now outweighed by the benefits of troop access. By January 2011 the US military had given up the fight. It announced the closure by 1 March of the 'social media office' it had opened in 2009 to coordinate its responses, recognizing that social media use was now ubiquitous and that using it appropriately was everyone's responsibility.

The battle, however, had been lost long before then. The interest in blogging had long been superseded by taking and posting military videos. Ubiquitous camera-phones and helmet-cams, together with the rapid rise of YouTube from 2005 and of specialist military-video sharing communities such as Militaryvideos.net, enabled any serving soldier to record, upload and share their experiences. As a result, a vast amount of unedited, raw, graphic and violent footage of patrols and firefights in Afghanistan and Iraq was now being produced and shared online. This was a remarkable, uncensored record of daily life as a soldier that was far more real than anything produced by the embedded journalists. Having little regard for official narratives or PR, it showed instead the unexpurgated boredom, confusion, routine and the random violence of soldiers' military experience. Some videos became famous. 'U.S. Soldier Survives Taliban Machine Gun Fire During Firefight' – a YouTube video of Ted Daniels being shot in Afghanistan uploaded in September 2012 – had received

34.5 million views by late 2017. Although it is undoubtedly more 'real' than mainstream media coverage of the conflict, its helmet-cam origins and editing highlight how it remains a selected and framed vision of war; one placing the viewer on the side of the US military whilst appealing to western viewers by inadvertently echoing contemporary video-game, first-person shooter, point-of-view (POV) aesthetics.

The authorities continued to emphasize the security risks of user-generated content. There were reports in 2007 of Taliban tapping British troops' mobile phone calls and ringing their families with threats, and warnings that Facebook's timeline could reveal soldier's routes or habits or that information on social media or a phone could be used against individual soldiers if they were captured in action. In September 2012, an Australian government 'Review of Social Media and Defence' warned that the Taliban were posing as 'attractive women' on Facebook in order to befriend soldiers online and gain crucial information that could be used against them. Increasingly photography was seen as a security risk too, with the revelation that in 2007 Iraqi insurgents had exploited the automatic geotagging feature of photos uploaded to Facebook to destroy four US AH-64 Apache helicopters when photos taken by soldiers when the fleet arrived were used to pinpoint their location for mortar fire.

Although the military emphasized the issue of security, underlying this was the continuing fear that soldiers would embarrass their governments. These fears had been realized in September 2005 when the story of NTFU.com broke. NTFU ('Nowthatsfuckedup') was an amateur pornography site created in February 2004 that began to offer free access to military personnel who couldn't use a credit card if they could prove they were in military employment. Soldiers began by sending benign photos but, as word spread, images of nude female soldiers or sexual activity and horrific trophy images of dead-Iraqi insurgents, civilians killed in roadblocks, severed heads, IED victims and so on came to dominate. As Zornick commented, 'The website has become a stomach-churning showcase for the pornography of war – close-up shots of Iraqi insurgents and civilians with heads blown off, or with intestines spilling from open wounds. Sometimes photographs of mangled body parts are displayed: Part of the game is for users to guess what appendage or organ is on display' (Zornick, 2005).

The use of dead Iraqis for entertainment and the sadistic glee with which the soldiers posted the images and commented approvingly on them risked another Abu Ghraib-style PR disaster: an eviscerated corpse captioned 'what every Iraqi should look like' was not the image the USA wanted to present to the world. In the end, the site's owner was prosecuted under Florida's obscenity laws and he agreed to hand over control of the website. Despite 30,000 military personnel being registered and despite too the nature of the images, an army investigation was closed by the end of September without any charges being brought.

Other cases would emerge in the following years. In Summer 2010 the US Army charged five members of a platoon with the murder of at least three Afghanis in Kandahar province and with taking body parts as trophies. The group called themselves 'The Kill Team' and staged combat situations to kill Afghan civilians, taking photos and physical mementos of their victims. The images became public at their trial. The most famous one shows the body of 15-year-old Gul Mudin. The soldiers had called for him to come closer. He'd raised his shirt to show he carried no weapons and the soldiers threw a grenade at him and opened fire with machine guns. They cut off his finger for a trophy and posed for photographs. Spc. Jeremy Morlock can be seen smiling whilst raising the dead child's head for the camera.

This 'war-porn' was common among the servicemen. As Mark Boal comments,

> The soldiers of 3rd Platoon took scores of photographs chronicling their kills and their time in Afghanistan. The photos, obtained by *Rolling Stone*, portray a front-line culture among U.S. troops in which killing Afghan civilians is less a reason for concern than a cause for celebration.

He adds,

> Among the soldiers, the collection was treated like a war memento. It was passed from man to man on thumb drives and hard drives, the gruesome images of corpses and war atrocities filed alongside clips of TV shows, UFC fights and films such as *Iron Man 2*. One soldier kept a complete set, which he made available to anyone who asked.

> Boal (2011)

The Pentagon desperately tried to find and suppress all the images to contain the scandal, even sending agents into the homes of relatives in the USA to find every possible copy. Their criminal investigation of possible homicides was, however, much less thorough, Boal says.

User-generated imagery continues to leak from the US military. In January 2012 video was posted on YouTube, LiveLeak and TMZ of a July 2011 incident, showing a line of US soldiers urinating on the dead bodies of Afghan Taliban soldiers and joking about it. The desecration of Muslim bodies caused outrage in Afghanistan and across the middle east, being seized on by the Taliban as evidence of US brutality and condemned as 'simply inhuman' by the Afghan President Hamid Karzai. An Afghan soldier who shot dead four French troops later that month claimed he did it because of the urination video. Then on 18 April 2012 images of US soldiers posing with the body parts of dead Afghan insurgents were leaked to the *LA Times*. At a time of increasing hostility to the USA in the country images of soldiers smiling whilst holding up the remains of a suicide bomber seriously strained the US–Afghan relationship, leading to condemnation by both governments.

The issue of US military social-media behaviour arose again in March 2017 with the discovery of the closed Facebook group 'Marines United', which hosted member-submitted, clothed and nude photographs of serving female marines (as well as of wives and girlfriends), often with their name, rank and duty station listed. The group had 30,000 members, mostly serving US marines, although it also included veterans and UK Royal Marines, and combined stalker-like photography and revenge-pornography with misogynist hate-speech, including rape and death-threats. It wasn't a new problem. The US military has had long-standing problems with sexual harassment and a Facebook group called 'F'N Wook' had already been discovered in 2013, containing images and comments systematically sexually denigrating female marines. Within days of the new discovery a criminal investigation into about 500 members had begun and it emerged that four branches of the military were now investigating image-sharing websites. By April it was reported that the images had spread across multiple platforms including Twitter, Snapchat, Tumblr, Instagram, Reddit, Pornhub and Anon-IB, and onto the Darknet, being offered for sale on the anonymous darkmarket 'AlphaBay'.

The problem of the digitally-empowered, individual wasn't confined to the US military. Saddam Hussein was captured by US forces on 13 December 2003 and was handed over to the interim Iraqi government on 30 June 2004 to face trial for crimes against humanity. His trial began on 19 October 2005 and on 5 November 2006, having been found guilty of the murder of 148 Shiites from Dujail in 1982, he was sentenced to death by hanging. He was executed at 6am local time on 30 December 2006 at the Iraqi–American military base Camp Justice in northern Baghdad. This was an important unifying moment for Iraq and a legitimizing event for the new government. Saddam's trial and execution were supposed to represent a break with the old regime, symbolizing the victory of the rule of law over arbitrary dictatorship and justice for the victims of Saddam. Official video of the lead-up to the execution was filmed and an edited, though soundless, version was shown on Al-Iraqiya and rebroadcast globally.

This official representation of Saddam's execution was undermined a day later when a grainy, stuttering, mobile-phone video of the execution shot by one of the guards began to circulate online. It not only showed Saddam's death in full; more importantly the sound-track exposed the chaos of the event and the sectarian abuse. Saddam was told to 'go to hell', guards chanted 'Moqtada', naming the radical Shia muslim cleric and leader of the Mahdi army militia, Moqtada al-Sadr, and shouted 'Long live Mohammed Baqir al-Sadr', referencing Moqtada's father-in-law, the Shia cleric killed by Saddam in 1980. At a time when Iraq was splitting down religious lines with Islamist groups trying to foment a civil war between Sunni and Shia, the attempt to demonstrate the execution was calm, respectful, dignified and an expression of the rule of law rather than sectarian revenge was destroyed by one of the new government's own guards.

As Roger notes, 'the Iraqi government misunderstood the uncontrollability of images in the information age' (Roger, 2013:134). Thinking they could enforce their own political message they found themselves instead in a 'remediation battle', with execution footage functioning as an 'image munition' (Roger, 2013:135), being widely shared by Sunnis and used to promote Saddam as a martyr to the insurgent cause. The government's embarrassment increased in early January 2007 when mobile footage of morgue workers showing off Saddam's body and wounds appeared online. Shot by the government's own workers, the video leaked and was posted on a pro-Baathist Arabic website under the headline 'New film of the late immortal martyr, President Saddam Hussein'.

A similar fate befell the new Libyan government in 2011 at the conclusion of the civil war. Muammar Gaddafi had fled Tripoli in August when it fell to opposition National Transition Council (NTC) militias, finding refuge in his hometown of Sirte. He was finally captured on 20 October when Sirte fell and his escaping convoy was hit by a NATO airstrike. A wounded Gaddafi was dragged from a drainage pipe in which he had taken refuge and was filmed by rebel fighters. Very quickly, images of him on a pick-up truck surrounded by shouting rebels appeared on Libyan TV and were carried worldwide. They were soon replaced by footage of his bloodied body with a head-wound under the headline 'Gaddafi killed', apparently having been killed by his captors. Although many Libyans were pleased about Gaddafi's death, the now-ruling NTC faced concern from western governments at chaotic mob-justice and from international rights organizations at a possible war-crime. The new government needed international support and legitimacy and wanted to demonstrate its break from the arbitrary killing of the former regime. Hence on 21 October the acting Prime Minister Mahmoud Jibril distanced NTC forces from the death, announcing Gaddafi had been accidentally killed soon after his capture: 'When the car was moving it was caught in crossfire between revolutionaries and Gaddafi forces in which he was hit by a bullet in the head' (BBC, 2011).

The problem was that there was no evidence of any regime resistance or gunfight after Gaddafi's capture, whilst new footage was emerging from rebel fighters of his abuse, mistreatment and final moments. Whilst the government were still claiming regime forces were effectively responsible for Gaddafi's death, new video appeared showing the convoy for his ambulance trip from Sirte to Misrata. As well as showing celebrating fighters driving along, it also shows the convoy stopping and crowds surrounding the ambulance where Gaddafi has just been shot in the head. The rebels throng round and cheer a man holding a gun. One fighter has his arm around him, shouting 'He's the killer and I am the witness that saw him. He's the guy who killed him' (Telegraph, 2011). As in Iraq, it was the new government's own soldiers and their personal video footage of the events that destroyed the credibility of official claims, undermining their own government on the world stage.

On 24 October, the NTC announced a new investigation into the circumstances of Gaddafi's death, but by then his treatment had become a wider story. After being filmed being dragged along the street in Misrata, Gaddafi's half-naked body had been dumped on a mattress in a refrigerated meat store and kept as a trophy, with Libyans queuing to see and to film his corpse. More footage of Gaddafi's abuse emerged on 25 October when a video was posted on the GlobalPost news-site showing Gaddafi being sexually assaulted after his capture with a knife or pipe shoved into his anus. In July 2012 more video emerged of Gaddafi's body being used as a ventriloquist doll by a rebel fighter, who moves his head as he speaks Arabic. Whilst abuse and the defiling of bodies isn't new in warfare, now the combat-zone is filled with digitally empowered individuals creating and sharing their own personal record of the events. The time when a government could single-handedly control information and shape and determine the narrative of events was over.

Just as social media has become an everyday technology for many, so it has become a normal part of the soldier's day too, being used for many of the same reasons – to communicate with friends, to document experiences, to update others on what they are doing, to express opinions, and to store and share photographs and videos. But even everyday images and information can cause problems for military public relations. In August 2010, the ex-Israeli Defence Force (IDF) soldier Eden Abergil became famous when photographs posted on her Facebook page of her posing smiling in front of blindfolded and bound Palestinians were picked up by bloggers and global media. A comment beneath the photo by one of her friends reads 'You look sexiest here', to which she replied, 'Yeah I know … I wonder if he's got Facebook! I have to tag him in the picture!' (Shabi, 2010).

For Abergil it was a normal Facebook post, documenting her daily life. She couldn't see what was wrong with the images, claiming there was no violence, abuse or contempt involved. The IDF called them 'shameful' and distanced itself from them, although 'Breaking the Silence', the veterans group campaigning to raise awareness about the occupied territories, pointed out that the practice was 'a widespread phenomena', with most soldiers taking photographs of their 'everyday life'. As in Abu Ghraib, however, images taken in a personal context take on new meanings in different, public contexts, hence many Palestinian commentators saw the photographs as representing the casual dehumanization of Palestinian lives and the lack of regard for their dignity and privacy by occupying Israelis. For them, the 'everyday' images exposed an underlying mindset towards the Palestinians, symbolizing an everyday mode of oppression and racist humiliation.

The IDF were in the news again in February 2013 when an Instagram photo posted by the 20-year-old IDF soldier Mor Ostrovski of a Palestinian boy's head in the crosshairs of a sniper rifle was discovered and publicized by Palestinian activists. Ostrovski claimed the image wasn't taken by him and closed his account, with the IDF issuing a statement that his actions were 'not in accordance with the spirit of the IDF and its values' (Greenwood, 2013).

The photograph's discovery, however, wasn't accidental. Online activists for both Israel and the Palestinians regularly scour the net for videos, comments and photos that support their views, with IDF soldiers' accounts being closely monitored. The picture was discovered by Ali Abunimah, a co-founder of the Electronic Intifada website, who said:

> When I looked at the picture of the child in the crosshairs, to me it really captured in a sense symbolically the way that the Israeli army and occupation views Palestinians – as potential targets.
>
> Knell (2013)

The Electronic Intifada website has found numerous examples of aggressive IDF-related posts online. In December 2012, they discovered posts by a 22-year-old soldier, Nisim Asis, who posted racist images on his Instagram page, including a picture of himself licking what is probably tomato ketchup off a knife with the caption 'Fuck all Arabs their blood is tasty' (Greenwood, 2013).

Social media continued to cause problems for the IDF. A few days after Ostrovski's image became famous, news broke of the IDF sending two soldiers to military prison for 14 and 21 days respectively after their artillery unit posted a video to YouTube of themselves dancing around a cannon doing 'The Harlem Shake', a popular meme of the time. The soldier who made the video and the commander who approved it were punished despite the video's positive reception and the appearance of the IDF in a positive light. At the end of the month the IDF issued a warning to its troops to stop uploading videos and pictures 'not appropriate to the spirit of the IDF' (Knell, 2013).

The warning didn't stop the problem. In early June 2013 four female Israeli recruits were disciplined by the IDF for 'unbecoming behaviour' when photographs emerged online of them posing half-naked with their rifles (Evans, 2013). Just over a week later a video appeared showing more half-naked female recruits pole-dancing around a rifle being held like a stripper's pole. One of the soldiers can be heard telling another to 'dance on the rifle like a slut' whilst another says she will post the video on Facebook and tag the others (Didymus, 2013). Another problem arose in July 2014 when an IDF sniper, David D. Ovadia, bragged on Instagram about his 'kills', posting an image of himself with his rifle, accompanied by the message '@sherii_elkaderi I killed 13 childrens today and ur next fucking muslims go to hell bitches'. The hacker-group Anonymous responded by taking over and deleting his Instagram account and hacking Mossad websites 'for the brave IDF sniper' and also took down the Israeli Ministry of Defence website (Simeon, David and Jamal, 2014). Ovadia confessed to the IDF to making the post, whose claims had no basis in reality, in order to harass and terrorize Palestinians on social media and was sentenced to 30 days in jail.

Russia has also been embarrassed by its troops on social media. Following Russia's annexation of Crimea in 2014, President Vladimir Putin admitted that pro-Russian separatists in Eastern Ukraine had links with Russian intelligence and were receiving training, intelligence and weapons from them, but he denied that the Russian Federation had deployed military forces there. This claim was disproved in July 2014 when Buzzfeed's Max Seddon found a Russian soldier posting photographs to Instagram that were geotagged in Ukrainian locations, whilst a BBC reporter discovered a Russian soldier posting images of troop convoys in Ukraine on the Russian version of Facebook, VKontakte. A year later, in September 2015, Russia's denial that it had sent ground troops into Syria to help support President Assad were similarly exposed when numerous images, some dating back to April, were discovered to have been posted on social media by troops in various locations in Syria.

Some social media leaks have been more humorous. On 18 June 2007, a video entitled 'German Navy Boats Crashing' was uploaded to YouTube showing two boats from the German Navy on manoeuvres off the Lebanese coast running into each other. The video went viral and was picked up by the major news organizations, compounding the Navy's humiliation. The short sequence was obviously shot by the sailors themselves and, as it consists of two different point-of-view shots, was also edited together before it was posted. The Navy confirmed the captain lost his job as a result of the accident.

The social media battlefield

New personal digital technologies and Web 2.0 platforms and services, therefore, have transformed the media ecology, massively-expanding who can produce and share material and turning every soldier into their own embed, producing information for their own, familial, networked or public consumption. Increasingly governments realize they can no longer control information about the combat zone or their soldier's lives and activities. The move from the predictable, known and more manageable broadcast era to a world in which everyone, including their own troops, can affect the public perception of their operations and combat zones is difficult for the military, but they have understood that they need to deal with the new challenges and to master the changed ecology.

As we've seen issues around security and the risk of embarrassment have led military authorities to attempt to ban or control mobile phones and social media. This remains an ongoing battle for the authorities. In May 2015, for example, *The People's Liberation Army Daily*, the Chinese military's official newspaper, announced troops were forbidden from using new wearable technology after a recruit received a smartwatch as a birthday present. The military recognized that wearables with internet access, location information and voice-calling functions represented an important security risk, from possible misuse, from inadvertently broadcasting sensitive information or from the risk of being hacked. 'The moment a soldier puts on a device that can record high-definition audio and video, take photos and process and transmit data, it's very possible for him or her to be tracked or to reveal military secrets', the paper said (BBC, 2015). The US military remains concerned with security too. In August 2015, it launched a new campaign for serving soldiers with a poster saying 'Loose tweets destroy fleets' (updating the World War II-era slogan 'loose lips sink ships'). The military warned especially about social media use leaking information about training, shortages, current or future deployment, internet speeds and the locations of core facilities and key personnel.

Most governments, agencies and militaries have now realized that social media offers new opportunities for propaganda. The authorities' voice may only be one among many but it is important to take part in those conversations, to present information positively and to counter opposing claims and criticism online and in real time. Many organizations have set up YouTube channels for officially produced videos, as well as creating accounts on Twitter, Facebook, Flickr, Instagram and even Pinterest. The IDF especially has embraced social media. A specialist social media team emerged out of 'Operation Cast Lead' (the 2008–09 Gaza conflict) when 25-year-old Aliza Landes began to use YouTube to share IDF videos and by August 2009 she had talked the IDF into giving her a dedicated budget. Through the unit's cross-platform approach, multi-media posting and creation of content for sharing, her successor Sacha Dratwa turned it into 'the most globally visible arm of the Israeli military' (Hoffman, 2012). By 2016 the IDF was active on 30 platforms and in six different languages. Even social media experts can be embarrassed, however, as Dratwa was in November 2012 when a photo he posted on Facebook showing his face covered in black paint and entitled 'Obama style' led to worldwide attention and accusations of racism.

For Dratwa social media allows Israel to get its narrative out in real time whilst also bypassing old media to provide information directly to supporters and critics: 'We believe people are getting information from social media platforms and we don't want them to get it from other sources – we are the ones on the scene, and the old media are not on the scene as are the IDF' (Hoffman, 2012). Social media serve, therefore, as a real-time, informational, propaganda battlefield. This was seen most clearly on 14 November 2012 when the IDF live-tweeted the launch of 'Operation Pillar of Defense' in Gaza and the death by drone strike of the Hamas commander Ahmed al-Jabari. His death was confirmed moments later by al-Qassam's Twitter feed. Over the following days the IDF posted a link to a video of the drone strike and a wanted poster covered with the words 'eliminated' and traded threats with Hamas, with @IDFSpokesperson tweeting 'We recommend that no Hamas operatives, whether low level or senior leaders, show their faces above ground in the days ahead', to which @AlqassamBrigade quickly replied, 'Our blessed hands will reach your leaders and soldiers wherever they are (You opened Hell gates on Yourselves)'. Meanwhile Hamas gave a running commentary on Twitter of its mortar and rocket attacks on Israeli targets and posted a YouTube video showing the launch of a Fajr 5 missile in the direction of Tel Aviv (Borger, 2012).

In December 2013, the IDF also launched another front in the social media war, creating a 'selfie squad' to boost their image and narrative. This is a trained combat camera unit that deploys alongside IDF troops as videographers able to film, edit and broadcast from the battlefield straight onto their social media platforms. It was a response in part to the Israeli B'Tselem movement who have distributed cameras to Palestinian volunteers to film human rights abuses by Israeli soldiers in the occupied territories. B'Tselem's successes include footage in July 2008 of a soldier firing a rubber-coated bullet at point-blank range at a blindfolded, handcuffed Palestinian, June 2012 footage of an Israeli border police officer kicking a Palestinian child in Hebron, the July 2013 detention of a five-year-old child in Hebron for stone-throwing, and a November 2013 incident in which an Israeli officer fired a tear-gas canister directly at the chest of the volunteer videographer. As one of the Israeli graduates explained, the squad's aim was to give the other side of the story: 'We are here to explain and to document for the entire world that we don't use force for bad' (Kalman, 2013).

Other nations have followed suit. In April 2015, the British Army created the 1,500-strong 77 Brigade, based in Hermitage in Berkshire to specialize in 'non-lethal warfare' – or social-media psyops. Inspired by the successes of Israel and the USA, its creation was a response to Russia's propaganda activities in the Crimea and the effective use of social media by the Islamic State. An army spokesman said: '77th Brigade is being created to draw together a host of existing and developing capabilities essential to meet the challenges of modern conflict and warfare. It recognizes that the actions of others in a modern battlefield can be affected in ways that are not necessarily violent' (MacAskill, 2015).

Today social media has become an established part of everyday life and of global conflict and ideological struggles. The military authorities can't control information and perception as easily as they did in 2003, but instead they devote considerable resources to directly crafting and pushing their own stories. Although the military journalist remains important in securing mainstream media coverage, militaries increasingly take on the role of media producers themselves. There is obviously a long history of this, with specialist media and propaganda teams working within the military through the twentieth century, for example, but now the military no longer need to rely on mass-media intermediaries to tell their story: the twenty-first-century military is a multi-media, multi-platform, direct-to-the-public media organization. Its control of its own internal information, however, was severely tested in 2010 by the new threat of Julian Assange's mass-leaking website, Wikileaks, the subject of the next chapter.

Key reading

For an overview of the digital revolution see Merrin (2014). Chapters 1 and 2 cover the rise of computing and networking and the digitalization of older analogue forms. For information about Abu Ghraib see Sontag (2004), Danner (2004), Hersh (2004, 2005, chapter 1), Baudrillard (2005), Kennedy (2007), Gourevitch and Morris (2008), Morris (2008) and Roger (2013). Further context for the Abu Ghraib imagery can be found in Gopal (2014, chapter 7), Rose (2004), Worthington (2007), Stafford Smith (2008) and Rosenberg (2016), which cover activities at the USA's Afghanistan's prisons, at the CIA 'black sites' and at Guantanamo Bay. Greenberg (2016) provides an excellent overview of the Bush administration's attempts to reshape the US legal system to support the War on Terror. The idea of 'Web 2.0' is covered in O'Reilly (2005) and the cultural consequences of this shift are detailed in Merrin (2014, chapters 3–5) as well as other texts such as Anderson (2006) and Shirky (2008). Information about military blogging and the US clampdown can be found in Wall (2005), Hockenberry (2005) and Burden (2006). Roger (2013) and Hoskins and O'Loughlin (2010) provide excellent discussions of recent war-porn, abuse-imagery and execution videos. For more information about the mobile phone and social media examples I discuss here follow the articles in the footnotes and also Google the specific cases. These – and many not covered here – are easy to find.

References

Anderson, C. (2006) *The Long Tail*, London: Random House Business Books.

Baudrillard, J. (2005) 'War Porn', *The International Journal of Baudrillard Studies*, 2(1), January [orig. 2004], at: http://www2.ubishops.ca/baudrillardstudies/vol2_1/taylor.htm.

BBC. (2004) 'In Quotes: Rumsfeld Faces Congress', *BBC News*, 7 May, http://news.bbc.co.uk/1/hi/world/americas/3694995.stm.

BBC. (2011) 'Libya's Col Gaddafi Killed in Crossfire, Says NTC', *BBC News*, 21 October, http://www.bbc.co.uk/news/world-africa-15397812.

BBC. (2015) 'China Imposes Smartphone and Wearable Tech Army Ban', *BBC News*, 13 May, http://www.bbc.co.uk/news/technology-32718266.

Benjamin, M. (2006) 'Salon Exclusive: The Abu Ghraib files', *Salon*, 16 February, http://www.salon.com/2006/02/16/abu_ghraib_10/.

Boal, M. (2011) 'The Kill Team: How U.S. Soldiers in Afghanistan Murdered Innocent Civilians', *Rolling Stone*, 27 March, http://www.rollingstone.com/politics/news/the-kill-team-20110327.

Borger, J. (2012) 'Israel and Hamas Deploy Twitter Feeds in Media War', *The Guardian*, 15 November, https://www.theguardian.com/world/2012/nov/15/israel-hamas-twitter-media-war.

Burden, M. C. (2006) *The Blog of War: Frontline Dispatches From Soldiers in Iraq and Afghanistan*, New York: Simon and Schuster.

CNN. (2004) 'Bush "Sorry" for Abuse of Arab Prisoners', *CNN*, 7 May, http://edition.cnn.com/2004/ALLPOLITICS/05/07/bush.apology/index.html?iref=mpstoryview.

Danner, T. (2004) *Torture and Truth: America, Abu Ghraib and the War on Terror*, New York: The New York Review of Books.

Department of Defense. (2006) 'Information Security/Website Alert', http://www.dtic.mil/whs/directives/corres/writing/SecDef_Msg_InfoSec.pdf.

Didymus, J. T. (2013) 'Video of Half-Naked, Pole-Dancing Female Israeli Soldiers Emerges', *Digital Journal*, 13 June, http://www.digitaljournal.com/article/352152.

Evans, B. (2013) 'Female Israeli Soldiers Disciplined for "Unbecoming Behaviour" After Posing for Pictures Dressed Only in their Underwear and Combat Fatigues', *The Daily Mail*, 3 June, http://www.dailymail.co.uk/news/article-2335015/Female-Israeli-soldiers-disciplined-unbecoming-behaviour-posing-pictures-dressed-underwear-combat-fatigues.html.

Gallagher, P. (2015) 'How Britain's Treatment of "The Hooded Men" During the Troubles Became the Benchmark for US "Torture" in the Middle East', *The Independent*, 20 February, http://www.independent.co.uk/news/uk/home-news/how-britains-treatment-of-the-hooded-men-during-the-troubles-became-the-benchmark-for-us-torture-in-10060242.html.

Greenberg, K. (2016) *Rogue Justice. The Making of the Security State*, New York: Crown.

Greenwood, P. (2013) 'IDF Soldier Posts Instagram Image of Palestinian Child in Crosshairs of Rifle', *The Guardian*, 18 February, https://www.theguardian.com/world/2013/feb/18/israeli-soldier-posts-instagram-palestinian.

Gourevitch, P. and Morris, E. (2008) *Standard Operating Procedure*, London: Picador.

Hersh, S. M. (2004) 'Torture at Abu Ghraib', *The New Yorker*, 10 May, http://www.newyorker.com/magazine/2004/05/10/torture-at-abu-ghraib.

Hersh, S. M. (2005) *Chain of Command*, London: Penguin.

Hersh, S. M. (2007) 'The General's Report', *The New Yorker*, 25 June, http://www.newyorker.com/magazine/2007/06/25/the-generals-report

Hochschild, A. (2004) 'What's in a Word? Torture', *New York Times*, 23 May, http://www.nytimes.com/2004/05/23/opinion/what-s-in-a-word-torture.html.

Hockenberry, J. (2005) 'The Blogs of War', *Wired*, 1 August, http://www.wired.com/2005/08/milblogs/.

Hoffman, S. (2012) 'The "kids" Behind IDF's Media', *Tablet*, 20 November, http://www.tabletmag.com/jewish-news-and-politics/117235/the-kids-behind-idf-media.

Hoskins, A. and O'Loughlin, B. (2010) *War and Media*, Cambridge: Polity.

Kalman, M. (2013) 'Israel Defence Forces Deploy 'Selfie Squad' to Boost Image', *The Guardian*, 27 December, https://www.theguardian.com/world/2013/dec/27/israel-defence-forces-selfie-squad-combat-image.

Kennedy, R. (2007) *Ghosts of Abu Ghraib, HBO Video*, at: https://www.youtube.com/watch?v=FGpaOp6_I7M.

Knell, Y. (2013) 'Israeli Army Ire Over Social Media Posts', *BBC News*, 2 March, http://www.bbc.co.uk/news/world-middle-east-21627500.

MacAskill, E. (2015) 'British Army Creates Team of Facebook Warriors', *The Guardian*, 31 January, https://www.theguardian.com/uk-news/2015/jan/31/british-army-facebook-warriors-77th-brigade.

Manovich, L. (2001) *The Language of New Media*, London: MIT Press.

Matheson, D. and Allan, S. (2009) *Digital War Reporting*. Cambridge: Polity.

Merrin, W. (2014) *Media Studies 2.0*, Abingdon, Oxon: Routledge.

Morris, E. (dir.) (2008) *Standard Operating Procedure*, Participant Media [DVD].

O'Reilly, T. (2005) 'What is Web 2.0: Design Patterns and Business Models For the Next Generation of Software', *O'Reilly*, 30 September, http://www.oreilly.com/pub/a/web2/archive/what-is-web-20.html?page=1.

Roger, N. (2013) *Image Warfare in the Age of Terror*, Houndmills, Basingstoke, Hamps: Palgrave Macmillan.

Rose, D. (2004) *Guantanamo Bay. America's War on Human Rights*, London: Faber and Faber.

Rosenberg, C. (2016) *Guantanamo Bay. The Pentagon's Alcatraz of the Caribbean*, Coral Gables, FL: Mango Media Inc.

Sealey G. (2004) 'Hersh: Children Sodomized at Abu Ghraib, on Tape', *Salon*, 15 July, http://www.salon.com/2004/07/15/hersh_7/.

Shabi, R. (2010) 'Anger over ex-Israeli Soldier's Facebook Photos of Palestinian Prisoners', *The Guardian*, 16 August, https://www.theguardian.com/world/2010/aug/16/israeli-soldier-photos-palestinian-prisoners.

Shachtman, N. (2009a) 'Army Orders Bases to Stop Blocking Twitter, Facebook, Flickr', *Wired*, 10 June, https://www.wired.com/2009/06/army-orders-bases-stop-blocking-twitter-facebook-flickr/.

Shachtman, N. (2009b) 'Military May Ban Twitter, Facebook as Security "Headaches"', *Wired*, 30 July, https://www.wired.com/2009/07/military-may-ban-twitter-facebook-as-security-headaches/.

Shirky, C. (2008) *Here Comes Everybody*, London: Allen Lane.

Simeon, A, David, M. B., and Jamal, I. A. (2014) 'IDF Sniper Admits on Instagram to Murdering 13 Gaza Children', *Counter Current News*, 31 July, http://countercurrentnews.com/2014/07/idf-sniper-admits-on-instagram-to-murdering-13-gaza-children/.

Sontag, S. (2004) 'Regarding the Pain of Others', *The New York Times Magazine*, 23 May, at: http://www.nytimes.com/2004/05/23/magazine/regarding-the-torture-of-others.html?_r=0.

Stafford Smith, C. (2008) *Bad Men. Guantanamo Bay and the Secret Prisons*, London: Orion Publishing Group.

Surowiecki, J. (2005) *The Wisdom of Crowds*, New York: Anchor Books.

Taguba, A. (2004) *AR 15-6 Investigation of the 800 Military Police Brigade*, http://nsarchive.gwu.edu/NSAEBB/NSAEBB140/TR3.pdf.

Telegraph. (2011) 'The Man Who Killed Muammar Gaddafi?', *YouTube*, 23 October, https://www.youtube.com/watch?v=LwIm_MLcdX8&bpctr=1472216537.

Wall, M. (2005) '"Blogs of War": Weblogs as News', *Journalism*, 6(2), pp. 153–72.

Worthington, A. (2007) *The Guantanamo Files*, London: Pluto Press.

Zornick, G. (2005) 'The Porn of War', *The Nation*, 23 September, http://disaffecteddems.proboards.com/thread/3583/war-george-zornick.

7 Transparent war

Wikileaks and the war logs

Post-invasion Afghanistan and Iraq

With the end of the Afghanistan campaign in December 2001 the first task of the War on Terror seemed accomplished. The Taliban had been overthrown and although Bin Laden had escaped, al-Qaeda's immediate capacity to strike against the west had been destroyed. In Washington, there was a sense of relief and triumphalism as the Bush administration began preparations for its next target, Iraq. With the war's conclusion, western media attention also turned away from the country. Reporting the complexities of the post-war situation was of less interest to western audiences than the war and over the following years most media organizations offered only a limited, sporadic coverage of Afghanistan. The success of the Taliban insurgency from 2006 returned the country to the news but it also made independent reporting difficult, forcing journalists to rely on military protection. Most reports thereafter focused on the military's activities and losses, offering little insight into the reality of the conflict or the experiences of the Afghan population. As a result, few in the west understood the problems of the country, the reasons for the ongoing conflict or what precisely western forces were fighting and dying for. The full story would take years to become public.

Hamid Karzai was sworn in as President on 22 December 2001, leading the Afghan Interim Administration, which, on 13 July 2002, became the Afghan Transitional Administration of the new Islamic Republic of Afghanistan. Although Karzai was a Pashtun, the government was dominated by non-Pashtun Northern Alliance warlords, alienating Afghanistan's ethnic majority. These warlords had been responsible for much pre-Taliban violence and were often hated by ordinary Afghanis. The southern Pashtuns especially were suspicious at the west returning them to power and remained hostile to the Karzai regime. Although the west presented the new government as a western-style democracy the reality was an unpopular, unrepresentative regime run through personal alliances, patronage, bribery and systematic corruption.

Ultimately, central government's influence barely extended beyond the major cities. Large swathes of Afghanistan had no experience of government and the Bush administration's initial preference for a light military footprint, reluctance to engage in 'nation-building' and preparations for Iraq meant that it failed to secure the country or establish Kabul's rule. As a result, tribal communities rarely recognized the invasion's legitimacy, whilst the power vacuum in the countryside was exploited by corrupt warlords and their police forces as well as by drug-traffickers and criminals, creating the conditions for later local Pashtun support for the returning Taliban.

Where it did operate, the USA was highly unpopular. It was exploited by the warlords and tribal leaders who used the USA's hunt for Taliban to settle old scores or increase their own power, ensuring many innocent Afghanis were killed, imprisoned or fled to join the Taliban. The suspects they rounded up were refused Geneva Convention rights and imprisoned and tortured at Bagram airbase (sometimes, until death), with many being sent to Guantanamo Bay. The US 'night-raids', where forces would break into compounds after tip-offs of Taliban, were especially hated by ordinary Afghanis, breaching cultural norms in a country where the home had a special sanctity, where laws of hospitality governed and where contact with female relatives was closely regulated. The USA's over-reliance on air power and indiscriminate and excessive use of force also led to increasing civilian casualties and anti-American sentiment until even President Karzai complained about the loss of Afghan lives.

On 20 December 2001, the UN Bonn Conference and resolution 1386 authorized a NATO-led International Security Assistance Force (ISAF) under a rotating national command to provide security for Kabul and the surrounding region. With the USA's attention shifting to Iraq, NATO took over the ISAF command on 13 August 2003. UN resolution 1510 on 13 October 2003 expanded ISAF's mission to cover the whole country and NATO began to deploy member-nation troops to the north in late 2003, then through the west, east and south in 2006. At its height, the ISAF's forces were 130,000 strong, with troops from 51 NATO and partner nations in Afghanistan, although its forces were inferior to the US military and suffered from a confusing command structure and inter-nation conflicts. NATO's major problem, however, was that it found itself facing a full-scale Taliban insurgency.

The Taliban had not been defeated in 2001, but had decamped to the tribal areas of western Pakistan. They had supporters there and were nurtured by the Pakistan intelligence agency, the ISI, as part of Pakistan's long-standing influence over Afghanistan, opposition to the pro-Indian Karzai government and strategy of 'security-in-depth' against an Indian invasion. With Pakistan support, the Quetta Shura – the Afghan Taliban organization in Pakistan – launched an Afghan insurgency in May 2003 (shortly after Donald Rumsfeld had declared major combat activity over, with the bulk of the country 'secure' (CNN, 2003)). The Taliban targeted government buildings, US bases and aid workers, adopting tactics from Iraqi insurgents including using suicide bombers for the first time and refining the use of Improvised Explosive Devices (IEDs). Although the USA increased its Afghan troop contingent, its focus was on Iraq and it took until the escalation of the insurgency in 2006 for western politicians to realize the extent of the problem in Afghanistan. By the time ISAF deployed British troops to Helmand province in 2006 they found themselves facing an organized and powerful enemy.

The wisdom of deploying Afghanistan's hated nineteenth-century enemies, the British, as peacekeepers in the Pashtun south was questionable. Their role defending local government structures, however, meant they were propping up corrupt, violent and unpopular warlords and police forces, antagonizing local Afghanis and driving many to join the resistance. The UK's opium crop eradication policies also proved unpopular with the local farmers, driving many to the Taliban who tolerated, but taxed, the produce. The under-manned, over-stretched, under-equipped and over-confident British forces bore the brunt of the southern Taliban and public insurgency, suffering especially from deadly roadside IEDs. After losing only five personnel from April 2002 to March 2006, the British would lose hundreds over the following years.

The insurgency intensified again in 2008 and in February 2009 President Obama authorized an Afghan-surge of 17,000 extra troops, with US marines being deployed to

Helmand to prevent the British mission collapsing. The USA, however, also struggled. Their counter-insurgency strategies failed as they lacked support among the local population and they couldn't prevent the Taliban returning to areas they had cleared. The Afghan National Army (ANA) proved an ineffective partner and cases of US troops being shot by ANA soldiers increased the mistrust. Fundamentally, the west failed to appreciate that, for most southern Afghanis, the Pashtun Taliban were preferable to the hated, predatory and corrupt US-Karzai administration, the local warlords and police, and the foreign military 'occupiers'.

The political situation was also deteriorating. Although Karzai had won the first democratic Presidential election on 9 October 2004, the second election on 20 August 2009 was marked by massive electoral fraud by Karzai, undermining western claims to be fighting for democracy and souring their relationship with the president. Karzai himself increasingly turned against the western forces, tapping into widespread anti-Americanism and describing the western presence as a 'foreign occupation' (Fairweather, 2014:373).

Afghanistan remained one of the poorest nations on earth, with its infrastructure decimated by decades of war. The post-war economic investment in Afghanistan was smaller than that poured into other nation-building efforts. A lot of pledged aid never arrived, whilst the international aid programme was poorly coordinated, followed western ideas of what was required and was marked by confusion and competition among agencies and significant security problems. After Iraq, the USA belatedly accepted the need for nation-building in Afghanistan leading to a new injection of funds, but most of the money was wasted due to corruption and profiteering by private companies and organizations. As Fairweather says, 'what is striking about the early years of the US reconstruction effort is how often it was American firms ripping off their own government' (Fairweather, 2014:120). Corruption by Karzai's government, the urban elites and the warlords added to this problem. In the meantime, the drugs trade had increased. By 2007 the narcotics industry was worth $3bn out of a gross national product of $8bn, with the record production of 8,200 tonnes of opium (supplying 93% of the world's illicit heroin supply) being twice that of 2002. Confused and ineffective policies by the USA and UK had only increased the power of the traffickers and support for the Taliban. The increasing opium harvest funded new insurgents every year through Taliban 'taxes'.

Increasing Taliban successes led Obama to agree in December 2009 to deploy 30,000 more troops to Afghanistan. The accompanying proviso that a military withdrawal would begin in September 2011 and be complete by December 2014 signalled, however, that the western powers were looking for a face-saving, stage-managed end to their commitments. By then the point of the invasion had long since been lost. Whilst the western powers claimed to be fighting the War on Terror, in truth they were mired in an ongoing civil war and regional proxy conflict begun decades before 9/11, against an enemy, the Taliban, who had only a tangential relationship with terrorism, in support of a corrupt, warlord-based government of limited democratic credentials that was unwanted by many ordinary Afghanis.

US troops (especially special forces, airpower and drones) remained in Afghanistan after the 2014 'withdrawal' under the terms of a bilateral security agreement, signed between the two governments on 30 September 2014. By the end of 2015 the war had cost 2,325 American and 454 British lives, with an estimated 31,000 Afghan civilians violently killed since 2001. In February 2016, Obama launched a 'mini-surge' of hundreds of troops into Helmand and in July he announced a delay to the US military withdrawal, saying 8,400 troops would remain there indefinitely. President Trump announced another troop surge in June 2017, bringing the total troop numbers there to 14,000 by the end of the year. Significant US military operations continue against the increasingly successful Taliban, resulting in more civilian casualties, whilst Taliban and other anti-government activities kill many more – in 2016, 4,561 people were killed in 1,340 terrorist attacks. By the end of 2017, after 16 years and over $1tn spent on military operations, the USA was losing the war in Afghanistan.

Throughout the conflict, the military losses were continually justified by western politicians and media through a caricatured presentation of the Taliban as a foreign body in Afghanistan, as a force of evil and as a threat to world peace, although all of these were highly debatable. For some, such as US special representative to Afghanistan and Pakistan Richard C. Holbrooke, the USA had been fighting 'the wrong enemy in the wrong country' all along (Gall, 2012:ix), with Pakistan, the USA's partner in the War on Terror, being the real regional supporter of the Taliban, radical Islamism and its violence.

Post-war Iraq fared worse than Afghanistan. The Bush administration thought Saddam was the problem, hence after removing him they naively believed they could hand the country over to the exiles and the people to create a stable, pro-western democracy. In reality, Iraq was broken and bankrupt before the invasion. The economy had collapsed after years of war and punitive UN sanctions and there was massive unemployment and mass destitution. There had been no investment in the infrastructure, the power grid was obsolete and public services had collapsed. The fall of the regime by 9 April led to widespread looting in the major cities. Government buildings and public facilities such as hospitals and schools were stripped and thousands of artefacts from the National Museum were stolen, whilst US troops looked on, doing nothing. Rumsfeld's dismissive response to the looting – 'freedom is untidy' – was widely condemned.

The Bush administration had no plans for the aftermath. They'd established the Office of Reconstruction and Humanitarian Assistance (OHRA) under Jay Garner in January 2003 to serve as a caretaker administration until a government was elected, but it lacked a budget and was poorly supported. It was quickly replaced by the Coalition Provisional Authority (CPA) on 21 April, with Garner himself being replaced on 12 May by L. Paul Bremer III. Bremer assumed the title of US Presidential Envoy and Administrator, with the CPA operating out of the heavily fortified 'Green Zone' in Central Baghdad.

The USA made two military mistakes, however. First, it assumed the regime's fall signalled the end of combat hostilities, failing to recognize that it wasn't 'mopping up' the 'dead-enders' of the old regime (as Rumsfeld described them), but facing a planned Baathist guerrilla war whose networks, arms caches and strategies had been organized before the invasion. Second, Rumsfeld's neo-liberal predilection for the cost-effective, light-touch of network-centric warfare meant there weren't enough troops in Iraq. Network-centric warfare could rapidly defeat an army, but couldn't secure the country, and without that security the CPA couldn't function.

The CPA also made two significant mistakes. First, on 16 May 2003, it issued CPA order #1 aimed at 'De-Baathification' – a strict banning of Saddam's Baathist party-members from posts that led to doctors, teachers and other key workers being made redundant. Secondly, on 23 May, CPA order #2 disbanded the army and security services, sending the message that the USA wasn't merely anti-Saddam, but anti-Iraq in considering itself the only legitimate military force in the country. It also disproportionately punished the Sunni community and significantly increased unemployment. Together these orders signalled that there would be no quick transition of power and that the USA was an occupying force with Bremer ruling as an imperialist Viceroy. — *almost like a*

"This is how we do it whether you like it or not!"

Recognizing the shift of power from the Sunni to the majority Shia, many newly unemployed Sunni servicemen joined the Baathist insurgency, which soon became a Sunni resistance movement. This was complicated by the arrival of foreign Sunni Islamists who flocked to Iraq as the new site of global jihad for a chance to fight the 'far enemy', America. Ironically, the jihadis the USA had claimed were in Iraq to justify the war arrived as a result of the invasion, with their passage being aided by neighbouring states such as Syria and Iran who wanted to destabilize the American operation. One of the most successful Sunni jihadi groups was run by Abu Musab al-Zarqawi whose first acts were the Baghdad bombings of the Jordanian embassy on 7 August 2003 and the UN head-quarters on the 19. Zarqawi's aim was the creation of an Islamic caliphate. His plan was to use extreme violence to cause chaos, repel the Americans, split Iraqis and unite the Sunni. His hatred of and attacks upon the Shia would open up another fault-line in post-war Iraq.

As car bombs, suicide bombings, IEDs, firefights, ambushes, murders and kidnappings became common, CPA members and the military rarely ventured out without an armoured convoy. The frightened, trigger-happy US soldiers used excessive violence, killing many innocent civilians, whilst the ongoing failure to find any WMDs added to the occupation's illegitimacy. From within their bubble in the Green Zone the CPA continued issuing declarations about the improved state of the country that had no relationship to reality. For the Iraqis, power, fuel and food shortages meant daily life was worse now than before the invasion, with the CPA's inability to generate electricity making them more unpopular than Saddam. Soon, even its own staff would joke that CPA stood for 'Can't Provide Anything'.

The Iraq War was an explicitly neoconservative project: whereas George Bush Sr's 1991 Gulf War was a conservative, restorative war, to return the middle east to the status quo, Bush Jr's war was transformative. As Patrick Cockburn says, 'It was nothing less than an attempt to alter the balance of power in the world' (Cockburn, 2006:1). From the start, the CPA was the expression of the same ideological worldview. Appointments were based upon personal connections and Republican and neoconservative credentials, with background and allegiance mattering more than expertise. Most of the CPA were young, inexperienced and unqualified. Few had worked abroad, spoke the language or knew the region and culture, although this didn't matter: as the aim of the CPA was to remake Iraq in America's image, they only had to know the *American* way of life.

Hence instead of simply restarting industry, the CPA attempted to transform Iraq's economy from a state-run and subsidized, centrally-planned economy into a US-style, modern, hi-tech, free market economy. The CPA's ambitious plans for privatization and the elimination of the state's role faltered, however, as the security-situation scared off foreign investors. The neo-liberal agenda was also evident in the preference for using private companies over government troops and workers. In a 'seedy gold-rush' (Cockburn, 2006:73) politically well-connected private American contractors poured into Iraq, taking over security and reconstruction. Corruption was endemic, with huge sums being siphoned off state contracts, and there were no background checks on security staff who were not subject to the rules of war (although, conveniently, nor did they count in casualty statistics).

The capture of Saddam on 13 December 2003 had little impact on the CPA's standing. Hostility towards the USA erupted in April 2004. The murder of four US security men in Fallujah on 31 March prompted a massive US retaliatory response on 4 April, with three battalions besieging the city, a centre of Sunni resistance. The bombardment fuelled Sunni anger, leading to more uprisings in Baghdad, Samarra and Ramadi. The USA met stiff resistance in Fallujah and the fierce US attack (during which 572–616 civilians died) stoked anger within Iraq and abroad. With the operation's failure, a negotiated ceasefire on 8 April effectively left the Sunni insurgents in control.

The USA faced a simultaneous uprising by the 'Mahdi Army', an Iranian-backed, 10,000-strong militant group controlled by the Shia cleric Moqtada al-Sadr. In late March Bremer closed al-Sadr's newspaper, leading al-Sadr to mobilize protests against the CPA. When special forces arrested his deputy, al-Sadr ordered attacks on 4 April. Fighting in Najaf, Sadr City and Basra was put down by the USA by the end of April but the Mahdi Army rose again in August. The Battle of Najaf, from 5 to 27 April, ended in a ceasefire and an October agreement by al-Sadr to end his military opposition but by the time it came into effect in November the USA was fighting the second Battle of Fallujah. Insurgent strength had grown there since April, with al-Zarqawi making it his operational base. The resulting battle, from 7 November to 23 December, saw the heaviest US urban fighting since Vietnam, leading to 95 US casualties with 560 wounded and between 1,200 and 1,500 insurgents killed. The USA was victorious but not decisively. Al-Zarqawi escaped, the insurgents weren't defeated and by 2006 they would control most of Anbar province.

The April uprisings were popular with the Iraqis. The Iraqi police and army units refused to fight their own countrymen on American orders, whilst the exposure in May 2004 of the Abu Ghraib abuse fuelled this growing anti-American sentiment. Meanwhile George Bush was pushing for a formal handover of power before the US presidential election in November. On 13 July 2003, the CPA had created the Iraqi Governing Council (IGC), whose job it was to draft the Transitional Administrative Law (TAL) that would govern Iraq until a constitution was approved. On 28 June 2004, when the IGC handed authority over to the Iraqi Interim Government (IIG) with Ghazi Mashal Ajil al-Yawar as President and Iyad Allawi as Prime Minister, the CPA was formally dissolved and the official US occupation ended. It wasn't the end of its military involvement, however: in 2004, it would lose 848 soldiers with 7,989 wounded and in 2005 another 846 would be killed and 5,944 wounded. Casualties would continue for many years after.

Elections on 30 January 2005 for an interim National Assembly were boycotted by many Sunnis. They led, on 3 May, to the IIG being replaced by the Iraqi Transitional Government (ITG) with Ibrahim al-Jaafari as PM. Its job was to write the constitution, which was ratified in a referendum on 15 October 2005. Elections were held on 15 December 2005 for the first permanent elected Iraqi government and a coalition government with Nouri al-Maliki as PM was eventually formed on 20 May 2006. This was not, however, the pro-western democracy the USA had envisaged. It was dominated by Iranian-backed Shia parties, was hostile to the Sunnis and was inefficient, bureaucratic and massively corrupt.

By this time a sectarian civil war had broken out. Al-Zarqawi's group, Tawhid w'al-Jihad, had pledged allegiance to al-Qaeda on 17 October 2004, changing its name to Tanzim Qaudat al-Jihad fi Bilad al-Rafidayn, or 'Al Qaeda in Iraq' (AQI). Following the advice of Abu Bakr Naji's 2004 online text, *The Management of Savagery*, al-Zarqawi used extreme violence to cause chaos. On 22 February 2006, his bombing of the Shia al-Askari shrine in Samarra led to a full-blown sectarian civil war. Over 1,300 mostly Sunni bodies were found in and around Baghdad over the following days and the next few years saw an explosion of murders, kidnappings and torture, attacks by Shia militias and Sunni insurgent groups, and the ethnic cleansing of neighborhoods as an ethnic division held in check by Saddam came apart.

Iraq's government wasn't neutral. The Iranian-influenced Ministry of the Interior's special police commandos operated as anti-Sunni death squads and in November 2005 the USA had already discovered a secret ministry torture bunker, freeing 173 Sunni prisoners who'd been tortured with electric shocks and drills. Al-Maliki's government would become known for its systematic anti-Sunni policies. Even Saddam's execution on 30 December 2006 fed this sectarianism as leaked mobile phone footage revealed he'd been subjected to abuse by Shia government guards on the gallows.

AQI and other groups merged into the Mujahideen Shura Council on 15 January 2006. Al-Zarqawi was killed in a US airstrike on 7 June 2006, being succeeded by Abu Omar al-Baghdadi who changed the group's name to the Islamic State of Iraq (ISI) on 15 October. By then, however, their violence, harsh laws, use of foreign fighters and attempts to enter the smuggling market had alienated the Sunni tribes in Anbar. Sheikh Sattar al-Rishawi created an alliance that became the Anbar Salvation Council and by January 2007 the tribesmen were in armed revolt against the jihadi insurgents in a movement known as 'the awakening'. The US military allied with the tribesmen, creating the official watch organization 'The Sons of Iraq', whilst implementing a 'surge' of 21,000 extra troops and beginning a series of successful major military operations against the insurgents from June 2007 that left ISI in disarray by 2008.

The highly-sectarian Iraqi government, however, was hostile to the Sons of Iraq (as an official, armed Sunni organization) and it arrested many of its leaders and disbanded the movement. It reneged on US promises to the Sunni community, persecuted Sunni politicians and repressed the tribesmen, many of whom were also killed by Shia militias. On 18 April 2010, al-Baghdadi was killed by the USA and under the influence of former Baathist officers ISI reformed around Iraqi fighters. Its new leader, Abu Bakr al-Baghdadi was announced on 16 May 2010. With the help of Sunni communities driven back to the jihadis by government repression, he would rebuild the Islamist insurgents into a major force. By 2014 his fighters would capture a huge swathe of territory across Syria and Iraq, including the city of Mosul. It would take three years, and the return of American military operations, for the Iraqi government to defeat 'Islamic State'.

As with Afghanistan, western media reporting of post-war Iraq was limited. The initial coverage was celebratory, with the speed of the war appearing to justify the invasion and its optimism, but the complexities of post-war Iraq was of less interest to the media or the public. Indeed, media organizations that had argued for the invasion, and who had donned fatigues and cheered on the military as humanitarian liberators of Iraq were in no position afterwards to be critical of the official line from Washington, London and the CPA. Hence, they were understandably muted about the failure to find WMDs, and were conflicted about the obviously escalating violence the war had created. It soon became too dangerous for journalists to leave the Green Zone but the violence had become too common and too confusing to report anyway. As Patrick Cockburn says, 'so many people were being killed in Iraq every day for so many reasons that the outside world had come to ignore the slaughter and Iraqis themselves were almost used to it' (Cockburn, 2006: 199–200). For years after, Iraq appeared in the news only as a catalogue of random violent incidents, with little context or explanation. As in Afghanistan, coverage often focused on patriotic support for 'our troops' who were doing a difficult job and eulogies for the casualties. What Wikileaks would expose, therefore, was the simultaneous failure of the mainstream media to report the reality of the wars, and the refusal of the authorities to inform us of this reality.

Julian Assange

The character of Wikileaks owes much to the background, motivations and political ideas of its founder, Julian Assange. His personal philosophy was formed from the influence of four phenomena: hacking, cypherpunk, online collaboration and leaking.

Born on 3 July 1971 in Townsville, Queensland, Australia, Assange developed an early interest in computing. Operating under the name 'Mendax', he became a key part of the 1980s Melbourne hacking scene. His group, the 'international subversives', focused on penetration and exploration, but they were also pioneering 'hacktivists' (hacking for political purposes), with Assange writing a program called 'Sycophant' to penetrate and attack US military computers. By 1991, Leigh and Harding claim, 'Assange was probably Australia's most accomplished hacker' (Leigh and Harding, 2011:42). His penetration of the Canadian telecoms company Nortel's master terminal led to his October 1991 arrest. Charged in 1994, he was found guilty in 1996 and sentenced to a $2,100 fine and $5,000 good behaviour bond.

'Hacking' had originated as a term for an inspired act or especially impressive solution to a problem. Applied to computing this meant a well-worked and minimally coded program, but the concept developed to include feats of computing such as system penetration and many hackers became interested in exploring how computers worked. Hacking moved, therefore, from an activity to a world-view and, as Stephen Levy explains, an 'ethics'. This hacker world-view revolved around a belief in the freedom of information and expression, the freedom to tinker, explore and access, a joy in learning and improving oneself, and a libertarian, individualistic anti-authoritarianism that favoured open, decentralized and transparent systems.

These hacker ideals were influential upon Assange, as well as upon the 'cypherpunk' movement. This emerged from the late 1980s based around John Gilmore, Eric Hughes and Timothy C. May, the email list they created in 1992 and a series of online texts including May's 'The Crypto Anarchist Manifesto' (1992) and 'Cyphernomicon' (1994) and Hughes' 'A Cypherpunk Manifesto' (1993). The cypherpunks were marked by a right-wing libertarianism that was suspicious of all government, hence their promotion of encryption to protect individual privacy. Recognizing that they couldn't trust anyone to create these technologies, they advocated producing their own: as Hughes said, 'cypherpunks write code' (Hughes, 1993). The cypherpunks' anti-statism was most-famously expressed in Jim Bell's 'Assassination politics' essay, which suggested secure, anonymous communications could be the basis for a 'lottery', with bets on when politicians would die acting as an incentive to assassinate them to claim the money. Others looked to transparency as a check on government, with John Young – whom Wikileaks later called 'the spiritual godfather of online leaking' (Greenberg, 2012:102) – setting up the website Cryptome in 1995 as a repository for leaked government information.

Assange contributed to the cypherpunk mailing list but whilst he shared their belief in privacy and suspicion of the state he rejected their individualism, retaining a left-wing faith in the collective and democracy. Where Young saw leaking as a means to protect the individual, Assange saw it as a means to protect democracy and as a social good. For Assange, collective action brought benefits to society. He was inspired especially by the 'Free Software Project': Richard Stallman's mobilization of online programmers to produce a non-proprietal operating system, completed as GNU/Linux in 1992. Assange liked both the crowd-sourced collaboration and the idea of opening up the source code so anyone could improve it. Eric Raymond's famous claim that 'given enough eyeballs all bugs are shallow' (Raymond, 2000), Assange realized, could be applied to politics: make power transparent and everyone could scrutinize and improve it. Following Young, he understood that leaking could achieve this.

Leaking had its own long history. The most famous precursor to Wikileaks was the *New York Times*' publication of the 'Pentagon Papers' from 13 June 1971. This was a 7,000-page, US DOD history of government involvement in Vietnam that demonstrated that four administrations had lied about the nature and extent of US intervention. It was photo-copied by one of its researchers, Daniel Ellsberg, and leaked to the press. Although the government attempted to stop publications, court action against the *Times* was derailed by spreading publication across fifteen other papers, whilst a lawsuit against Ellsberg collapsed in 1973. The leaks were widely seen as an important journalistic success in exposing government lies and the truth about the Vietnam War.

By late 2006, Assange's plans and ideas were coming together. He registered Wikileaks.org on 4 October and on 10 November and 3 December published two, linked blog essays, 'State and Terrorist Conspiracies' and 'Conspiracy as Governance'. Here he describes the world as dominated by authoritarian governments and corporations who maintain their power through conspiratorial links. Cut these, he says, and you throttle their communications and reduce their ability to think and act. In a further post on 31 December he suggests leaking as a way to disrupt this conspiracy: 'In a world where leaking is easy, secretive or unjust systems are nonlinearly hit relative to open, just systems', he says: 'leaking leaves them exquisitely vulnerable to those who seek to replace them with more open forms of governance' (Assange, 2006).

As he explained in an email to potential supporters,

> We have come to the conclusion that fomenting a world wide movement of mass leaking is the most cost effective political intervention available to us. New technology and cryptographic ideas permit us to not only encourage document leaking, but to facilitate it directly on a mass scale. We intend to place a new star in the political firmament of man.
>
> Greenberg (2012:131)

Coupling liberal–democratic ideas of the consent of the governed and checks on authority with cyber-libertarian ideas of freedom of information and anti-authoritarianism and a socialist vision of a more participative democracy, Assange realized that secure, anonymous leaking together with the internet's ability to publish beyond state jurisdictions could increase political transparency and accountability and lead to more open and better government.

Assange made the remarkable scope of his ambitions clear in another email to potential supporters, which asked, 'Are you interested in being involved with a courageous project to reform every political system on earth – and through that reform move the world to a more humane state?' (Greenberg, 2012:96). John Young rebuffed Assange's approach (posting all of his communications on his website), but Assange found new friends in December 2007 in the Chaos Computer Club annual congress in Berlin. There he met key supporters, Jacob Appelbaum and Daniel Domscheit-Berg, and contacts such as the Swedish web-hosting company PRQ.

Wikileaks

Wikileaks posted its first material on 28 December 2006 – a 'secret decision' by a Somali rebel leader, culled from traffic passing through the TOR network to China. The site had some early successes. Its August 2007 publication of the April 2004 Kroll Report, exposing widespread government corruption in Kenya, made the *Guardian*'s front-page, whilst its November 2008 publication of a report about Kenyan death squads, was widely praised, winning Wikileaks the Amnesty International 2009 Media Award at a ceremony in London on 3 June. A defining moment for Wikileaks came in January 2008 when the Swiss bank Julius Baer began legal action against it after it posted leaked information about its operations. The complexity of Wikileaks' organization, ownership and registration, however, placed the site beyond the court's jurisdiction. The court ruling on 18 February that Dynadot should block Wikileaks' domain name had no effect on Wikileaks' global mirror sites and led to opposition from civil liberties groups who got the ruling overturned on the 29th. Baer's attempted injunction against the site backfired as, following the 'Streisand effect', the publicity drew global attention both to Wikileaks and the documents they wanted to suppress. Baer dropped all actions by 5 March.

By 2009 Wikileaks was an established, if minor, leaking site. It had fulfilled Assange's hopes of exposing the powerful, proven itself beyond state control and attracted media support as a freedom of speech cause, but problems remained. So far Wikileaks had failed to have the revolutionary impact Assange wanted. Its control of publication prevented broader public participation, Assange's hopes of crowd-sourcing analysis to an army of visitors who would find and promote stories had failed, whilst few of its leaks attracted mainstream media attention. It lacked a revenue stream, relying upon public donations, and its future depended on a supply of quality submissions. Luckily it had been noticed by Bradley Manning.

Private First Class Manning was a US intelligence analyst in Iraq with access to the military networks SIPRNet (Secret Internet Protocol Router Network) and JWICS (Joint Worldwide Intelligence Communications System). He had enlisted in August 2007 out of patriotism and was sent to Iraq in October 2009. As Leigh and Harding comment, he quickly became disillusioned by the reality of the occupation: 'The more he read, the more alarmed and disturbed he became, shocked by what he saw as the official duplicity and corruption of his own country' (Leigh and Harding, 2011:21–22). The turning point came on 27 February 2010 when he was asked to look at 'anti-Iraqi literature' found on 15 civilians arrested by the Iraqi police. Instead of insurgent propaganda, he discovered a scholarly political critique of the Prime Minister and cabinet corruption. He informed his superior officer but, 'he didn't want to hear any of it. He told me to shut up, and explain how we could assist the FPs (Federal Police) in finding *more* detainees' (Leigh and Harding, 2011:84).

Realizing that he was helping identify political opponents who would very likely be tortured and killed, Manning decided he couldn't be part of this system. He copied information from the military networks onto CD, with the aim of exposing the truth of the war and encouraging 'worldwide discussion, debates and reforms' (Leigh and Harding, 2011:83). Unlike the left-wing, anti-statist Assange, Manning was motivated by a fierce sense of American patriotism and a high moral integrity. What they shared, however, was the hacker worldview – as Manning argued, 'It's public data. It belongs in the public domain. Information should be free' (Leigh and Harding, 2011:30) – as well as a belief in the importance of information to democracy. In order to make an informed decision as to whether they supported a war carried out in their name the public needed to *know* what this war was.

Manning contacted Assange in November 2009, passing the material he had copied to him accompanied by a 'readme' file saying, 'This is possibly one of the more significant documents of our time, removing the fog of war, revealing the true nature of 21st century asymmetric warfare. Have a good day' (Greenberg, 2012:40). The contact was anonymous but Manning confessed what he had done in online chats with Adrian Lamo from 21 to 25 May 2010, being arrested on the 26th after Lamo contacted the military. In a 2013 statement for his pre-trial hearing Manning repeated his reasons for contacting Wikileaks, saying, 'I believe that if the general public, especially the American public, had access to the information … this could spark a domestic debate on the role of the military and our foreign policy in general … as it related to Iraq and Afghanistan' (Manning, 2013).

The first material Assange released was an AH-64 Apache helicopter video from 12 June 2007 in Baghdad, showing the crew opening fire on a group of people, killing 10 civilians and two Reuters journalists. To maximize its impact, Assange asserted editorial control. The 39-minute video was posted online as an original source whose veracity could be checked, whilst Wikileaks edited together a 17-minute version entitled 'Collateral Murder', which it presented at the National Press Club in Washington on 5 April 2010 (Sunshinepress, 2010a; 2010b). The video was horrific. The crew's desperation to fire and casual pleasure in the deaths appalled viewers: 'Oh yeah, look at those dead bastards', said one soldier. 'Nice', the other replied. In the most distressing sequence an Iraqi who'd stopped to help the wounded is killed, whilst his children in the minivan are wounded.

Manning later explained that what he found most alarming was 'the seemingly delightful bloodlust the aerial weapons team seemed to have', their dehumanization of the individuals on the ground and their lack of value for human life. Their hope that a wounded man will pick up a weapon so they can reengage, he says 'seemed similar to a child torturing ants with a magnifying glass'. They assume everyone in the area is a threat and show no remorse when told about the children they'd hit. More than anything, Manning says,

> I wanted the American public to know that not everyone in Iraq and Afghanistan were targets that needed to be neutralized, but rather people who were struggling to live in the pressure cooker environment of what we call asymmetric warfare.
>
> Manning (2013)

The video didn't have the impact Assange had hoped for. Wikileaks had shifted into being a hybrid activist media organization and there was suspicion at the organization's motives and at the editorializing that had added a title and commentary to the video. Assange was surprised by the lack of interest from the media and public, whilst Reuters refused to offer support due to the ongoing legal investigation into their reporters' deaths. The US government defended the aircrew, saying they acted lawfully within the rules of engagement, and, in what would become a familiar tactic, shifted the media discussion onto Wikileaks itself.

Realizing he needed to find a way to reach a mass audience, Assange changed his publication strategy for the 'war logs', deciding to partner with what he disparagingly called 'the mainstream media' ('msm'). Assange agreed to a multi-jurisdictional alliance between Wikileaks and the *Guardian*, the *New York Times* and *Der Spiegel* with simultaneous, joint publication to spread the legal risk and prevent political pressure upon any one organization. As well as political credibility and influence and a huge, daily audience, the newspapers brought with them a range of professional skills, including data analysis (building a searchable database out of the information), foreign affairs expertise and the journalistic ability to identify key stories and evaluate and verify information. Although he had formed Wikileaks to overturn the existing informational structures that he saw as too close to power, Assange pioneered here a new alliance of 'old' and 'new' media worlds, one locating Wikileaks within the journalistic rubric, allowing it to receive 'accreditation and attention' for its leaks and greater influence than it would have had from posting the material online (Benkler, 2011a).

It wasn't an easy alliance, with Assange's personality, insistence on secrecy and his own leaking of copies causing problems. The biggest issue was over the redaction of identifying information where the journalists' desire to protect people ran up against Assange's hacker insistence that information should be free. Although he eventually agreed to redactions, the journalists were alarmed by his initial dismissive reaction to the issue.

The 'Afghan War Logs', published from 25 July 2010, was a set of US Army field reports, constituting 91,731 documents running from January 2004 to December 2009. It was the largest unauthorized publication of confidential government information in history. Its revelations included the exposure of a US special forces 'kill or capture' squad, 'Task Force 373' (a code-name for Joint Special Operations Command, or JSOC), as well as other units working to a secret target hit-list that required no evidence and had no judicial review. The logs also revealed repeated patterns of civilian casualties, with hundreds killed or wounded and not reported, along with trigger-happy troops who ignored 'escalation of force' (EOF) procedures and were liable to fire first. The *Guardian*, for example, identified clusters of excessive use of force against civilians by certain British units. As Leigh and Harding comment, 'These events and hundreds like them, together constitute the hidden history of the war in Afghanistan, in which innocent people were repeatedly killed by foreign soldiers' (Leigh and Harding, 2011:124).

The logs also demonstrated the unreported complexity of the Afghan conflict. They highlighted the USA's belief that Pakistan was hampering attempts to stop the Taliban and that elements in the country were actively supporting the insurgency. There were hundreds of reports of border clashes between US and Pakistan troops and Afghan Army and Pakistan troops and about Iranian and Pakistan intelligence agencies running riot, supporting particular warlords and engaging in combat missions. Along with escalating deaths from IEDs, and numerous friendly-fire incidents, what emerged from the logs was a picture of chaos unreported by western media or governments. As the *Guardian* editorial on 25 July 2010 concluded,

> These war logs – written in the heat of engagement – show a conflict that is brutally messy, confused and immediate. It is in some contrast with the tidied-up and sanitized 'public war', as glimpsed through official communiqués as well as the necessarily limited snapshots of embedded reporting.
>
> Leigh and Harding (2011:126)

The US response was angry. Defense Secretary Robert Gates' statement that Wikileaks was 'morally guilty' of putting lives at risk and endangering national security, was picked up by the right-wing press, with *The Times* claiming in a front-page article on 28 July that 'Hundreds of Afghan lives have been put at risk by the leaking of 90,000 intelligence documents because the files identify informants working with NATO forces' (Brooke, 2011:177). There was no evidence anyone was ever hurt from the leaks, but as Brooke points out, 'This was speculative blood, unlike the real blood revealed to be on the hands of Task Force 373'. The speculative blood, however, became the story. Admiral Mike Mullen, chairman of the joint chiefs of staff, for example, claimed in a Pentagon news conference on 29 July, '. . . the truth is they might already have on their hands the blood of some young soldier or that of an Afghan family' (Leigh and Harding, 2011:113). As Leigh and Harding note, 'This slogan – "blood on their hands" – was in turn perverted from a speculation into a fact, endlessly repeated, and used as a justification for bloodlust on the part of some US politicians' (Leigh and Harding, 2011:113). 'Particularly repellant', they say, 'was hearing the phrase being used by US Generals who, as the Wikileaks documents revealed, had gallons of genuine blood on their own hands' (Leigh and Harding, 2011:113). In October Robert Gates admitted to the chair of the Senate Armed Services Committee that 'the review to date has not revealed any sensitive intelligence sources and methods compromised by the disclosure' (Levine, 2010).

On 22 October 2010, the same papers began publication of the 'Iraq war Logs' – 391,832 US Army field reports from January 2004 to December 2009. A significantly bigger leak than the Afghan logs, the reports immediately gave the lie to US General Tommy Franks' claim at a 2002 news conference, 'We don't do body counts'. In fact, the army did, recording 66,081 civilian deaths from 2004 to 2009 (a period excluding the invasion phase and its immediate aftermath). If anything, the logs underplayed the numbers. Whilst Iraq Body Count estimated 1,200 civilian deaths from the battles over Fallujah the field reports listed 0, whilst many innocent deaths – such as the Reuters journalists killed in the 'Collateral Murder' incident – were recorded as 'enemy combatants'. Iraq Body Count used the logs to revise up their estimate of civilian deaths in Iraq from 2003 to 2010 as between 99,383 and 108,501 people. Handily, the USA even categorized the deaths, recording 31,780 killed by IEDs and 34,814 as sectarian killings, for example. By December 2009, in addition to the 66,081 civilian deaths, there had been 15,196 Iraqi Security Force personnel, 23,984 'enemy combatants' and 3,771 coalition troops killed.

The logs exposed the level of sectarian abuse. The Shia army and police force, Leigh and Harding note, continued 'to arrest, mistreat and murder its own citizens almost as if Saddam had never been overthrown' (Leigh and Harding, 2011:131). The USA was complicit in this, failing to investigate hundreds of reports of abuse, torture, rape and murder by Iraqi police and soldiers whose conduct went unpunished. As Nick Davies notes, the USA had a 'formal policy' of ignoring torture allegations, recording 'no investigation necessary' and handing reports over to the units implicated in the crimes (Leigh and Harding, 2011:132). As Manning had claimed, the logs showed the USA handed captives over to the special police commando 'Wolf Brigade' torture squads and were even present during their operations.

There were other revelations too – about US troops killing insurgents trying to surrender and killing civilians at checkpoints; about British abuse of Iraqi detainees; about continual friendly-fire incidents; and about Iranian intervention in Iraq – but the Iraq war logs were primarily about the numbers. In the sheer volume of incidents they recorded, the logs provided the most detailed picture of the conflict, exposing the chasm between the official image of Iraq and the reality. They demonstrated that Afghanistan and Iraq were not largely successful and completed missions complicated only by minor insurgencies but rather complex, chaotic and failing countries with the government, non-state actors and neighbouring nations embroiled in a multi-sided, multi-faceted civil war. The logs showed that western governments knew this and that mainstream journalism had been unwilling or unable to report the truth.

With the publication of the Iraq logs, the US military information media management system swung into action again, turning the story once more onto Wikileaks itself rather than the wars and the details revealed. The potentially criminal acts of the coalition forces were overlooked again in favour of claims that it was Wikileaks that had blood on its hands.

Information warfare

Wikileaks also had another cache of information from Manning: 251,287 diplomatic cables (around 2,000 books' worth of material) from 274 US consulates and embassies, dating from December 1966 to February 2010. Assange again arranged a media deal, this time also bringing in Spain's *El Pais* and France's *Le Monde*, with simultaneous publication from 28 November 2010. With a volume of information that couldn't have been collected and leaked in the pre-digital era, 'Cablegate' eclipsed the war logs as 'the biggest leak in history'. The *Guardian's* editor Alan Rusbridger said, 'it was a fruit machine. You just had to hold your hat under there for long enough' (Leigh and Harding, 2011:182).

The cables were mostly honest assessments by embassy staff highlighting the geopolitical interests and preoccupations of the USA. There were some sensitive revelations, such as a June 2009 directive revealing the USA had been illegally spying on the UN, but few of the cables were as embarrassing as the authorities feared. Some had a positive effect, with the USA's frank comments about the greed and corruption of the ruling regime in Tunisia improving the USA's reputation and contributing towards the protests from 18 December that led to the 'Arab Spring' and the overthrow of the Ben Ali regime on 14 January 2011. As Leigh and Harding note, 'Assange may have regarded the US as his enemy, but in this case he had unwittingly helped restore American influence in a place where it had lost credibility' (Leigh and Harding, 2011:249).

Despite being the least-damaging leak, the USA's response to Cablegate was the most virulent. The Vice-President Joe Biden described Assange as a 'high-tech terrorist', the State Department condemned the newspaper's 'reckless and dangerous action', and Pete King, Chair of the Homeland Security Committee, described the publication as 'treason' and Wikileaks as 'a foreign terrorist organization'. 'Wikileaks presents a clear and present danger to the national security of the US', he said, calling for the death penalty for Assange (MacAskill, 2010; Leigh and Harding, 2011:202). Sarah Palin even called for Assange's assassination, although (as Wikileaks revealed in October 2016) this had already been considered, with Hillary Clinton asking in a meeting on 23 November, 'Can't we just drone this guy?' (Wikileaks, 2016). State Department spokesman P. J. Crowley claimed the cables' publication had caused 'substantial damage', claiming 'hundreds of people have been put at risk' through their publication, although in June 2011 the department admitted it had no evidence of harm to any individual and an official later told congress the leak was 'embarrassing but not damaging' (Hosenball, 2011). This tough stance was designed to bolster attempts to shut Wikileaks down and charge the leakers. The Justice Department had opened a criminal investigation against Wikileaks, demanding Twitter turn over account information of key Wikileaks members.

The informational battle over the legitimacy, motives and meaning of Wikileaks escalated into an online informational war. On publication of the cables Wikileaks came under a massive botnet DDoS (distributed denial of service) attack from the right-wing, former military, patriot hacker, 'the Jester' ('th3j35t3r'). Wikileaks moved to Amazon's

EC2 cloud services platform but had to move again when Senator Lieberman pressured Amazon to withdraw its facilities, highlighting, Benkler argues, the free-speech limitations of 'a public sphere built entirely of privately-owned infrastructure' (Benkler, 2011a:340). Free, online software used by Wikileaks was removed by the US company Tableau Software whilst EveryDNS removed Wikileaks' domain-name registration and every associated email address. Wikileaks survived by moving to a Swedish ISP with servers housed in a nuclear-proof bunker.

Following a letter from the State Department on 27 November, on 3 December US payment company Paypal suspended Wikileaks' account. On 6 December, the Swiss bank Post Finance closed Assange's account and froze $31,000 of assets and Mastercard stopped processing payments to Wikileaks. On the 7th, Visa and the Bank of America followed suit, and on the 21st Apple removed an app containing links to the cables. As Benkler points out, 'the implicit alliance' of private infrastructure companies and government 'was able to achieve extra-legally much more than law would have allowed the state to do by itself' (Benkler, 2011a:342).

The government response prompted an internet backlash – a global, decentralized, hacker-libertarian protest defending Wikileaks. John Perry Barlow of the Electronic Frontier Foundation announced on Twitter, 'The first serious infowar is now engaged. The field of battle is Wikileaks. You are the troops' (Beckett and Ball, 2012:65), and the internet rallied round this attack on free speech, creating mirror sites for Wikileaks. The hacker collective Anonymous extended their 'Operation Payback' to defend the site. With the aid of two botnets they launched a DDoS attack against Visa, Mastercard, Paypal, Amazon and Post Finance (taking down most of their targets, but not Paypal and Amazon). Wikileaks supporters pointed out Mastercard and Visa still allowed organizations such as the Ku Klux Klan to use their services, whilst Anonymous published a manifesto declaring, 'We support the free flow of information' (Leigh and Harding, 2011:207–08).

The biggest threat to Wikileaks proved to be personal. Assange's fractious personality and relationships with other media organizations led to rifts with his newspaper collaborators, whilst allies such as Domscheit-Berg left after criticizing Assange's despotic leadership style and 'cult of stardom'. Commentators had already noted Assange's male chauvinism and 'restlessly predatory attitude towards women' (Leigh and Harding, 2011: 145–46) and after visiting Sweden in August 2010 Assange became the subject of two sexual assault allegations. A European arrest warrant was issued in December 2010 and Assange was arrested by British police on 7 December, appearing in court on the 14th and being bailed on the 16th. Fearing extradition from Sweden to the USA he fled bail and since 19 June 2012 he has been living in the Ecuadorean embassy in London, being granted asylum on 16 August. The allegations were a PR disaster for Wikileaks. Assange was personally discredited, whilst his legal costs escalated just as the US economic blockade was hitting hard. Wikileaks stopped accepting submissions and by October 2011 it had lost 95 percent of its revenue. Key personnel left and new competitors emerged, such as Domscheit-Berg's 'Open Leaks' in December 2010.

Assange's self-imprisonment has limited his activism, although Wikileaks has survived and remained an important cause and force. In the following years, it published secret files on Guantanamo Bay in April 2011, emails from Stratfor in February 2012, emails from Syrian politicians in July 2012. It returned to the news in 2016 when it published Russian-hacked emails from the US Democratic National Committee and Hillary Clinton and when it published 'the biggest ever leak of secret CIA documents' (MacAskill and Thielman, 2017), entitled 'Vault 7', in March 2017. Bradley Manning fared worse. Charged with 22 offences, including aiding the enemy, he was found guilty of 17 charges in a political show-trial lasting from 3 June to 30 July 2013. On 21 August 2013 Bradley – now Chelsea – Manning, was sentenced to 35 years in prison: a deterrent sentence designed to warn off future leakers. On 17 January 2017, the outgoing President Obama commuted Manning's sentence to the seven years served, freeing her on 17 May 2017.

Wikileaks' significance

Whilst Wikileaks may not have led to the reform of 'every political system on earth', as Assange hoped, it remains of significance for a number of reasons.

First, Wikileaks is significant because of its continuities with what has gone before, as a journalistic organization. Benkler is critical of the mainstream media's complicity in the 'political attack' on Wikileaks and the framing of the events as an equivalent threat to the US system as global terrorism rather than as legitimate journalism (Benkler, 2011a:331). For him, Wikileaks is indisputably a journalistic organization as its entire purpose is 'the gathering of information for public dissemination' (Benkler, 2011a:360). Beckett and Ball similarly argue Wikileaks is an example of 'the new forms of journalism' that are emerging in the digital-era and reshaping 'the nature of news itself' (Beckett and Ball, 2012:152). Sifry agrees, pointing out that in verifying and publishing its material Wikileaks has done what every other newspaper or media outlet who has received a leak has done. Hence 'if Wikileaks can be prosecuted and convicted for its acts of journalism, then the foundations of freedom of the press in America are in serious trouble' (Sifry, 2011:140).

Second, Wikileaks shares the traditional aim of journalism of holding power to account for the collective good: as Assange says, 'We exist to protect the public interest. We are a new spy network: the public intelligence agency' (Brooke, 2011:58). Indeed, Beckett and Ball point out, Wikileaks admits its journalistic inspiration, as 'Assange asserts that Wikileaks is part of the tradition of radical journalism that seeks not just to reveal uncomfortable or disturbing facts, but also to change society and shift power' (Beckett and Ball, 2012:156–7). For Alan Rusbridger, Assange is a 'new media baron' (Leigh and Harding, 2011:4); one who, with the war logs, achieved one of the greatest journalistic scoops in history.

But if Wikileaks is just another journalist organization why was it so controversial? Benkler explains this best:

> while WikiLeaks came under attack, the traditional media that published exactly the same cables were spared. To the political establishment, the threat lay not in the content of the published cables but in the new organizational and technical model that WikiLeaks represents: a networked fourth estate, distributed across the world instead of being housed in traditional media enterprises. That is what provoked the panic response, and it is precisely that response that makes the 2010 WikiLeaks releases this generation's version of the Pentagon Papers.
>
> Benkler (2011b)

Wikileaks is significant, therefore, because it represents a new mode of publishing, as the world's first stateless media organization. It exploits the unique affordances of digital technology, taking advantage of the new world of digital insecurity that allows anyone to easily copy, remove and distribute unthinkable quantities of information at zero cost and thus leak sensitive material. It also takes advantage of technologies that allow anonymous submission, and hosting services and mirror sites that enable it to publish globally, beyond national legal jurisdictions. Wikileaks is independent of government regulatory frameworks, commercial necessities, advertisers and corporate or lobby-group interests. Coupled with its libertarian philosophy of information and pro-democratic desire to expose the powerful, Wikileaks represents a genuine threat to all authorities and the systems upon which they rely.

Third, Wikileaks' significance lies in highlighting the failure of mainstream journalism to inform the public and hold power to account. Although the deteriorating security situation in Afghanistan and Iraq made reporting difficult, the limitations of journalism run deeper than this, encompassing the failure adequately to question the basis for the wars, the uncritical war coverage, the refusal of journalists in the aftermath to address their own culpability and complicity, and the failure to adequately report the ongoing conflicts or question their government's position. Wikileaks demonstrated the organizational, technical, commercial and political failure of mainstream newspapers to report the reality of these wars. As Beckett and Ball argue,

> The Afghan and Iraq leaks showed that America had not been honest about the conduct of the war and its failure. It lied about the use of torture, the civilian deaths and the lack of success. Other mainstream journalists have told parts of that story but only Wikileaks was able to tell it so comprehensively and irrefutably.
>
> Beckett and Ball (2012:80)

Fourth, therefore, Wikileaks also exposes the limits of contemporary democracy. If democratic governments go to war in the name of their publics, putting them at risk of retaliation for their actions, then – exactly as Chelsea Manning argued – the public must be able to make an informed decision about supporting these wars. As well as keeping this information from the public, once it was leaked the US government partnered with politically sympathetic newspapers to attack Wikileaks and with private infrastructural companies to remove it from the internet. As Benkler argues, with Wikileaks probably protected by the First Amendment on freedom of speech, these partnerships gave the government 'an extralegal pathway to suppress information that it would have no authority to censor directly' (Benkler, 2011b).

Fifth, Wikileaks is important as an expression of a broader 'global transparency movement' (Sifry, 2011:85) and philosophy of open governance. Part of the post-2008-crash anti-politics, this movement is suspicious of the political establishment and representative democracy, demanding a more transparent, participative politics. As Sifry argues, it is based on 'one core idea' – that when information about what governments are doing is made available the public can better watch over them, question them and 'foster real accountability' (Sifry, 2011:186–7). Ultimately, he says, 'more information, plus the internet's power to spread it beyond centralized control, is our best defense against opacity and the bad behaviour it can enable' (Sifry, 2011:187).

Sifry argues we live in a new networked age where 'elemental changes in the economics of information, connectivity and time have occurred' (Sifry, 2011:49). New technologies are opening up government whilst the public are refusing their formerly passive roles and emerging as active players, monitoring what governments and politicians are doing. His examples include the Sunlight Foundation, GovTrack.us, Mysociety.org and

TheyWorkForYou, all of which aim at making public-financed data and information about public servants available, and new initiatives such as Ushahidi, Ipaidabribe and Wiki-crimes, which crowd-source knowledge to hold authorities to account.

Politicians, Sifry says, have paid lip service to openness, but Wikileaks showed them what transparency meant, 'by exposing systemic details of how America actually conducts its foreign and military policies' (Sifry, 2011:138). Wikileaks also helped expose the hypocrisy of western governments regarding the freedom of information. On 21 January 2010 Hillary Clinton delivered her speech, 'Remarks on Internet Freedom', in which she denounced authoritarian countries for trying to censor the net and defended the free flow of information, stating, 'We stand for a single internet where all humanity has equal access to knowledge and ideas' (Clinton, 2010). She was less happy about the free flow of US government information and although she argued that this information was illegally obtained and disclosed and that its publication was tearing apart 'the fabric of the proper function of responsible government' (Sifry, 2011:142), others saw it as aiding the democratic process.

Sixth, Wikileaks exposed the Afghan and Iraq Wars. The war logs were limited in what they revealed, their content was obscured by the attacks on Wikileaks itself, and it would be several years before their picture of the failure of the War on Terror would be publicly accepted. Nevertheless they played an important role in undermining the simplistic image of these conflicts presented by western governments and the sympathetic right-wing press. The scale of hostility directed at Wikileaks was indicative of how much the authorities didn't want these conflicts questioned or their failures exposed. At a time when patriotism and anti-terrorism dominated public discourse, the war logs showed these conflicts were chaotic, violent, counter-productive and unwinnable. It was a picture that would eventually prove correct.

Finally, Wikileaks revealed a new type of war – or rather it exposed an existing war and the new ways in which it was being fought. This was an ongoing, real-time ideological war over information itself: about the image of the world, of governments and of their policies and about what the public should be allowed to access or know. It was a war that successfully positioned Wikileaks as the enemy for informing the public and as having 'blood on its hands' for exposing the huge death-toll of the War on Terror. As Beckett and Ball argue, however, 'The threat is not Wikileaks and the new networked news, it is still secrecy and the abuse of power' (Beckett and Ball, 2012:159); a point Heather Brooke also makes: 'free speech is not the great danger for humanity. Concentration of power is' (Brooke, 2011:230). Established power may have the upper-hand in this war but transparency has its own weapons. As Benkler observes, 'reporting based on documents leaked securely online using multiple overlapping systems to reach the public and evade efforts at suppressing their publication is here to stay' (Benkler, 2011a:350).

Wikileaks, however, remains an unsatisfying standard-bearer for the transparency movement. Its critics point out that this anti-authoritarian, pro-transparency organization is run as an opaque, unaccountable, personal dictatorship by its founder and owner, Julian Assange, and that its main targets have remained the USA rather than more authoritarian countries. More recently, in the run-up to the 2016 election Assange posted hacked emails from the Democratic National Convention (DNC) and Hillary Clinton and appeared to develop a positive relationship with Donald Trump. In November 2017 *The Atlantic* magazine published leaked Twitter direct messages between Assange and Donald Trump Jr that took place months before the election. Whatever his dislike for Clinton, Trump remains a right-wing authoritarian, whilst the DNC emails themselves originated with Russian hackers who were involved in a Putin-ordered informational war against the USA that aimed to undermine faith in the democratic process and aid Trump's election. These were not actors working in the spirit of that radical, democratic, open politics Assange claims to support. For many Assange supporters, the allegations about Trump were more important than the sexual assault allegations, in representing a political rather than a personal betrayal.

Ultimately, however, Wikileaks' most important contribution has been to encourage other whistleblowers and leakers. Within a few years of Wikileaks' exposure of the USA's foreign activities – and following in the same patriotic and pro-democratic spirit as Chelsea Manning – Edward Snowden would warn the American public about the USA's domestic surveillance activities.

Key reading

As I have suggested here, it's useful to have some knowledge of what happened in Afghanistan and Iraq after the invasion-phase operations. To understand the political situation in Afghanistan from December 2001 onwards look at Fairweather (2014), Gopal (2014), Burke (2011), Fergusson (2011), Gall, S. (2012), Abbas (2014) and Gall, C. (2015). To understand what happened in Iraq after Bush's declaration of 'mission accomplished' look at Burke (2011), Cockburn (2006) and Packer (2006). A special mention must go to Chandrasekaran (2006), whose account of the activities and failures of the Coalition Provisional Authority is essential reading. The best introductions to Wikileaks are found in Leigh and Harding (2011), Beckett and Ball (2012), Sifry (2011), Brooke (2011), Benkler (2011a) and Gibney (2013). Stoekley (2014) covers the trial of Chelsea Manning in detail whilst Greenberg (2012) also provides an excellent background to issues around encryption, leaking and the cypherpunk movement that inspired Wikileaks.

References

Abbas, H. (2014) *The Taliban Revival*, London: Yale University Press.

Assange, J. (2006) 'Sun 31 Sec 2006: The Non-Linear Effects of Leaks on Unjust Systems of Governance', *IQ.Org*, http://web.archive.org/web/20071020051936/http://iq.org/.

Beckett, C. and Ball, J. (2012) *Wikileaks: News in the Networked Era*, Cambridge: Polity.

Benkler, Y. (2011a) 'A Free Irresponsible Press: Wikileaks and the Battle Over the Soul of the Networked Fourth Estate', *Harvard Civil Rights – Civil Liberties Law Review*, https://dash.harvard.edu/bitstream/handle/1/10900863/Benkler.pdf?sequence=1.

Benkler, Y. (2011b) 'The Real Significance of Wikileaks', *The American Prospect*, 10 May, http://prospect.org/article/real-significance-wikileaks.

Brooke, H. (2011) *The Revolution Will Be Digitised*, London: William Heinemann.

Burke, J. (2011) *The 9/11 Wars*, London: Penguin.

Chandrasekaran, R. (2006) *Imperial Life in the Emerald City*, London: Bloomsbury.

Clinton, H. (2010) 'Remarks on Internet Freedom', U.S. Department of State, 21 January, Washington, https://www.state.gov/secretary/20092013clinton/rm/2010/01/135519.htm

CNN. (2003) 'Rumsfeld: Major Combat Over in Afghanistan', *CNN*, 1 May, http://edition.cnn.com/2003/WORLD/asiapcf/central/05/01/afghan.combat/.

Cockburn, P. (2006) *The Occupation*, London: Verso.

Fairweather, J. (2014) *The Good War*, London: Penguin Random House.

Fergusson, J. (2011) *Taliban*, London: Corgi Books.

Gall, C. (2015) *The Wrong Enemy: America in Afghanistan 2001–14*, New York: Mariner Books.

Gall, S. (2012) *War Against the Taliban*, London: Bloomsbury.

Gibney, A. (dir.) (2013) *We Steal Secrets*, Jigsaw Productions/Global Produce [DVD].

Gopal, A. (2014) *No Good Men Among the Living*, New York: Metropolitan Books.

Greenberg, A. (2012) *This Machine Kills Secrets*, London: Virgin Books.

Hosenball, M. (2011) 'U.S. Officials Privately Say Wikileaks Damage Limited', *Reuters*, 18 January, http://www.reuters.com/article/us-wikileaks-damage-idUSTRE70H6TO20110118.

Hughes, E. 1993. 'A Cypherpunk's Manifesto', 9 March, http://www.activism.net/cypherpunk/manifesto.html.

Leigh, D. and Harding, L. (2011) *Wikileaks. Inside Julian Assange's War on Secrecy*, London: Guardian Books.

Levine, A. (2010) 'Gates: Leaked Documents Don't Reveal Key Intel, but Risks Remain', *CNN*, 17 October, http://edition.cnn.com/2010/US/10/16/wikileaks.assessment/.

MacAskill, E. (2010) 'Julian Assange Like a High-Tech Terrorist, Says Joe Biden', *Guardian*, 19 December, https://www.theguardian.com/media/2010/dec/19/assange-high-tech-terrorist-biden.

MacAskill, E. and Thielman, S. (2017) 'Wikileaks Publishes "Biggest Ever Leak of CIA Documents"', *Guardian*, 7 March, https://www.theguardian.com/media/2017/mar/07/wikileaks-publishes-biggest-ever-leak-of-secret-cia-documents-hacking-surveillance.

Manning, C. (2013) 'Bradley Manning's Personal Statement to Court Martial: Full Text', *Guardian*, 1 March, https://www.theguardian.com/world/2013/mar/01/bradley-manning-wikileaks-statement-full-text.

May, T. C. (1992) *The Crypto Anarchist Manifesto*, https://www.activism.net/cypherpunk/crypto-anarchy.html.

May, T. C. (1994) *Cyphernomicon*, http://www.cypherpunks.to/faq/cyphernomicron/cyphernomicon.txt.

Packer, G. (2006) *The Assassin's Gate: America in Iraq*, London: Faber and Faber.

Raymond, E. (2000) *The Cathedral and the Bazaar*, http://www.catb.org/esr/writings/cathedral-bazaar/cathedral-bazaar/.

Sifry, M. L. (2011) *Wikileaks and the Age of Transparency*, London: Yale University Press.

Stoekley, C. (2014) *The United States vs Pvt. Chelsea Manning*, London: O/R Books.

Sunshinepress. (2010a) 'Wikileaks Leaked Video of Civilians Killed in Baghdad – Full Video' *YouTube*, 3 April, https://www.youtube.com/watch?v=is9sxRfU-ik.

Sunshinepress. (2010b) 'Collateral Murder – Wikileaks – Iraq', *YouTube*, 3 April, https://www.youtube.com/watch?v=5rXPrfnU3G0&t=32s.

Wikileaks. (2016) 'Hilary Clinton on Assange...', *Twitter*, 3 October, https://twitter.com/wikileaks/status/782906224937410562.

8 Drone war

Telepresent assassination

Changing war

In 1985 Orson Scott Card published his teen-novel *Ender's Game*. Set in the future, it told the story of a child, Andrew 'Ender' Wiggin, and his training as a commander for Earth forces in their interplanetary war against the alien insect race, the Formics, or 'buggers'. At the book's end, he celebrates passing his final examination – a video-game simulation in which he destroys the alien's home world – only to be told that it wasn't a simulation and that he'd actually been remotely controlling Earth's forces: 'Ender, for the past few months you have been the battle commander of our fleets. This was the Third Invasion. There were no games, the battles were real, and the only enemy you fought were the buggers' (Card, 2002:296–7). The future Card described came sooner than he'd imagined. The dream of remote, telepresent control of weaponized drones became real in 2001. By 2010 the USA was training more drone pilots than fighter and bomber pilots combined.

Variously known as an 'unmanned aerial vehicle' (UAV), 'remotely-piloted vehicle' (RPV), 'remotely-piloted aircraft' (RPA) or 'unmanned aerial system' (UAS), the 'drone' has become 'the defining weapon of America's seemingly endless global "war on terror"' (Woods, 2012:23) and one of the most important military technologies of the early twenty-first century. It has transformed conflict, raising questions about what it is, how it is fought, who fights it and where it is fought, as well as issues around the ease and ethics of remote killing, its role within democracies and the future consequences of its global proliferation. This chapter offers an introduction to the drone, exploring its history, development and deployment, the benefits of unmanned conflict and the issues it raises.

The history of drones

The US *Department of Defense Dictionary of Military and Associated Terms* (JP1-02) defines an 'Unmanned Aerial Vehicle' (UAV) or 'drone' as: 'A powered aerial vehicle that does not carry a human operator, uses aerodynamic forces to provide vehicle lift, can fly autonomously or be piloted remotely, can be expendable or recoverable, and can carry a lethal or nonlethal payload'. It notes that 'ballistic or semi-ballistic vehicles, cruise missiles and artillery projectiles are not considered unmanned aerial vehicles' (Joint Chiefs of Staff, 2001).

Despite appearing to be cutting edge technologies, drones are not new. They are usually contextualized within the history of remote-controlled vehicles and weapons. In 1898

Nikola Tesla publicly demonstrated small, radio-controlled boats, although the US military was uninterested in the idea. There were other demonstrations of remote control pre-World War I, including Leonardo Torres-Quevedo's remote-controlled trike, shown in Madrid in 1904, Gustave Gabet's remote-controlled torpedo, demonstrated in the Seine in December 1909, and Jack Hammond and Benjamin Miessner's automatic, light-following three-wheeled cart, the 'electric dog', designed in 1912 and tested in 1915.

World War I saw the military experiment with a range of remote control technologies, including the land torpedo, a small, armoured tractor carrying explosives. Most never saw service, although the French Crocodile Schneider Torpille Terrestre Type B was field-tested in 1916. There were also attempts to create aerial torpedos, such as the British Aerial Target (1917) and the USA's Hewitt-Sperry Automatic Airplane (1917), and Kettering Bug (1918), though their late development and unreliability meant they weren't used in combat. The most successful device was the German FL-7, a remotely controlled boat of explosives designed to defend the coast, which was successfully deployed against the British HMS Erebus on 28 October 1917.

By World War II the technology had improved. From 1940 the Germans deployed their wire-controlled 'Goliath' land torpedos against tanks and produced the first precision-guided munition, the FX-1400 bomb ('Fritz'), which was directed after release by a joystick with radio-link steering. The aerial torpedo was revived in the British GB-1 glider bombs and the failed 'Operation Aphrodite', which attempted to create radio-controlled explosive-packed US bombers. The most successful use of remote control was for anti-aircraft training targets. The British de Havilland Queen Bee and Airspeed Queen Wasp target planes were developed in the 1930s, but the US OQ-2 Radioplane (or 'Dennymite') was the first mass-produced UAV, in use from 1939. The noisy, insect-like flight and unintelligent service of these targets gave rise to the term 'drones'.

The only major US post-war drone project was the jet-powered Ryan Aeronautics Firebee drone in use by 1948 and fitted with cameras by 1955. The shooting down of Gary Powers' U2 spy plane over Russia on 1 May 1960 exposed the vulnerabilities of manned planes and with satellite technology limited the US explored the drone's surveillance capabilities. From February 1962 Ryan contracted to modify the Firebee, creating the Ryan Model 147A Fire Fly and the 147B Lightning Bug reconnaissance drones. The Bug was first used on 20 August 1964 to monitor Chinese activities and eventually flew 3,435 missions over Vietnam, mostly for surveillance although it was also part of the 'Igloo White' electronic battlefield system. As satellites became more reliable the Bug was discontinued after the war. The only US post-Vietnam project, the 1975 Lockheed MMQ-105 battlefield target-assessor drone, was cancelled in 1987 due to reliability issues and lack of DoD interest.

It was Israel that revived the drone. It had used them in 1969 to monitor Egyptian, Syrian and later Jordanian troops and used Firebee and Chukar drones as decoys in the 1973 Yom Kippur War. In 1973, it began its own programme, creating the Tadiran Mastiff reconnaissance UAV. This was the first modern military drone as it was the first equipped with cameras transmitting real-time imagery to its operators: the modern drone, therefore, is defined not by remote control, but by *telepresence*. In the 1982 Lebanon War Israel sent drones to gather the electronic frequencies of the Syrian Integrated Air Defence System (IADS), then sent them back with fake signals as decoys to attract missiles, allowing follow-on planes to destroy the reloading, unmasked batteries. The USA recognized the value of Israel's drones and the Navy worked with their aircraft industry to produce the Pioneer drone, which was used for bombardment-reconnaissance in the 1991 Gulf War.

When surrendering Iraqis waved a white flag at a Pioneer it marked, Singer says, 'the first time in history that human soldiers surrendered to an unmanned system' (Singer, 2009:57).

The modern US drone programme was created by the Israeli inventor Abraham Karem. He had moved to California in 1977 and in 1984 his company, Leading Systems Incorporated, had secured a DARPA contract to create the Amber UAV. His company was bought by Hughes Aircraft who sold it to General Atomics (GA) in 1991. The Amber became the Gnat 750 and in February 1994 the CIA deployed it over Bosnia. The CIA and GA redeveloped the Gnat into the Predator by 1995 and flew it that Spring over Yugoslavia. Two were shot down by Serb air defences, they were susceptible to poor weather and couldn't be flown by satellite link but they could operate for 24 hours and the 159 missions they flew over Bosnia in their first year produced intelligence that helped push Serbia towards the Dayton Peace Accords.

1990s US military budget cuts made the drone attractive, its video-game technology aided youth recruitment, and its removal of risk fitted the post-Gulf War desire for zero-casualty warfare. The 1993 'Battle of Mogadishu' exposed again the political cost of troop losses and the UAV offered a solution to this. With renewed government investment, Predators improved, with a new 'nose' added, allowing satellite data transmission and piloting. In 1998 the USAF took command of drones, reactivating the Vietnam-era 11 Reconnaissance Squadron, at Creech US AFB, Nevada, with video and sensor analysts based at Langley AFB, Virginia, now designated 'Distributed Ground System One'. Predators and Global Hawks were used for reconnaissance in the 1999 Kosovo War and continued to evolve, adding a GPS signal and a sensor-ball laser-designator lighting up targets for air strikes. On 8 February 2000, the Senate Armed Services Committee ordered that by 2010 one third of aircraft attacking behind enemy lines and one third of all combat vehicles be unmanned.

It was terrorism that ensured the drone's success. The CIA's Counterterrorism Centre established the Bin Laden Issue Station ('Alec Station') in 1996 and tracing him had become a priority after the 1998 African embassy bombings. On 7 September 2000, a joint Pentagon-CIA 'Afghan Eyes' project used drones to try to find him. On the 25th a Predator sent back images of a tall figure at the Tarnak Farms complex near Kandahar. Washington soon realized that if drones could find Bin Laden then they could also kill him. The Predator first successfully test-fired a Hellfire missile on 23 January 2001. Inconclusive results from a CIA test-bombing of a simulated compound in June 2002 and debates about the legality of assassination meant the proposed strike was cancelled.

9/11 changed everything. The weaponized drone was immediately deployed in Afghanistan. The Bush administration waived legal and technical doubts and embraced the drone. The annual defence budget rose 74 percent to $515bn from 2002 to 2008 and a significant amount went into unmanned systems. Just as World War II spurred the development of computing, so the War on Terror spurred the development of drones.

Box 8.1 US drones

The smallest US drones are hand-launched 'miniature UAVs' such as AeroVironment's RQ-11 Raven. In service since May 2003, this Intelligence, Surveillance and Reconnaissance (ISR) drone is 3ft in length, weighs 4.2 pounds and has a wingspan of 4.5ft. It can fly for up to 90 minutes up to 500ft above ground level, with a range of 6.2 miles and a top speed of 60mph. The most famous drone is General Atomics'

MQ-1 Predator, which is 27ft long, with a wingspan of 48.7ft. Costing $4.03m each in 2010, this is a multi-role, medium altitude, long-endurance UAV, armed with two laser-guided AGM-114 Hellfire missiles, operating for up to 24 hours at up to 25,000ft, with a range of 675 miles and a maximum speed of 135mph. As well as a colour nose camera for the pilot, the Predator includes a rotating 'ball' carrying a Raytheon 'multi-spectral targeting system' (sensors providing target-acquisition, tracking, range-finding and laser-designation of the missiles or airstrikes), real-time, full-motion video-cameras, still cameras, infra-red night-vision cameras and electronic surveillance equipment.

The USAF is replacing the Predator with the more powerful Predator B, or MQ-9 Reaper. Introduced in 1997 and costing $16.9m each, the Reaper is 36.1ft long with a 65.7ft wingspan. It can operate for 14 hours at up to 50,000ft, with a cruising speed of 194mph and a range of 1,151 miles. It carries up to 4 Hellfire missiles and has improved sensors and electronic warfare surveillance pods. The largest US UAV is Northrop Grumman's RQ-4 Global Hawk, a huge, high-altitude surveillance drone, 47.5ft long, with a 130.9ft wingspan and a cost of $222.7m. It can operate for over 32 hours, with a cruising speed of 357mph, a range of 14,154 miles and sensors capable of surveying 40,000 square miles of territory per day.

The US drone programme

The USA has developed the most extensive drone programme, pioneering their incorporation into front-line military strategy. Their primary use has been for intelligence, surveillance and reconnaissance (ISR), though weaponized drones have been used as a target designator for airstrikes, as support for ground troops and for targeted killings on and away from the battlefield. The smallest drones are operated at unit level, but most ISR drones are best understood as a network of systems comprising (1) a forward-operating team, (2) a remote Ground Control Station (GCS), (3) the Distributed Common Ground System (DCGS) for intelligence analysis, and (4) feeds sent to the intelligence and command structure.

US drones are portable, being delivered to, assembled on and launched from airbases around the world. They have a large forward-operating team who arm and maintain them, including a pilot overseeing launch and landing. Once airborne, the drone uses a GPS satellite receiver to establish its location and connect to the GCS which takes over its flight. The GCS are located on permanent airbases in Nevada, New Mexico and California, although that can be delivered by plane to any theatre. Each GCS operates four Predator or reaper drones, with a pilot, sensor operator and mission intelligence coordinator ('MC') backed by a large technical support team. Sitting opposite real-time screens the pilot uses a joystick and real-time camera feed alongside a sensor operator, whilst the MC interfaces with the base commanders. Missions are flown for 'customers' such as the CIA, and the GCS team may know little about their mission.

Drone feeds are sent to the Distributed Common Ground System: installations housing intelligence analysts who survey the real-time data. Langley AFB formed DGS-1, but by 2015 there were 27 sites worldwide, including in Germany and South Korea, with the CIA and Air Force Special Operations Command (AFSOC) running their own analysts. By 2015, the USAF said, the DCGS 'operational tempo' included 'more than 50 ISR sorties exploited, over 1,200 hours of motion imagery reviewed, approximately 3,000 Signals Intelligence (SIGINT) reports produced, 1,250 still images exploited and 20 terabytes of data managed daily' (Department of the Air Force, 2015). Their analysts work 12-hour days, monitoring real-time battlefield streams, communicating via headsets with pilots and instant messaging commanders on the airbase and in the field. The DCGS is considered so important that the drones and analysts are classified as separate 'weapons systems' (Woods, 2016:183).

Drone feeds are sent to others in the command structure. Generals have access to feeds, whilst the President has watched live operations. The CIA have feeds for their missions and the NSA accesses feeds from drones carrying their electronic surveillance equipment. Their 'Gilgamesh' pod, used by Joint Special Operations Command (JSOC) can locate sim-cards or handsets, whilst the CIA's 'Shenanigans' pod vacuums up data from wireless routers, smartphones, tracking devices or other electronic devices within range. The intelligence is used to guide special forces missions or direct strikes.

The USA's drone network grew rapidly after 9/11. Whereas in October 2001 it could only field one Predator above Kandahar, by 2014 it could fly 63 simultaneous 'combat air patrols', 24 hours a day, over half a dozen countries, deploying 3,000 drone operators and 5,000 analysts. Just as the drone system has expanded, so has their global deployment. Weaponized drones were first used in the Afghanistan War, although their first mission on 7 October 2001 was a disaster when a confused command structure led to a premature Predator strike on a convoy, allowing Taliban leader Mullah Omar to escape an adjacent building. Despite this, CIA-run Predators proved successful, beginning a 'targeted killing programme' (Woods, 2016:45) (TKP) tracking down senior al-Qaeda and Taliban militants and guiding the airstrike that killed al-Qaeda's military commander Mohammed Atef in Kabul in November 2001. On 3 November 2002, the TKP was expanded beyond the battlefield for the first time when a drone killed the al-Qaeda leader Qaed Salim Sinan al-Harethi in Yemen.

Drones were next used in the 2003 Iraq war, although their impact was limited as the USAF's reconnaissance drones were unarmed and only seven armed CIA Predators were deployed. Drones proved more useful in the war's aftermath, being embraced by the elite JSOC counter-insurgency (COIN) forces, who used their ISR capabilities against insurgents. JSOC got their own drone fleet in October 2005 and pioneered a new 'pattern of life analysis' (Woods, 2016:78), integrating multiple intelligence assets to build up a picture of insurgent activity over time that was used to direct special forces operations. In August 2006 JSOC established its own drone video feed analysis unit, 'DGS: Special Ops' at Hurlburt Field, Florida. Within two years it was analysing half the UAV video being collected in Iraq and Afghanistan. Drones played a key role in the killing of the al-Qaeda in Iraq (AQI) leader Abu Musab al-Zarqawi. Hundreds of hours of predator time and thousands of hours of analyst time went into finding him and, after an intelligence breakthrough, drones tracked him to a house in Hibhib on 7 June 2006 and gave a positive ID for an F-16 airstrike. By then the USA had also refined the battlefield use of armed drones as close air support for ground troops in combat.

Drones continued to be used in Afghanistan, although until 2004 only the CIA's drones were armed. The most important development was the opening of a new front inside Pakistan. The Taliban and al-Qaeda had fled to the Federally Administered Tribal Areas (FATA) of northern Pakistan and in 2002 drones crossed the border to monitor the pro-Taliban Haqqani network and provide ISR for CIA/Pakistan ISI capture operations. Following a new Taliban insurgency into Afghanistan in May 2003, in June 2004

Pakistan's President Musharraf secretly allowed the CIA's drones access to Pakistan military sites, including Shamsi airfield in Balochistan and Shabaz airfield near Jacobabad to strike targets within the FATA.

Given conservative opposition within Pakistan, the ethics of a government allowing a TKP against its own citizens and the dubious legality of a US civilian intelligence agency engaging in clandestine military operations, both Pakistan and the USA denied the programme existed and hid evidence. Pakistan journalists who investigated the drone strikes were threatened, beaten, tortured or killed. The freelance journalist Hayatullah Khan, for example, was found dead in June 2006 after publishing photographs exposing drone strikes (whilst his widow was later killed by a bomb in November 2007 after blaming the ISI).

Most of the attacks were within Waziristan, an area housing 800,000 people who are among the poorest on earth, leading to many innocent deaths. The first civilian casualties were 14-year-old Irfan Wazir and his 8-year-old brother Zaman, killed on 17 June 2004 whilst sitting next to the Taliban militant Nek Mohammed who was visiting their father. By late 2014 CIA drones had killed over 2,000 people in 'the longest sustained bombing campaign in US history' (Woods, 2016:93). By 2006 Pakistan began expressing official anger at the drone strikes, partially under domestic pressure from the loss of life, but also because the ISI was trying to cultivate relations with a 'good' Taliban to gain influence in any post-US Afghanistan. US pressure, however, forced more concessions, leading to an increase in CIA strikes and an increase in their lethality too as they moved from launching 'personality strikes' on identified individuals to 'signature strikes' based on a 'pattern of life' analysis and likely militant affiliation that killed scores in single bombings. What began, therefore, as a joint policing operation morphed into a systematic, intensive, unilateral bombing campaign within Pakistan.

With its resources concentrated in Iraq, the USA struggled with the renewed Taliban insurgency from 2006. Its response, using airpower, increased civilian casualties, with 2,100 killed during 2008 alone. President Obama was determined to refocus US effort on the 'AFPAK' (Afghanistan/Pakistan) security situation and he reduced airstrikes by a quarter and increased drone ISR missions supporting targeted military and special forces kill-or-capture operations against Taliban militants. Obama will be remembered, however, for his embrace of drone strikes, 'overseeing more strikes in his first year than Bush carried out during his entire presidency' (Bureau of Investigative Journalism, 2017a). His first strike, three days into his presidency, on 23 January 2009, was reported as killing 'foreign militants' but in reality, 9–11 of the 15 people killed were civilians. Another strike the same day killed 5–10 people, all civilians. Although Obama was 'disturbed' by these deaths (Klaidman, 2012), having vowed to change the Muslim world's perception of the USA, 'there were ten times more air strikes in the covert war on terror during President Barrack Obama's presidency, than under his predecessor, George W. Bush' (Bureau of Investigative Journalism, 2017a).

The aim of the drone strikes also changed. From February-August 2009 the CIA carried out multiple strikes on the Tehrik-e-Taliban Pakistan (TTP), a militant group with no connection to the Afghanistan Taliban and whose target was the Pakistan state. They also aided a Pakistan military operation against them in South Waziristan in October 2009, despite the TTP not being in an armed conflict with or threatening the USA. The strikes brought more civilian casualties and increased pressure from tribal leaders. By June 2012 at least 2,500 people had been killed in Pakistan by drones, with at least 482 of them civilians. Pakistan complained the attacks contravened international law, though in reality they were government sanctioned. On 16 May 2009 there was the first confirmed use of the 'double-tap' bombing method (developed in Vietnam), where drones returned to launch a second strike at anyone aiding the victims on the basis that they were related militants. Those killed included villagers and healthcare workers helping the injured and naturally curious children.

Obama expanded drone use beyond the official War on Terror battlefields. In February 2012, a reported drone strike in the Philippines targeted Abu Sayyaf and Jemaah Islamiyah leaders and in June 2012 a drone hit a convoy of trucks in Northern Mali, killing al-Qaeda in the Islamic Maghreb (AQIM) militants. With the emergence of al-Qaeda in the Arabian Peninsula (AQAP) in June 2009, Yemen's president Saleh gave the USA full access for COIN drone and special forces operations. Their cruise missile strike on 17 December 2009 against a claimed AQAP training camp at Al-Majalah was a disaster, killing 41 civilians, including 21–22 children and wiping out entire families, and the USA and Yemen conspired to hide the deaths. There were regular JSOC drone strikes, air strikes and ground operations from 2011, although it was a joint JSOC/CIA drone operation that eventually succeeded in killing the AQAP-linked preacher (and US citizen) Anwar al-Awlaki, along with another US citizen, Samir Khan, on 30 September 2 011.

With AQAP-proxy Ansar Al-Sharia seizing territory following the 2011 Arab Spring Yemen uprising, the USA replaced Saleh and coordinated a major Yemeni offensive backed with airstrikes, drone strikes and JSOC forces to retake the south. Yemen's human rights minister Hooria Mashhour was critical of the resulting civilian deaths, including the 17 killed on 12 December 2013 in a wedding convoy, writing in the *Washington Post* in January 2014:

> More often than not, US drone strikes leave families bereaved and villages terrified. Drones tear at the fabric of Yemen society … Our president may reassure the United States of his support for drone strikes but the reality is that no leader can legitimately approve the extrajudicial killing of his own citizens.
>
> Mashhour (2014)

How can Yemenis believe in their fledgling democracy, he asks, when the parliamentary vote banning drone flights after the wedding party deaths is 'ignored by democracy's greatest exponent'?

US Yemeni drone strikes continue to the present, through the ongoing civil war. The Bureau of Investigative Journalism documents a minimum of 154 strikes from May 2011 to September 2017, killing a minimum of 615 people, including at least 68 civilians. Woods suggests a higher figure of 500–1,400 deaths by September 2014, including at least 120 civilians.

US drones were also employed in the multi-state NATO military intervention in the Libyan civil war, with Obama claiming the 14 September 2001 'Authorization for the Use of Military Force' covered this expansion of the battlefield. From 21 April to 20 October 2011 the USA launched 145 Predator strikes against Gaddafi's forces, using drones for block-by-block targeting surveillance for airstrikes and to monitor Gaddafi's Sirte hideout. It was a drone missile that stopped Gaddafi's fleeing convoy on the 20 October and allowed him to be captured and killed.

Obama also extended drone use to Somalia, a nation with no government or media organizations, enabling the Pentagon and CIA to operate without scrutiny. The USA has been involved in covert military operations inside Somalia since 2001 against al-Qaeda in

East Africa and Islamist groups such as al-Shabaab. In December 2006, their intervention escalated in support of the Ethiopian invasion to overthrow the al-Qaeda-linked Islamic Courts Union. Following the withdrawal of Ethiopian troops in January 2009 the USA has continued to aid the Federal Government of Somalia with naval bombardments, airstrikes, JSOC ground operations as well as drone strikes from Camp Lemonnier in Djibouti. The first lethal strike was on 23 June 2011. The BIJ claims there have been 36 drone strikes till the end of 2016 killing 242–418, including up to 12 civilians.

Meanwhile, the drone campaign continued in Afghanistan and Pakistan. In 2010 there was an 'Afghan surge' of troops and an expansion of targeted JSOC raids supported by AFSOC drones. With US special forces operating with impunity (following, as Wikileaks revealed, a 'kill list'), there was growing Afghan government concern at the increasing civilian casualties. US–Pakistan relations also deteriorated with the ongoing CIA operations, the growing civilian deaths, the impact of US operations on their preferred Taliban allies and a diplomatic spat over the killing of two Pakistan men by the CIA contractor Raymond Davis on 27 January 2011. On 17 March – the day after Pakistan released Davis – the CIA retaliated with a drone 'signature strike' against a meeting of militants at Datta Khel, killing 44 and injuring 10. The meeting was a legal tribal 'jirga' held to resolve a mining dispute and almost all the dead were civilian elders. The USA disputed this, with an anonymous official claiming they were 'a large group of heavily armed men, some of whom were clearly connected to al-Qaeda and all of whom acted in a manner consistent with AQ-linked militants' (Woods, 2016:232).

Drone strikes were halted, but resumed in April against Islamabad's objections. The CIA were eventually forced to abandon their Pakistan bases in December 2011, moving Pakistan operations to a Saudi airbase following the 2 May 2011 special forces raid that killed Bin Laden and the 26 November 'Salala incident' when US helicopters killed 28 Pakistani security forces at a border check-point. The US-Pakistan relationship has fluctuated since. Drone strikes continue but have tailed off in recent years with 13 in 2015, three in 2016 and four from January–September 2017. By then, the BIJ estimated there had been 429 drone strikes in Pakistan, killing 2,514–4,023 people, of whom 424–969 were civilians (including up to 207 children).

Drones, therefore, have become a vital part of an expanded, diffuse, permanently-on, low-visibility, low-footprint, secretive, global US 'War on Terror', predominantly conducted using special forces and the CIA, and flown from up to 60 bases worldwide, including the USA, Afghanistan, Burundi, Burkina Faso, Cameroon, Chad, Djibouti, Ethiopia, Italy, Iraq, Kuwait, Mauritania, Niger, Pakistan, Philippines, Qatar, Saudi Arabia, the Seychelles, Somalia, Tunisia, Turkey, Uganda, the United Arab Emirates and Uzbekistan. From September 2014, the USA began drone strikes in Syria as part of a campaign against Islamic State (IS) and also returned to Iraq for the first time in three years, providing assistance, including drone operations, to the Iraqi government. Under a Bilateral Security Agreement with the Afghan government special forces continue to carry out COIN ground and drone operations from Afghan bases.

President Trump has increased drone use, especially in Yemen, which he declared a zone of 'active hostilities'. From March–April 2017 the USA launched 80 drone and airstrikes, leading to an increase in civilian casualties. Trump also reversed Obama's restriction on CIA use of drones in Pakistan, leading to two strikes in 2017 and increased attacks in Afghanistan (with more than 600 airstrikes from January-April 2017). In March 2017, the Trump administration loosened restrictions on drone strikes, removing Obama's May 2013 Presidential Policy Guidance rule requiring the 'near-certainty' of no civilian deaths. It allowed the Pentagon and CIA to conduct strikes against IS without presidential approval and allowed strikes such as the one that killed UK citizen Sally Jones in June 2017 whilst fleeing Raqqa possibly along with other civilians, including her 12-year-old son. Trump is determined to give the CIA and special forces more autonomy and allow expanded, less legally-restricted, anti-Islamist operations.

The benefits of drones

The most important benefit of drones is the removal of the combatant's body from the battlefield. By March 2009, 70 Predators had been lost, due to poor weather, equipment failure or pilot error, with four shot down – all without a single pilot killed. Drones allow the projection of power without vulnerability, promising to realize that dream of zero-casualty warfare the USA has pursued since the Gulf War, perfecting the Baudrillardian model of 'non-war' that is unilaterally imposed on the other without any possibility of response. This has made drones popular as a 'risk-free' (Woods, 2016:xiv) form of warfare, avoiding a public backlash at troop commitments and losses. Drones are cheaper than aircraft, helping strained military budgets and are popular with the public: a February 2012 US poll found 83 percent support for their use. They have taken on the ideological role played by 'smart-bombs' in 1991, as a 'patriotic' technology, precisely eliminating 'evil' whilst saving western lives.

Drones are ideal technologies for the 'post-conflict' battlefield and the War on Terror. Their ISR capability is invaluable against an amorphous and secretive enemy, whilst they also allow targeted attacks against individuals and small groups. This combination is powerful: as Lt General Dave Deptula commented, 'you have on one aircraft all the Find, Fix and Finish capability that had always been separated on different systems. You can now observe and take an action to engage in a matter of minutes, seconds quite frankly, which you never had before' (Woods, 2016:41). Drones can reach remote areas inaccessible to conventional forces, are near-silent, are difficult to detect, can conduct missions that would be too dull or dangerous for manned aircraft and are capable of long-endurance operations, building intelligence over time. They are less visible and provocative and leave a smaller military footprint than conventional forces, making military operations easier, especially in politically sensitive regions.

Drones are claimed to be 'the most precise weapons in the history of warfare' (Woods, 2012:24). John Brennan, Obama's counter-terrorism advisor, described them as an 'exceptionally precise and surgical' weapon, causing next to no collateral damage (Dilanian, 2011). They are a perfect response to networked terrorism, allowing a 'pin-point war' by taking out key nodes and damaging the network. Drones arguably reduce loss of life as they are less destructive than the attritional carpet-bombing that marked earlier wars. As Woods says, from 1965 to 1973 the USA released 4.7m tons of ordnance in half a million sorties over Cambodia and Laos, killing at least 100,000 civilians. Drone deaths are troubling but they are 'a far remove from the indiscriminate bombing of earlier wars' (Woods, 2016:xv). Drone defenders point to their successes. Drone attacks have decimated the al-Qaeda, Taliban and Islamic State leaderships, stripping key personnel and forcing them to change their organizations, communications and OPSEC, reducing their ability to operate and hence their international threat. As a Pentagon official said, 'We denied the enemy the ability to move. We denied him the ability to communicate. Prior to that they could do a thousand things. Now they could do one thing. Hide' (Woods, 2016:277).

For all their benefits, however, drones remain controversial technologies. In particular, issues around their use can be grouped into three areas: (1) military issues, (2) legal and political issues, and (3) moral issues.

Military issues

One issue worrying the military is deskilling. US DoD unmanned aircraft rose from 5–60 percent from 2005 to 2012, leading to a 'UAV pilot crunch' and a need to train more drone operators. Fearing deskilling, the USAF insisted only trained pilot fly drones, but this isn't necessary and drone pilots are not being trained to fly aircraft. Ongoing developments in AI allowing autonomous drone flight will extend this deskilling by eradicating the pilot completely. More broadly, as Singer notes, unmanned technologies threaten the entire military culture as 'there is nothing inherently military about the ability to punch a keyboard and move a joystick around' (Singer, 2009:371). For these new 'cubicle warriors', he says, '"going to war" doesn't mean shipping off to some god-forsaken place to fight in a muddy foxhole but a daily commute in your Toyota Camry to sit behind a computer screen and drag a mouse' (Singer, 2009:329). When office drones drive robotic drones, Singer says, 'there is a striking convergence between military tasks and civilian life' (Singer, 2009:371).

This leads to the new problem of combining military and family life. As Gary Fabricius says, 'You are going to war for twelve hours, shooting weapons at targets, directing kills on enemy combatants, and then you get in the car, drive home, and within twenty minutes you are sitting at the dinner table talking to your kids about their homework' (Singer, 2009:347). Drone operators also complain their 'combat' experience isn't valued, that they face worse promotion prospects and that they have less official recognition in being denied combat medals.

Although critics argue drones represent a disconnected warfare, others claim the opposite is true, claiming post-traumatic stress disorder (PTSD) from the intimate connection to those on the ground and the deaths caused. Woods is more careful, suggesting that whilst traditional PTSD doesn't fit, the unique telepresent relationship may lead to a 'prolonged virtual combat stress' (Woods, 2016:187), with operators haunted by what they've seen on the screens. A December 2011 US government-commissioned report found half of drone pilots reported 'high operational stress', with a third suffering 'burnout' and 17 percent 'clinically depressed' (Benjamin, 2013:96). Chamayou, however, disputes such 'trauma', arguing operators don't have 'direct personal experience' of a threat to their lives and although they may witness death, 'they are the *authors* of that death, that injury, that threat' (Chamayou, 2015:110–11). Chamayou argues claims of PTSD are designed to publicly rehabilitate drone killing, serving 'to apply a layer of humanity to an instrument of mechanized homicide' (Chamayou, 2015:108). He quotes a psychological study of pilots that failed to diagnose a single case of PTSD, arguing instead that their stress is a result of poor working conditions, including the 12-hour shifts, alternating day, night and overnight shifts, the commute, tiredness, disillusionment and boredom. As he concludes: 'warfare becomes telework with shifting timetables, and the symptoms its agents present are all connected with this' (Chamayou, 2015:110).

The technical limitations of UAVs have also caused military concern. On 3 October 2001 – at exactly the point when the US military was embracing armed drones – an Office of Operational Test and Evaluation report concluded 'The Predator UAV is not

operationally effective', succumbing too easily to bad weather and with poor performance in target location and communications (Cockburn, 2015:71). Drones struggled
for a long time with the quality of their camera images, leading to problems of
analysis and identification. The move to HD video and improved cameras has helped
but problems of analysis remain, compounded by the daily volume of information
being captured today.

Drones are also highly vulnerable technologies, with many crashing due to equipment
failure or pilot error. In 2015, 20 large drones were lost or sustained over $2m in damages –
the worst annual toll at the time – including 10 Reapers that dropped out of the sky following
sudden electrical failures. The problems continue: in August 2017 two Predators were lost
within four days when taking off from Incirlik airbase in Turkey. Drones can also be shot
down. A Predator was lost on 17 March 2015, for example, when it was hit by Syrian air
defences near Latakia, whilst their vulnerability to aircraft was demonstrated on 23 December
2002 when a Stinger-armed Predator was easily destroyed by an Iraqi MIG-25. It will be
decades before an AI drone will succeed in a dogfight.

Even the precision of drones can be questioned. Although the targeting is precise, the
munitions need not be: the USA has twice reduced the lethality of its missiles to limit
local casualties. Precise targeting also says nothing about the validity of the target.
'Signature strikes' can misinterpret innocent behaviour as hostile, whilst 'personality
strikes' can be based on misidentification. Human intelligence (HUMINT) can be flawed
or deliberately wrong whilst even signals intelligence (SIGINT) can make mistakes,
targeting phone signatures on the assumption that they belong to particular people. The
US policy of accepting a certain level of civilian deaths if a 'high-value target' (HVT) is
killed, also renders moot questions of precision.

These problems of identification mean drones may be less successful than is claimed.
Critics point to a 2 percent HVT hit rate, suggesting most killed are not leadership targets.
In reality 'signature strikes' dominate and most of those killed are simply described as
'militants'. Given the similarity of civilians and militants errors are inevitable and as the
real status of victims may never be established it is certain that the numbers of 'militants'
killed inflates the drone's efficacy and value. Where drones have been successful, militants
have learnt to adapt, developing OPSEC counter-measures, forcing the drone programme
to rely more on HUMINT. Bin Laden's Abbottabad compound, for example, had no
internet or phone lines, relying on a human courier, whilst most militant groups eschew
mobile phones as too dangerous.

The use of drones also has the military effect of radicalizing more people and creating
'blowback'. For those affected, the fear of living under drones and the death and
destruction caused creates a hatred of the west and a desire for revenge that aids local
militant recruitment. As Benjamin says, 'While violent extremists may be unpopular, for a
frightened population they may seem less of a threat than an omnipresent, hovering enemy
that at any moment could choose to eliminate one's loved ones with a Hellfire missile'
(Benjamin, 2013:206–07). This blowback includes inspiring western terrorists. Faisal
Shahzad, who tried to detonate a car bomb in Times Square, NYC, on 1 May 2010 cited
child drone victims in Pakistan as his motive, whilst Michael Adebolajo, one of the killers
of Lee Rigby, near the Royal Artillery Barracks in Woolwich, London, on 22 May 2013
gave an impromptu mobile phone press conference immediately after the murder implicitly blaming drones for civilian Muslim deaths. Video footage of child drone victims and
funerals are popular on Islamist websites, providing powerful imagery that helps radicalize and recruit people for their cause.

Hence, whilst drones appear as a wonder weapon for the west, there is a reputational cost for their use. As Chamayou says (echoing Baudrillard's critique of the Gulf War), as an attempt to eradicate all 'reciprocity', erasing the traditional principles of military bravery and sacrifice, 'a drone looks like the weapon of cowards' (Chamayou, 2015:17). In traditional cultures where bilateral laws of hospitality, respect, 'face' and honour govern relationships this remote-controlled, faceless technology whose operators are so frightened of the loss of their own lives and who exist in a purely unilateral and protected relationship with those they kill appears disrespectful, cowardly and weak. Instead of projecting power without vulnerability, the drone projects western pusillanimity.

Legal and political issues

Drones change the nature of warfare and this has legal and political implications. As Chamayou argues, drones represent a shift from conventional, attritional warfare to 'a militarized manhunt' based on 'targeted assassinations' (Chamayou, 2015:32). War becomes 'hunting warfare', moving from mass destruction to a game of hide and seek where 'the primary task is no longer to immobilize the enemy but to identify and locate it (Chamayou, 2015:33–4). This is a 'preventive' warfare, designed to destroy threats in advance: a 'prophylactic elimination' of individuals that is closer to a policing function (Chamayou, 2015:34). The drone, therefore, is a tool of state assassination, hence, Chamayou says, what we call warfare today 'takes the form of vast campaigns of extrajudicial executions' (Chamayou, 2015:35). Turning war into assassination, therefore, raises key legal questions. Drone attacks constitute a shoot-first policy stripping individuals of their rights under international law and killing them without trial.

The USA explicitly banned assassinations after revelations about the activities of its intelligence agencies. President Ford signed Executive Order (EO) 11905 on 18 February 1976 stating 'no employee of the United States Government shall engage in, or conspire to engage in, political assassination' (Ford, 1976). This was reinforced by EO 12036 in January 1978 and EO 12333 in December 1981, but later presidents tried to find loopholes. By 1998 the government wanted to loosen restrictions to deal with both Saddam Hussein and Bin Laden, but the prohibition was secure enough to make plans to weaponize a Predator to kill Bin Laden in 2001 politically sensitive. The USA was also a vocal critic of Israel's sniper-based TKP. After 9/11 Bush dropped this criticism and signed a memorandum of notification (MON) clearing the way for the CIA to kill any terrorist who appeared on a 'high value target list' (Woods, 2016:49).

The justification was that this wasn't assassination, it was 'war' (Woods, 2016:50) – a claim best expressed by the legal advisor to the State Department, Harold Koh, in a speech on 25 March 2010:

> as a matter of international law, the United States is in an armed conflict with al-Qaeda, as well as the Taliban and associated forces, in response to the horrific 9/11 attacks, and may use force consistent with its inherent right to self-defense under international law. As a matter of domestic law, Congress authorized the use of all necessary and appropriate force through the 2001 Authorization for Use of Military Force (AUMF). These domestic and international legal authorities continue to this day. As recent events have shown, al-Qaeda has not abandoned its intent to attack the United States, and indeed continues to attack us. Thus, in this ongoing armed conflict, the United States has the authority under international law, and the responsibility to its citizens, to use

force, including lethal force, to defend itself, including by targeting persons such as high-level al-Qaeda leaders who are planning attacks.

Koh (2011)

These attacks, therefore, are not 'extrajudicial killings', as 'a state that is engaged in an armed conflict or in legitimate self-defense is not required to provide targets with legal process before the state may use lethal force'.

International law, however, recognizes only two types of war: international armed conflict (between nation-states) and non-international armed conflict (or civil wars). Bush's War on Terror was neither. The Afghan War fitted these criteria, but problems began with the November 2002 drone killing of AQAP leader Al-Harethi in Yemen who was not on a battlefield and there was no attempt made to capture him. The strike also killed the US citizen, Kamel Derwish, who was protected by the fifth amendment of the US Constitution, which says no person shall be 'deprived of life, liberty, or property, without due process of law'.

In fairness, the current conception of warfare in international law appears anachronistic in an age of global, non-state terrorist actors with global political aims and globally coordinated activities, and a right of national 'self-defence' seems reasonable here. This would require, however, a careful, subtle and agreed refashioning of international norms and this didn't happen. Instead Bush and the neoconservatives illegally expanded the concept of 'armed conflict' to a global battlefield and jettisoned all domestic and international laws that clashed with their aims. As Geoffrey Robertson, QC, argues, the USA replaced international law with 'the punishment of the Red Queen' – 'sentence first, trial later' – denying victims the presumption of innocence, right to a trial and right to life (Robertson, 2012:27). The drone TKP, therefore, was part of a suite of US extrajudicial policies including the removal of POW Geneva convention rights, extraordinary rendition, Guantanamo Bay's military jurisdiction and the use of EIT torture.

By 2009 the USA had a list of 367 'prioritized targets' on what was effectively a kill list whose basis was beyond all legal, journalistic or public scrutiny. There was no due process for a potential victim or their family to challenge the information on which a targeting decision was made. In effect, Robertson says, 'What the Pentagon is doing is secretly sentencing people to death for an unproven crime' (Robertson, 2012:26). Although post-9/11 fears of terrorism may make this targeting acceptable to us, the reality is globally prosecuted, extrajudicial state assassinations that set a dangerous precedent for any nation to kill anyone, anywhere in the world they define as a 'terrorist' threat. Defenders claim drone strikes are among the most regulated in the military, with a complex 'kill chain' of decision-making and 'judge advocates' advising the commanders, but critics argue the legal advisors aren't there to protect the victims but are instead present to give the TKP a legal immunity and veneer of respectability.

The US drone killing of its own citizens has proven especially controversial. In April 2010, the government put Anwar al-Awlaki, a joint US-Yemen citizen on a target list, rejecting in December his father's legal challenge to this decision. The government argued the courts had no right to oversee the executive's war-time targeting decisions, argued al-Awlaki's father had no legal standing to bring the case and invoked state secrets privilege. In effect, Jameel Jaffer, Deputy Director of the ACLU said, 'If the court's ruling is correct, the government has unreviewable authority to carry out the targeted killing of any American, anywhere, whom the president deems to be a threat to the nation' (Benjamin, 2013:136).

On 30 September 2011, al-Awlaki was killed by a Predator. His 16-year-old son, Abd al-Rahman, and his teenage friends were also killed by a drone on 14 October. The USA claimed al-Awlaki was a senior al-Qaeda leader but although his YouTube videos certainly inspired many terrorists, "no evidence of al-Awlaki's relations with al-Qaeda beyond his propaganda writings has ever publicly emerged' (Woods, 2016:139). A US Justice Department memo, leaked in February 2013, justified killing US citizens if there was an 'imminent threat' of attack, if they were engaged in 'armed conflict' with the USA and if their capture was unfeasible. The government, however, gets to define these terms and it remains difficult to see how a middle-aged YouTuber and a teenager having lunch 8,000 miles away constituted an 'imminent threat' to the most powerful nation on earth.

Dozens of westerners have been killed by drone strikes since 2002, often with the aid of their own country's intelligence agencies. The UK has targeted and killed citizens involved with Islamic State. On 21 August 2015, it carried out its first targeted killing outside of military operations when a drone killed Reyaad Khan. In April 2017, a parliamentary Intelligence and Security Committee (ISC) report concluded Khan had 'posed a very serious threat to the UK' as a recruiter and attack planner with a group that was at war with the west (MacAskill, 2017). US and UK drones also killed UK citizen Mohammed Emwazi on 12 November 2015. Better known as 'Jihadi John', he had appeared in IS videos beheading hostages. PM David Cameron described his killing as 'self-defence' and although few would mourn his death it still represented state execution without trial, in an area in which the government lacked parliamentary authority to operate, using a justification that stretched the definition of national self-defence.

The US government's justification of 'armed conflict' and 'self-defence' to defend killing 'militants' in the FATA, Afghanistan, Iraq, Yemen and Somalia, are similarly questionable. Few of these militants, many of whom own little more than a rifle, are an 'imminent threat' to the USA and nor are they in 'armed conflict' with the USA as their primary objectives are mostly local. Moreover, no 'pattern of life' analysis or 'signature strike' can discern the motives and threat-level of people who it can't even identify with certainty as militants. Like the Cold War, therefore, the War on Terror is primarily a global, ideological conflict fought to established preferred regional political authorities and oppose an assumed enemy ideology.

Other aspects of the drone programme also contravene international law. The 'double-tap' method violates legal protections for medical and humanitarian personnel and for the wounded: as UN special rapporteur Christof Heyns argues,

> Where one drone attack is followed up by another in order to target those who are wounded and hors de combat or medical personnel, it constitutes a war-crime in armed conflict and a violation of the right to life, whether or not in armed conflict.
>
> (Woods, 2016:161)

It is also illegal to target non-combatants in armed conflict but the USA has deliberately hit large gatherings such as weddings and funerals, calculating that the death of a HVT is worth a proportion of civilian deaths. Critics have also suggested that Obama's expansion of drone strikes was a way to avoid his predecessor's problems with extraordinary rendition, torture and detention without trial: ironically, his kill-don't-capture policy clears up the human rights issues of what to do with POWs.

There are also legal questions about the CIA's activities. Lethal force can only be used in a warzone by uniformed personnel but the CIA are a civilian agency fighting without uniform or insignia and engaging the enemy contrary to the laws of warfare. The same problem applies to civilian personnel employed by private defence contractors in drone maintenance, operation and intelligence analysis. The vast increase in ISR material and need to combat IS created a huge demand for more video and photographic analysts and this was largely filled by contracted civilians. Although they don't operate the drones, they work at the heart of a defined 'weapons system' and are intimately involved in strikes with their identifications – and mistakes – having lethal consequences. As civilians, they are not subject to the Uniform Code of Military Justice and are not accountable for war crimes or violations of rules of engagement. Once again, therefore, digital technologies transform the 'who', 'what' and 'where' of conflict, scrambling existing concepts and laws.

Drones also raise important political questions, especially about the functioning of democracies in wartime. The US drone programme has been kept secret, with politicians using the courts to block legal and public scrutiny. When the ACLU tried to get information about the CIA's drone use it argued, and the courts agreed, that 'the fact of the existence or non-existence' of the programme was classified (Benjamin, 2013:59). Consistent oversight only began after eight years when the Senate Intelligence Committee introduced 'monthly in-depth oversight meetings' (Woods, 2016:203), although there was no way to check CIA claims whilst JSOC operations have even less oversight. Most committee members have also expressed their support for drones and their investigations of strikes have remained limited.

The mainstream US media has proven equally unable or unwilling to interrogate administrations about the legality of drone strikes, largely because it shares the public's support for them in keeping American lives safe. Hence it rarely challenges official statements, repeating government claims regarding the number of 'militants' killed. Benjamin points out that when a drone killed the TTP leader Baitullah Mehsud on 7 August 2009 there was no mention of the deaths of his wife, father-in-law and eight others, or of the 204– 321 victims of the 15 prior attempts to kill him: 'All the American public heard was that justice had been done and that the evil Mehsud was dead' (Benjamin, 2013:107). When John Brennan, Obama's counterterrorism advisor, briefed the press in June 2011 that there hadn't been a civilian casualty for nearly a year due to 'the exceptional proficiency, precision of the capabilities that we've been able to develop' (Woods, 2016:252), his claims went unchallenged until a BIJ investigation identified at least 45 civilian deaths, and likely many more, in that period. Whilst it is true that journalists face problems investigating drone strikes, as they occur in remote, hard-to-reach and often unsafe places, and the sites are often sealed off by Pakistan security forces, nevertheless, as Woods points out, investigations are still possible and 'sustained efforts by lawyers, academics, NGOs and journalists have uncovered extensive details of many of these people who have died' (Woods, 2012:24).

These failures have important implications for democracy. Although democratic societies still need to conduct secretive operations, when this morphs into a secret war beyond public knowledge, legislative permission and legal oversight this is dangerous. A healthy democracy requires sufficient transparency to enable the public to understand the reality of their wars, so that governments can be held accountable and they can decide whether they support the action conducted in their name. This is especially important as they risk being held accountable themselves for their government's actions. Islamists often justify killing western civilians because, as Bin Laden explained, 'the American people have the ability and choice to refuse the policies of their government and even to change it if they want' (Maher, 2016:56). With drones, therefore, the public risk being attacked for operations they know little or nothing about.

Central to the problem is the growing disconnection between western populations and war. Whereas a major operation such as the 1991 Gulf War needed to mobilize domestic and international support, drones reduce the scale and cost of military actions and hence *don't* require the public to support them and function better *without* media coverage. Whilst the Gulf War turned war into something to watch on television with no risk for the domestic population, today even that level of participation isn't required. Whereas in 1991 'smart-bomb' videos entranced the population, today drone strike videos are noticeably rare, demonstrating how we have moved from a model of produced, public military engagement to one of deliberately produced disengagement.

The result is a disconnection from drone strikes. As Benjamin says, 'Although thousands of people have been killed and maimed by drone attacks, the American public has yet to see photographs or video from the aftermath of these strikes' (Benjamin, 2013:157). Hence, 'while the US military is engaged in more and longer conflicts than ever in our history, fewer people are involved, touched, concerned, or engaged' (Benjamin, 2013:151). The outcome, Singer says, is a public that is 'more disinvested in and delinked from its foreign policy than ever before in a democracy' (Singer, 2009:318). The UK public are no better informed or engaged. The Chair of the ISC report into the Reyaad Khan killing criticized the lack of access to ministerial decisions and material that made government scrutiny difficult.

Drone strikes also risk undermining the international order. Military operations against another nation usually constitute a highly visible and provocative breach of sovereignty. Being near-invisible and having a small military footprint, however, drones lower the threshold of military action, making breaches of sovereignty and warfare easier than ever. As such they appear an attractive, less risky option for politicians who would baulk at full-scale warfare and its commitments and consequences. Although some nations such as Yemen and Pakistan have given permission for drone strikes, this is irrelevant. As Benjamin points out, 'it doesn't really matter whether a government consents or not', as they cannot give legal consent to the killing of their own citizens (Benjamin, 2013:142). By 2013, however, Pakistan was publicly complaining of the US breach of international law and on 15 March the UN agreed drone strikes breached its sovereignty.

For the USA, however, sovereignty didn't matter: in a 'War on Terror' the idea of a battlefield as lines on a map was outdated as the enemy wasn't contained within or constrained by these. Hence, Chamayou argues, drones represent a shift from traditional, geographically identified and limited battlefields to a new conception of 'mobile micro-spaces', with the 'battlefield' reduced to a 'kill-box' the size of a human body. Here the whole world becomes a hunting ground, pursuing to find and fix targeted individuals (Chamayou, 2015:54–55). The 'rights of conquest' are thus replaced by 'the rights of pursuit' – 'a right of universal intrusion or encroachment' that tramples underfoot 'the principal of territorial integrity classically attached to state sovereignty' (Chamayou, 2015:53).

Moral issues

The most controversial issue around drones has been that of civilian casualties. Although drone defenders point to a lower rate of 'collateral damage' than traditional bombing, their secrecy and ease of use, together with the post 9/11 political climate that prioritized the eradication of any perceived threat above international laws, and specific targeting polices that work against people on the ground, have led to significant civilian casualties. The BIJ have recorded 4,413 confirmed US drone strikes from 2002 to October 2017 in Pakistan, Yemen, Afghanistan and Somalia, killing 6,826–9,930 people, of whom 753–1,488 were civilians, including 262–331 children. The real figures are much higher as the Afghan data is only from 2015 onwards, and the BIJ don't list strikes in other warzones such as Libya, Iraq and Syria.

The BIJ also accuse government officials of underestimating or denying civilian deaths. In July 2016, they took issue with the government's claim of 64–116 non-combatant deaths from January 2009 to the end of 2015, saying that the number they had recorded was 'six times higher than the US government's figure' (Bureau of Investigative Journalism, 2017b). The Trump administration continues these policies, claiming a drone strike on an Aleppo mosque on 16 March 2017 killed 'dozens' of terrorists without any civilian casualties, despite on-the-ground sources saying 'at least 47 civilians had also died in the strikes' (Gibbons-Neff, 2017). Whilst the USA claims it doesn't mean to kill civilians, this is only partially true. Although many are accidental, the USA has accepted and even pursued civilian deaths under certain circumstances.

International law allows the use of force necessary for war, provided it doesn't target or cause excessive incidental harm to civilians. The USA, however, has a particular approach to the definition of 'civilian'. First, its identification of 'militants' based on a 'pattern of life' analysis is problematic as combatants and non-combatants are so similar: as the Pakistani barrister Mirza Shahzad Akbar comments, 'since every man in Waziristan has a turban and a gun, every one of them is likely a CIA target' (Woods, 2012:24). Second, the USA defines enemy combatants as 'all military-age males in a strike zone' (Woods, 2012:24). Press reports of 'militants' killed, therefore, don't refer to *actual* militants, but to *all* males over 16 – or *looking* over 16 through drone cameras – within *any* area the USA has designated a combat zone, displaying behaviour that can't be immediately proven to be innocent. The USA says that the dead can be reclassified with evidence, but few in these remote and poor regions have the capacity to pursue such claims.

Hence the numbers of civilians killed by drones is undoubtedly higher than recorded. Other US targeting policies feed into this, including the varying 'non-combatant casualty value' (NCV), which defines acceptable death rates and which justifies strikes on large gatherings as acceptable if a HVT is present. Similarly, the 'double-tap' method increases civilian casualties by designating anyone helping drone victims as enemy combatants, as does the CIA's deliberate targeting of buildings in Pakistan's tribal areas, which, though described as 'militant compounds', are often simply civilian houses or contain civilians.

In his speech on 10 December 2009 accepting the Nobel Peace Prize, President Obama claimed US conflicts were fought to higher moral standards: 'I believe the United States of America must remain a standard bearer in the conduct of war. That is what makes us different from those who we fight' (Woods, 2016:240). A 'War on Terror' fought to avenge nearly 3,000 civilians, however, had long proven itself indifferent to other nation's civilian suffering. Indeed, the psychological basis of this war promoted killing anyone as a preventative act potentially saving US lives: as US military intelligence analyst Daniel Hale admitted:

> If you are of the mentality that American lives are more important than the lives of the people in the country that you're conducting a strike in, if the only people who

have to die in a drone strike are Afghans or Yemenis or Pakistanis, who cares if a kid also gets killed in a strike? You still got the bastard that was gonna kill an American.
                                                                                    Woods, (2016:245)

As we saw with the civilian deaths in the Afghanistan War, the western indifference to civilians killed by drones thousands of miles away leads again to the conclusion that, for all our public defence of individual freedom and human rights, the lives of those in developing countries count less for us than those in the west. Would a HVT strike against a wedding party really be made if the civilians were American or British and white, or if they were *our* families? Our entire set of values here is distorted: whilst we regularly condemn the sexual abuse of children, drone programmes that blow them up with Hellfire missiles barely enter the public consciousness. A good example of this hypocrisy was the western political, media and public treatment of Malala Yousafzai, a young Pakistani activist defending female education, who was shot in the head by the Taliban on 9 October 2012. The western response – her UK medical treatment and sanctuary – was presented in the media as proof of the superiority of our political system over a force so evil that they'd shoot a teenage girl in the head, but nowhere was there any reflection on the western drone programmes that were shooting at children with missiles.

Many see drones as creating a moral distance from death, with remoteness devaluing the lives of those on the ground for both domestic populations and for drone operators. Distanced from risk, watching targets on a screen for days at a time from a chair in a box, thousands of miles away, these operators appear to realize that unilateral, 'Nintendo war' critics had seen arriving in the 1991 Gulf War. Pilots have made these comparisons, with one saying, 'It's like a video-game. It can get a little bloodthirsty. But it's fucking cool' (Singer, 2009:332). Another reported the divine omniscience of this technological distancing: 'Sometimes I feel like a god, hurling thunderbolts from the air' (Martin and Sasser, 2010:3).

But bodily remoteness doesn't equate to psychological distance. Many pilots have spoken about the close connection built up over time with both ground troops and militants, a point confirmed by USAF Major General Jim Poss who argues, 'there's not many places in modern warfare where you get more intimately connected with the target than you do with unmanned aerial vehicle warfare' (Woods, 2016:175). From sensor operators and analysts spending hundreds of hours building up intelligence to commanders, Generals and even the President able to observe real-time feeds and communicate with troops in theatre there is a greater connection with the battlefield than ever before. The key here is *telepresence*. The McLuhanist, global electronic extension of the drone operator's central nervous system and senses ensures their minds and psyche are fully present in the combat zone: 'Physically we may be in Vegas', Air Force Major Shannon Rogers told *Time* magazine in 2005, 'but mentally we are flying over Iraq. It feels real' (Benjamin, 2013:96).

Hence the claims of some operators to have experienced guilt at drone operations – paradoxically drones are an intimate form of killing. For Gregory, it is not the remoteness of 'video-game war' that produces drone strikes with civilian casualties, but this telepresence. The sensors form part of a techno-cultural system, he argues, that is imposed upon another geographical region, transforming '"their" space' into '"our" space' (Gregory, 2011:201). The drones patrol a space that is always *American territory*, with all movement in this zone needing to be monitored and controlled. Telepresence thus heightens the perception of threat and delegitimizes indigenous life, movement and culture. The moral distance that removes the value of life on the ground is a function of this electronic hyper-presence.

The moral issues of drones, therefore, extend beyond deaths to include the impact on the everyday life of those living within these controlled spaces, as the defining weapon of the War on Terror is itself experienced as a weapon of state terror. The 2012 Stanford International Human Rights & Conflict Resolution Clinic report 'Living Under Drones' argues 'US drone strike policies cause considerable and under-accounted for harm to the daily lives of ordinary citizens, beyond death and physical injury':

> Drones hover twenty-four hours a day over communities in northwest Pakistan, striking homes, vehicles, and public spaces without warning. Their presence terrorizes men, women, and children, giving rise to anxiety and psychological trauma among civilian communities. Those living under drones have to face the constant worry that a deadly strike may be fired at any moment, and the knowledge that they are powerless to protect themselves. These fears have affected behavior. The US practice of striking one area multiple times, and evidence that it has killed rescuers, makes both community members and humanitarian workers afraid or unwilling to assist injured victims. Some community members shy away from gathering in groups, including important tribal dispute-resolution bodies, out of fear that they may attract the attention of drone operators. Some parents choose to keep their children home, and children injured or traumatized by strikes have dropped out of school. Waziris told our researchers that the strikes have undermined cultural and religious practices related to burial, and made family members afraid to attend funerals. In addition, families who lost loved ones or their homes in drone strikes now struggle to support themselves.
>
> Cavallaro, Sonnenberg, and Knuckey (2012: vii)

The mere presence of drones, capable of striking at any moment, terrorizes individuals, families and communities. As Benjamin says, 'Residents I met with said they had a hard time sleeping, that many people suffer from depression and PTSD, and that there is a widespread use of anti-depressants and anti-anxiety medications. They also reported a spate of suicides, something they said never existed before' (Benjamin, 2013:119). Hence, in the name of human rights and democracy, the USA wages a permanent terror-inducing drone war against some of the poorest and weakest people on the planet whilst presenting these same technologies 'as the cost-free magic wand that can eliminate terror' (Benjamin, 2013:122).

The future of drones

Superiority in weapons systems doesn't last: technologies proliferate and become standard in conflict. This is already visible in drone warfare. The only direct beneficiary of US technology to date is the UK. The UK had developed the BAE Phoenix UAV in the 1980s, then leased the Israeli Elbit Hermes-450 and, in collaboration with Elbit and Thales, developed the Watchkeeper, but in 2004 the RAF entered into an agreement with the USA to fly US-owned drones using personnel embedded in Nellis AFB, Nevada, under British authority, and employing stricter rules of engagement. In 2007 the UK stood up its own Reaper fleet based at first at Creech AFB then, from April 2013, at RAF Waddington, near Lincoln. By October 2013 UK Reapers had flown over 50,000 hours in Afghanistan, averaging three sorties a day, and they had launched over 418 drone strikes.

Today every major military power is either developing or investing in drone technology. Israel has retained its drone reputation, with several companies, including Aeronautics Defense Systems, Elbit Systems and Israel Aerospace Industries producing for the export market. Advertised as 'combat-proven', their drones have proven highly popular. The Hermes-450 surveillance UAV has been bought by the UK, Azerbaijan, Botswana, Brazil, Columbia, Cyprus, Macedonia, Georgia, Mexico and Singapore.

China unveiled its first model drone in 2006 and in November 2010 displayed 25 different drones in a trade show. By 2013 the PLA had a range of available drones including CATIC's ASN series UAVs, such as the ASN-15, and ASN-209; Reaper-style drones such as CADI's Yilong/Wing Loon 'Pterodactyl', CASC's CH-4, ALIT's CH-91 and CASIC's WJ-600; Global Hawk variants such as CAIC's 'Soaring Eagle', and stealth drones such as the RQ-170-style 'Wing Blade' and X74-B-inspired Anjian 'Dark Sword' drone. Importantly, China sees its drones as an important export. Its CAIG Wing Loon drones have reportedly been purchased by Saudi Arabia, Egypt, Kazakhstan, Nigeria, Uzbekistan, Pakistan and the UAE. In June 2017 China started commercial production for export of its CH-5 Rainbow, which rivals the MQ-9 Reaper in performance at half the cost. In time, China are expected to become the world's biggest drone manufacturer. Their low cost and willingness to deal with nations unable to source US/Israeli technology make them 'a highly affordable capability for a host of international customers' (Moss, 2013).

Military technologies also proliferate through theft and copying. Pakistan gave China access to the crashed Black Hawk stealth helicopter used in the Bin Laden raid, for example, whilst its PLA cyberwarfare units have systematically penetrated and exfiltrated secret information from western military and aerospace systems for their own use. On 4 December 2011 Iran claimed to have downed a US RQ-170 UAV over its territory by disrupting its control signals and GPS-spoofing it to force it to land (although others claim it merely crashed). On 10 November 2014 Iran claimed to have test-flown a reverse-engineered drone and on 1 October 2016 it unveiled its Saeqeh ('Thunderbolt') UAV based on the RQ-170. This is part of a broader Iranian drone programme that includes the armed Shahed-129 that has been deployed in Syria, with two shot down by US planes bear their Al-T'anf outpost in June 2017.

Drones don't have to be complex technologies and smaller states and non-state actors are also developing less sophisticated versions. In the 2016–17 battle for Mosul, for example, Islamic State sent remote-controlled commercial drones packed with explosives against Iraqi forces. Clearly, US dominance of drones is coming to an end, but the problem isn't simply the proliferation of drones but the proliferation of the specific *ways* the USA had used them. The USA has established the parameters of seemingly acceptable drone use, including a disrespect for international law, human rights, due process, sovereign borders and civilian casualties. Such practices may not have bothered us in the west when imposed upon a foreign 'other', but our disinterest in these issues will change when, as is entirely possible, we start becoming their targets.

As drones have proliferated, so too have anti-drone counter-measures. Security considerations are increasingly important. As the RQ-170 incident showed, there is a risk drones could be hacked and downed, or even turned against their own forces. In December 2009, a Predator video feed was found on an insurgent's laptop after an unencrypted satellite downlink was hacked with $26 'Skygrabber' software. The Predator itself wasn't affected, but in October 2011 US Predator and Reaper drone cockpits were found to be infected with a virus and keylogger software. Both state and non-state actors are undoubtedly exploring anti-drone cyberwarfare techniques.

Another emerging phenomenon are anti-drone defences. In January 2013, the German firm Rheinmetall Defence unveiled a 50KW laser that could shoot down drones; in November 2014, China demonstrated an anti-drone laser able to shoot down a 'small aircraft' in five seconds, whilst in the same month the USA demonstrated an anti-drone laser on the USS Ponce. It had been tested in 2009 at the naval weapons system at China Lake, California, where it destroyed 'threat representative' UAVs, and again in July 2012 on board the destroyer USS Dewey, where it successfully shot down three UAVs. In October 2017, it was reported that UK-designed 'Zapper' anti-UAV defences were successfully defending US bases against IS drones. Combining radar, radio-jamming, and video-tracking and thermal software, the technology can detect and jam drone signals, causing them to crash.

Whilst the short-term future will see the global proliferation of more and smaller, cheaper, more powerful and increasingly lethal drones, the longer-term future is even more worrying. Peter Singer's *Wired For War* (2009) placed drones as part of an emerging robotic 'Revolution in Military Affairs', and although his attribution seemed erroneous at the time for what remained human-controlled technologies, ongoing developments in artificial intelligence (AI) suggest he may one day be right. Already, developments in AI have led to autonomous UAVs that pilot themselves. The US X-47B Autonomous Unmanned Combat Air System (UCAS), for example, only requires human sensor and weapon operators, whilst BAE have produced a UK version called Taranis, flight-tested in 2013 and capable of independent operation. Even sensor and weapons operators may be dispensable: improvements in software such as facial recognition systems and biometric systems analysing the body and its movements will have important drone applications and we can envisage the development of 'pattern of life' analysis algorithms to reduce human error (and legal culpability).

Fully autonomous lethal robotic drones are only a matter of time. With weapons systems becoming too fast, too small, too numerous and too complex for humans to direct we can expect further military deskilling and redundancy. Drones will morph into many forms – tiny individual robots, robotic swarms, large motherships – all working together, algorithmically controlled and responding instantly to electronic sensors and their signals, removing the human even more from conflict. Unless, however, we match these developments with the creation of civilian drones, then the human body will remain in place as the victim of these technologies.

Key reading

The literature on drones has massively expanded in recent years and there are now many texts to choose from. Singer (2009) was the first book on drone warfare and is still required reading, although he places them within a broader concept of robotics that isn't always relevant here. The best general overviews of drone war are provided by Woods (2016) and Benjamin (2013) and I would recommend these as the starting point. Cockburn (2015) and Gusterson (2016) are also strong accounts. The most insightful book on drones, however, is by Chamayou (2015), which is essential reading about the underlying philosophy and politics of drone war. There is a huge amount of information available online regarding drones but I would especially recommend the Bureau of Investigative Journalism (2018), which tracks US drone strikes, and Drone Wars UK (2018) which tracks the UK's drone programme.

References

Benjamin, M. (2013) *Drone Warfare: Killing by Remote Control*. London: OR Books.

Bureau of Investigative Journalism. (2017a) 'Obama's Covert Drone War in Numbers: Ten Times More Strikes Than Bush', *The Bureau of Investigative Journalism*, 17 January, https://www.thebureauinvestigates.com/stories/2017-01-17/obamas-covert-drone-war-in-numbers-ten-times-more-strikes-than-bush.

Bureau of Investigative Journalism. (2017b) 'Obama Drone Casualty Numbers a Fraction of those Recorded by the Bureau', *The Bureau of Investigative Journalism*, 1 July, https://www.thebureauinvestigates.com/stories/2016-07-01/obama-drone-casualty-numbers-a-fraction-of-those-recorded-by-the-bureau.

Bureau of Investigative Journalism. (2018) 'Drone Warfare', *The Bureau of Investigative Journalism*, https://www.thebureauinvestigates.com/projects/drone-war.

Card, O. S. (2002) *Ender's Game*, London: Orbit Books.

Cavallaro, J., Sonnenberg, S. and Knuckey, S. (2012) *Living Under Drones: Death, Injury and Trauma to Civilians From US Drone Practices in Pakistan*, September, Stanford, New York: International Human Rights and Conflict Resolution Clinic, Stanford Law School; NYU School of Law, Global Justice Clinic, https://law.stanford.edu/wp-content/uploads/sites/default/files/publication/313671/doc/slspublic/Stanford_NYU_LIVING_UNDER_DRONES.pdf.

Chamayou, G. (2015) *Drone Theory*, London: Penguin.

Cockburn, A. (2015) *Kill Chain: Drones and the Rise of High-Tech Assassins*, London: Verso.

Department of the Air Force. (2015) 'Air Force Distributed Common Ground System', *U.S. Air Force Fact Sheets*, 13 October, http://www.af.mil/About-Us/Fact-Sheets/Display/Article/104525/air-force-distributed-common-ground-system/.

Dilanian, K. (2011) 'U.S. Counter-Terrorism Strategy to Rely on Surgical Strikes, Unmanned Drones', *Los Angeles Times*, 29 June, http://articles.latimes.com/2011/jun/29/news/la-pn-al-qaeda-strategy-20110629.

Drone Wars UK (2018) *Drone Wars UK*, https://dronewars.net.

Ford, G. (1976) 'Executive Order 11905: United States Foreign Intelligence Activities', *The American Presidency Project*, 18 February, http://www.presidency.ucsb.edu/ws/?pid=59348.

Gibbons-Neff, T. (2017) 'Civilian Deaths from US-led Airstrikes Hit Record High under Donald Trump', *The Independent*, 25 March, http://www.independent.co.uk/news/world/americas/us-politics/donald-trump-civilian-deaths-syria-iraq-middle-east-a7649486.html.

Gregory, D. (2011) 'From a View to a Kill: Drones and Late Modern War', *Theory, Culture and Society*, 28(7–8), pp. 188–215.

Gusterson, H. (2016) *Drone: Remote Control Warfare*, London: MIT Press.

Joint Chiefs of Staff. (2001) 'Department of Defense Dictionary of Military and Associated Terms', 12 April 2001 (as Amended through 31 October 2009), Joint Publication 1-02, Joint Staff, http://jitc.fhu.disa.mil/jitc_dri/pdfs/jp1_02.pdf.

Klaidman, D. (2012) 'Drones: The Silent Killers', *Newsweek*, 28 May, http://www.newsweek.com/drones-silent-killers-64909.

Koh, H. H. (2011) 'The Lawfulness of the U.S. Operation Against Osama bin Laden', *Opinio Juris*, 19 May, http://opiniojuris.org/2011/05/19/the-lawfulness-of-the-us-operation-against-osama-bin-laden/.

MacAskill, E. (2017) 'Briton Killed in Drone Strike on Isis "Posed Serious Threat to UK"', *The Guardian*, 26 April, https://www.theguardian.com/uk-news/2017/apr/26/briton-killed-in-drone-strike-on-isis-posed-serious-threat-to-uk-reyaad-khan.

Maher, S. (2016) *Salafi-Jihadism: The History of an Idea*, London: Penguin.

Martin, M. J. and Sasser, C. W. (2010) *Predator*, Minneapolis: Zenith Press.

Mashhour, H. (2014) 'Hooria Mashhour: The United States' Bloody Messes in Yemen', *The Washington Post*, 14 January, https://www.washingtonpost.com/opinions/hooria-mashhour-the-united-states-bloody-messes-in-yemen/2014/01/14/c21dfcec-7653-11e3-b1c5-739e63e9c9a7_story.html?utm_term=.0ea9bffbb140.

Moss, T. (2013) 'Here Come … China's Drones', *The Diplomat*, 2 March, https://thediplomat.com/2013/03/here-comes-chinas-drones/.

Robertson, G. (2012) 'Trial by Fury', *New Statesman*, 18 June, pp. 25–7: http://www.newstatesman.com/lifestyle/lifestyle/2012/06/drone-attacks-against-human-rights-principle-book.

Singer, P. W. (2009) *Wired For War: The Robotics Revolution and Conflict in the Twenty First Century*, London: Penguin Books.

Woods, C. (2012) 'Games Without Frontiers, War Without Tears', *New Statesman*, 18 June, pp. 23–5, http://www.newstatesman.com/politics/politics/2012/06/drones-barack-obamas-secret-war.

Woods, C. (2016) *Sudden Justice: America's Secret Drone Wars*, London: C. Hurst and Co. (Publishers) Ltd.

9 Ambient war

Cyberwar everywhere

Defining 'cyberwar'

The term 'cyberwar' has now entered common usage, being widely employed by governments and militaries, specialist commentators and academics, the technical and popular media and by the public too. The term's proliferation doesn't mean, however, that the concept is well understood and, indeed, its success owes much to how vague the idea is and how easy it is to apply to different phenomena. Even among experts there is no simple, single, agreed definition of 'cyberwar' or 'cyberattacks' and the terms are often applied to a range of activities, many of which wouldn't fit into conventional definitions of 'war' or even 'force'. Consider the definition offered by former US presidential Special Advisor on cybersecurity, Richard A. Clarke, who says, 'Cyberwarfare is the unauthorized penetration by, on behalf of, or in support of, a government into another nation's computer or network. Or any other activity affecting a computer system, in which the purpose is to add, alter or falsify data, or cause the disruption of or damage to a computer, or network device, or the objects a computer system controls' (Clarke and Knake, 2010:228). Although this appears precise, on closer inspection the emphasis on state actors, the question of what 'another nation's computer or network' actually includes, the suggestion that even minor acts of falsification or disruption constitute 'warfare', and the omission of acts such as espionage that, at a certain level, may be more serious and aggressive than others included here, all point to the problems of defining a complex and continually developing phenomenon.

Perhaps the best starting point is simply to say that 'cyberwar' and 'cyberattacks' are terms used to refer to a range of politically motivated computer-based actions that mostly involve either disruption (such as through denial-of-service attacks) or penetration (whether for espionage, vandalism, more serious damage to a system, or damage to systems and objects that are connected to and controlled by that computer). More complex questions, such as who does this, why, how serious it is, and whether it constitutes an act of war, should, for the moment, be left aside. The best way to begin to answer them is to consider first of all how computers and 'cyberspace' came to be recognized as constituting a new 'fifth domain' of warfare, joining land, sea, air and space (Harris, 2014:xxi).

The origins of cyberwar

The 1983 film *WarGames* told the story of high-school student and computer-geek David Lightman (Matthew Broderick), who, whilst searching for games, unwittingly hacks into a military supercomputer and launches 'global thermonuclear war', causing panic at NORAD (North American Aerospace Defense Command) who interpret it as a real nuclear assault. The films plays on many fears, including of the USSR, mutually assured destruction (MAD), artificial intelligence and children knowing more about computers than adults, but it warns especially about the dangers of 'hacking'. Relaxing at Camp David, President Reagan watched the film on 4 June 1983. Troubled by what he'd seen, four days later he interrupted a meeting to ask the chair of the Joint Chiefs of Staff, General John Vessey, if it could happen. After investigating the problem, Vessey was forced to report, 'Mr. President, the problem is much worse than you think' (Kaplan, 2016:2).

The problem of computer security had been recognized early. In April 1967, as ARPA was designing the first computer network, RAND computer scientist Willis Ware delivered his conference paper 'Security and Privacy in Computer Systems', which highlighted the dangers of resource-shared computing, especially for military and defense systems, arguing that 'deliberate attempts to penetrate such computer systems must be anticipated' (Ware, 1967:1). In October 1967, he participated in an ARPA task-force, organized under the DoD Defense Science Board, to study and make recommendations regarding computer security. Published in February 1970 and mostly written by Ware, the final report, 'Security Controls for Computer Systems', expanded upon his fears, focusing especially on the threat of 'deliberate penetration', including 'active infiltration' of systems and 'passive subversion' tapping communication lines (Ware, 1979: Part A, Section III). On 29 October 1969, however, the first nodes of the ARPAnet had been connected. With ARPAnet scientists primarily concerned with its successful operation and expansion, the security issues raised by Ware were largely ignored. In 1980, however, Ware was approached at RAND by two script-writers, Lawrence Lasker and Walter F. Parkes, who wanted advice on their new film *WarGames*, specifically about what a hacker would be able to do. Ware told them what was possible.

Vessey's report to Reagan about US vulnerabilities was based on the fact that the Cold War enemies had a long history of intelligence activities against each other's communications and were already inside each other's systems. Moreover some, such as Pentagon scientist William Perry, had already realized communications could be actively disrupted or interfered with, in an early recognition of what would become 'information warfare'. In response, on 17 September 1984 Reagan signed National Security Decision Directive (NSDD) –145 entitled 'National Policy on Telecommunications and Automated Information Systems Security': the first US attempt to secure its computing systems. The project failed, however, when its attempt to give the NSA the role of overseeing computing infrastructure led to a civil liberties backlash.

Around the same time as *WarGames*, the concept of 'cyberspace' was emerging. The idea of spatial, navigable, experiential, electronic, virtual realities had appeared in 1960s sci-fi, in the work of Philip K. Dick and Daniel Galouye especially, and the idea reappeared in the early 1980s in Vernor Vinge's 1981 novella 'True Names' and the 1982 film *Tron*. Inspired by arcade games and the Walkman, William Gibson coined the term 'cyberspace' in his 1982 short story 'Burning Chrome', employing the idea at length in his 1984 novel *Neuromancer*. Gibson imagined an electronic realm called 'the matrix' – 'a graphic representation of data abstracted from the banks of every computer in the human system' – into which the 'disembodied consciousness' could be projected (Gibson, 1995:11; 67; 12). Over the next decade the 'cyberpunk' genre would flourish. Central to almost all of its stories were hackers, electronic penetration and disruption, cybercrime and attacks on computing systems.

The real world was slower to understand the possibilities of connected computing. One wake-up call was the 'Cuckoo's Egg' in 1986 when Clifford Stoll at the Lawrence National Laboratory traced a minor system intrusion back to a West German hacker, Markus Hess, who was accessing US military systems to sell information about Reagan's Strategic Defense Initiative' (SDI) to the KGB. The 'Morris Worm' was another warning. Written by Cornell University graduate student Robert Morris and launched onto the internet on 2 November 1988 the worm crashed an estimated 10 percent of the global network. Morris was the first person prosecuted under the USA's 1986 Computer Fraud and Abuse Act (CFAA), itself introduced in response to the hacking fears following *WarGames*.

By the late 1980s the USA had developed the idea of 'Command and Control Warfare' (C2W) and they used it in the 1991 Gulf War, penetrating Saddam's C+C networks and using satellites to acquire microwave signals so they knew all his command instructions. The political and military authorities, however, were more interested in physical, kinetic abilities, so this remained an underdeveloped area. The NSA Director, Mike McConnell, was more receptive to the idea, being inspired especially by a speech in the 1992 film *Sneakers*

> The world isn't run by weapons anymore, or energy, or money. It's run by little ones and zeroes, little bits of data. It's all just electrons ... There's a war out there, old friend. A world war. And it's not about who's got the most bullets. It's about who controls the information. What we see and hear, how we work, what we think ... it's all about the information!
>
> Kaplan (2016:31)

The script was by Lasker and Parkes, the writers of *WarGames*.

As we saw in chapter 3, the early 1990s were a time when the US military was actively re-theorizing warfare to include computing and the concept of information. Arquilla and Ronfeldt's 1993 essay 'Cyberwar is Coming' and Alvin and Heidi Toffler's *War and Anti-War* were highly influential and the concept of 'information warfare' (IW), first officially introduced in a 1992 DoD Directive, was beginning to take off. In the following years the Air Force, Army and Navy would each develop their own IW units and doctrines, in an attempt to control this new field. The USAF was first, transforming its Electronic Warfare Centre into the Air Force Information Warfare Centre in 1993 and establishing 'the military's first cybercombat unit' (Zetter, 2014:207), the 609 Information Warfare Squadron, in 1995. Although policy documents such as the USAF's 1995 white paper *Cornerstones of Information Warfare* discussed both informational attack and defence, the latter was taken less seriously, despite hacker intrusions into DoD sites from April–May 1991 and into the USAF Rome Laboratory from March–May 1994. Ultimately, however, it was an act of physical terrorism – Timothy McVeigh's bombing of federal offices in Oklahoma City on 19 April 1995, killing 168 people – that next foregrounded the issue of cybersecurity.

Following the Oklahoma bomb, President Clinton signed PDD-39, 'US Policy on Counterterrorism', which established a committee to investigate the protection of 'critical infrastructure' (CI). Over half the resulting report focused on computer security and the need for cyber-protection of the eight, key, networked CI assets they identified:

telecommunications; electrical power, gas and oil; banking and finance; transport; water supply; the emergency services; and government systems. Their report led to Executive Order 13010 on 15 July 1996, which established the President's Commission on Critical Infrastructure Protection in 1996. The resulting Marsh Report, published on 13 October 1997, warned of the dangers to CI from hackers, identified private sector ownership of CI as a key vulnerability and called for a massive effort to protect the USA. The report concluded, 'We should attend to our critical foundations before we are confronted with a crisis, not after. Waiting for disaster would prove as expensive as it would be irresponsible' (Zetter, 2014:136).

Military interest in cyberwar was also increasing. On 9 June 1997, the USA launched a cyberwar exercise called 'Eligible Receiver', tasking an NSA 'red team' to infiltrate DoD networks using only commercially available technologies. Intrusion turned out to be 'absurdly easy' (Kaplan, 2016:69). The two-week test was over within four days, as the NSA team penetrated the entire defense establishment network, leaving markers to demonstrate their access and even interfering with communications: 'They intercepted and altered communications, sent false emails, deleted files and reformatted hard-drives' (Kaplan, 2016:69). Eligible Receiver was another proof-of-concept of both the possibilities of penetration and of actual disruption and damage.

Meanwhile, external hacking continued. The USA had initially believed the 1994 Rome Lab incursions were an act of 'infowar' from a foreign power, until they traced the penetration back to a house in North London and arrested 16-year-old Richard Pryce, alias the 'Datastream Cowboy', on 12 May 1994. Similarly, on 21 June 1996, 21-year-old Cardiff hacker Mathew Bevan, alias 'Kuji', was arrested for hacking into Griffiss AFB Research Laboratory in New York, in search of UFO files. Then on 3 February 1998, the Airforce Information Warfare Centre detected intrusions at several bases. which they codenamed 'Solar Sunrise'. The US government thought it might be an Iraqi attack, conducted in response to the USA's contemporary military strikes, until it was traced to two 16-year-olds in San Francisco, 'Makaveli' and 'Stimpy', who (along with an 18-year-old Israeli hacker, 'The Analyzer') had been competing to hack the Pentagon to show off their power. *WarGames* had become real.

If children could do this, then so too could a nation state. This was immediately confirmed with 'Moonlight Maze', a codename given to a series of sophisticated intrusions into military facilities, first detected in March 1998, hunting for specific information on aircraft design and microchips. A phony 'stealth aircraft' website was used as a 'honey pot', installing a bug that enabled the USA to track the attacks to Russia – an attribution inadvertently confirmed by a Russian General when a US delegation visited Moscow to investigate in April. 'Moonlight Maze' was the first recognized nation state attack and had no official end date as the hacking continued long after its discovery.

Although the USA established a new group to oversee military security, Joint Task Force – Computer Network Defense (JTF-CND), in operation from 10 December 1998, the problem of securing the critical infrastructure remained intractable. Clinton's attempts, including the PDD-63 'Critical Infrastructure Protection' on 22 May 1998, and the *National Plan for Information Systems Protection* published on 11 February 2000, floundered due to private companies' hostility to government regulation, their reluctance to spend on security, the military's preference for spending on physical assets, the government's lack of willingness to assert control and civil liberties fears over government intervention.

The USA's offensive cyber-capabilities were next employed in the late 1990s against Serbia. The Pentagon's cyberwarfare Unit J-39 responded to Serb-TV's organization of public demonstrations against NATO in 1997 by turning off the five key transmitters that served 85 percent of the population whenever TV urged the public to protest, and during the 1999 Kosovo War J-39 hacked into the Serbian phone network which ran the air defense lines and military telecommunications, using them to feed false information to the air defences so that they missed low-flying aircraft. The USA also employed a range of cyber-operations to sow mistrust among Milošević's forces and even engaged in prank calls to his home to annoy him and his family as part of a broader info- and psyops. The USA considered the campaign a success, although limited the tools they used, partially due to legal concerns (such as over hacking into Milosovic's bank accounts), but also to prevent disclosure of its capabilities or an equivalent response back. Although the physical informational war against Serbia's command, control and communications infrastructure (including roads, bridges and energy stations) was effective, the cyber-operations weren't fully integrated into the military plans and weren't valued by the commanders. As Kaplan concludes, 'few of America's senior officers evinced the slightest interest in the technology's possibilities' (Kaplan, 2016:119).

On 1 April 2000, the USA began reorganizing its cyber-capabilities, changing JTF-CND to JTF-CNO – 'computer network operations' – to explicitly include offensive capabilities under the US Space Command. The new NSA head Michael Hayden was unhappy, however, at the proliferation of IW and cyberwarfare units and wanted the NSA to become the primary cyber-organization, combining espionage, security and warfare. In 2000, he reorganized the NSA into three divisions, 'Global Response', dealing with everyday political issues, 'Global Network', to exploit digital communications, and 'Tailored Access Operations' (TAO), a secret, elite hacking force that would carry out CNE ('computer network exploitation'), including global penetration, espionage and cyberattacks using stockpiled 'zero day' exploits. The incoming President, George W. Bush, showed only a limited interest in cybersecurity. His 16 October 2001 Executive Order 13231, 'Critical Infrastructure Protection in the Information Age', established a White House office headed by Richard A. Clarke, who oversaw *The National Strategy to Secure Cyberspace* (signed by Bush on 14 February 2003), but he later chose not to replace Clarke and the issue fell down the agenda again. 9/11 appeared to reinforce this, as cyberattacks suddenly seemed less important than physical ones, but in reality, the terrorist attacks led to the secret expansion of the NSA and their cyber-capabilities.

Following revelations about the abuse of surveillance by the intelligence agencies in the 1970s, the 1978 Foreign Intelligence Surveillance Act (FISA) placed restrictions on domestic spying on US citizens. Now, realizing that 80 percent of the world's digital communications passed through US territory, the NSA wanted to access these communications, including analysing domestic communications and 'contact-chaining' US-linked phone numbers. After 9/11 they began operating under an 'expansive' interpretation of what was legal, but an executive order signed by Bush on 4 October 2001 gave them more powers, legalizing the bulk collection of phone and internet 'metadata' and the right to spy on communications where one node was domestic. The NSA set up the Metadata Analysis Centre (MAC), initially code-named 'Starburst' but renamed 'Stellar Wind' on 31 October, and from then, warrantless mass surveillance took off. An NSA 'Special Operations Group' worked with US companies, setting up surveillance equipment at their facilities and installing it on their networks and servers, forming a powerful state-private partnership with global reach.

The USA passed a series of acts to legalize the NSA activities, including the October 2001 Patriot Act, the August 2007 Protect America Act (amending FISA) and the July 2008 FISA Amendments act. The latter acts legalized bulk collection and provided cover for the participating companies, allowing the development of new surveillance programmes such as 'Prism' that would later be exposed by Edward Snowden. This surveillance was closely connected to CNE as Snowden also revealed the NSA was hacking into undersea cables, stealing data passing to the companies' data centres and hacking into US and foreign companies through its TAO unit. Surveillance, therefore, was only part of a broader NSA CNE-capability that formed a central part of the USA's informational and cyber warfare systems. It would be offensively deployed in this role in Iraq.

The USA had compromised Iraq's secure military network prior to the 2003 invasion, using it to send emails to military officers, encouraging them not to resist US forces. Cyber-operations were limited, however, until they realized they needed better strategies to defeat the post-invasion insurgency. When General Keith Alexander became head of the NSA on 1 August 2005, he upgraded the NSA systems with the 'Turbulence' programme, part of which, the 'Real-Time Regional Gateway' (RTRG), could be of value in Iraq. As Harris explains:

> With access to the telecommunications networks running in and out of Iraq, the NSA began scooping up and storing every phone call, text message, and email sent in and out of the country. It was a key pillar of the new strategy: collect all the data, then use it to map out the networks of terrorists and insurgents.
>
> Harris (2014:14–15)

Alexander deployed 6,000 NSA staff to Iraq, using the RTRG to pull real-time intelligence from hacked and intercepted insurgent communications. This cut the time from collecting to acting on intelligence 'from sixteen hours to *one minute*' (Kaplan, 2016:159), enabling special forces to act immediately and, by surprising insurgents, collect more equipment to analyse.

Again, surveillance and CNE were linked. The NSA manipulated enemy communications, sending texts to insurgent fighters and roadside bombers leading to their capture or killing, and infiltrated 'Obelisk', the US name for AQI's network of servers and websites, spreading malware to gather more intelligence and spread anti-AQ propaganda. They were able to target and kill individual propagandists by locating their IP addresses and computers and developed the tool 'Polarbreeze' to tap wirelessly into nearby computers in internet cafes. Combined with the 'surge' and the 'awakening', the cyber-operations were a huge success. Harris concludes: 'This was the most sophisticated global tracking system ever devised, and it worked with lethal efficiency' (Harris, 2014:23).

'The age of cyberwar'

In a September 2007 speech, Michael Wynne, the secretary of the USAF, announced, 'Tell the nation that the age of cyberwarfare is here' (Graham, 2008). In addition to US activities in Iraq, Israel was also demonstrating its capacities. On 6 September 2007, it launched 'Operation Orchard', an airstrike on an unfinished Syrian nuclear reactor. The strike was prepared by Israel's cyberwarfare group, Unit 8200, who hacked the Syrian air defences to feed false signals to the radar operators, enabling the Israeli planes to strike without response. Although the USA and Israel's actions were secret, cyberwar had already become public that April when Estonia suffered 'Web War One'.

The night of 26–27 April 2007 saw the beginning of several days of rioting in Estonia, with protesting ethnic Russians clashing with security forces over the government's decision to move 'The Bronze Soldier of Tallinn', a Red Army soldier statue commemorating the Russian dead of World War II. On the 27th, as the riots began, botnets launched a series of distributed denial of service (DDoS) cyberattacks at Estonian targets, including the parliament, ministries, banks, newspapers and broadcasters, impacting upon government and military communications, public records, bank accounts, the mass media, shops and telephony. DDoS attacks are a simple, but effective, disruptive technique, overloading servers with requests and causing computer systems to crash. Lasting until May, and hitting one of the most 'wired' countries in the world, the attacks involved over a million computers and beyond publicly known websites also hit servers running the telephone network, credit card verification system and the internet directory, demonstrating considerable sophistication. Western experts came to defend Estonia and the DDoS forced a decision from NATO that the attacks didn't trigger a military response under the treaty obligations.

Estonian researchers claimed the command and control servers were in Russia and that the computer code had been written on Cyrillic-alphabet keyboards, but there was no final proof for this attribution. Although Russian involvement seemed obvious, there was a dispute over the extent of state involvement or knowledge. Patriotic hackers or criminal groups such as the Russian Business Network (RBN) were also suggested as responsible, although both could have been acting under the direction or with the support of the Russian intelligence agencies. One ethnic Russian was found guilty of the cyberattacks in Estonia in January 2008, whilst in March 2009 a Commissar of the pro-Kremlin Nashi movement claimed to have organized the attacks with sympathizers, pointing out that simply sending internet requests to servers wasn't illegal. Few entirely believe these claims and ultimately Clarke's conclusion is the best we have: 'Did the Russian government security ministries engage in cyberattacks on Estonia? Perhaps that is not the right question. Did they suggest the attacks, facilitate them, refuse to investigate or punish them? And in the end, does the distinction really matter?' (Clarke and Knake, 2010:16).

Russia was suspected again in June–July 2008 when Lithuania suffered a series of web defacements after a ban on Soviet symbols. Over three hundred private and official Lithuanian websites were hit, with hackers posting the Soviet hammer-and-sickle and five-pointed star as well as profane messages. More significant were the cyberattacks that accompanied the 2008 Russo-Georgian War; a conflict centred upon Russian support for the two breakaway Georgian provinces, Abkhazia and South Ossetia. After a Georgian-separatist ceasefire broke down, Russian troops began engaging the Georgians on 7 August. Russia advanced into Georgia on the 10th, overwhelming its forces, and a ceasefire on the 12th effectively left the provinces independent, under Russian control. What was important here was that Russia launched simultaneous air, ground, naval and cyber military operations.

The DDoS attacks began before the ground campaign (demonstrating foreknowledge of it), crashing the Georgian president's website from 20 July. The attacks escalated when the ground war began, with botnets linked to Russian criminal groups taking down government and mass-media websites and disrupting official communications. A second phase of attacks went further, hitting an expanded target list that included financial institutions, businesses, educational institutions and western media, as well as involving web defacements and spam email campaigns. The overload to the banking system triggered a protective mechanism that cut Georgia off from the world financially. 'Without access to European settlement systems, Georgia's banking operations were paralyzed. Credit card systems went down as well, followed soon after by the mobile phone system' (Clarke and Knake, 2010:20).

The cyberattacks acted as a force multiplier for the ground attacks, being intended, Kaplan says, 'to confuse, bewilder or disorient the enemy and thus weaken, delay, or destroy his ability to respond to a military attack' (Kaplan, 2016:164). But they also played a significant psyops function, silencing and isolating Georgia, demonstrating to Georgians that their government wasn't in control of the country's communications and financial systems, and taking the war to every individual, impacting on their everyday life. There were attempts at defence, blocking traffic from Russia and hosting Georgian websites in other countries (such as the USA, Estonia and Poland), but the attacking botnets used servers from around the world, including Canada, Turkey and even Estonia, making defence difficult. Again, Russia was clearly responsible, although final attribution remained difficult and although western computer scientists linked the attacks to the Russian intelligence apparatus, the Kremlin refused responsibility, describing them as a populist reaction.

More Russian-based, RBN-linked Botnet DDoS attacks were suffered by Kyrgyzstan from 18 January 2009, removing 80 percent of the country's internet capacity. There were numerous explanations for the attacks, most focusing on Kyrgyzstan hosting a US airbase as part of the War on Terror, but as relations with Russia were good at the time one of the most interesting explanations was that the Kyrgyzstan president had *requested* the attacks to remove the online organizational and propaganda capabilities of the Kyrgyzstan opposition.

Cyber-operations had become a useful tool for the Russian government, whether as a psychological weapon or as a military adjunct, although most commentators accept that they held back from more destructive attacks. The question remained, therefore, as to how dangerous cyberattacks could be. Investigating the possibility of a real-world impact, the USA carried out the 'Aurora test' at the Idaho National Laboratory on 4 March 2007, setting hackers the task of accessing and damaging a specified generator. Within three minutes, with 21 lines of malicious code delivered via the internet, the $1m, 27-ton generator was thrown out of synch and reduced to 'a smoldering, lifeless mess of metal and smoke' (Zetter, 2014:130). If the attackers hadn't paused to assess each phase of damage, the attack 'could have achieved its aim in just fifteen seconds' (Zetter, 2014:163). The hackers opened and closed the key circuit breaker so fast that the safety systems didn't see it: 'Workers in the operations centre who monitored the grid for anomalies and weren't told of the attack before it occurred never noticed anything amiss on their monitors' (Zetter, 2014:163–4).

Later demonstrations confirmed the possibility of remote damage. In 2009 researchers at Sandia National Lab showed how they could remotely cause components at an oil refinery to overheat, with potentially dangerous consequences. Some suggest this physically destructive capability has already been employed, claiming that in June 1982 – a year before *WarGames* – the CIA discovered the KGB had stolen software and a control system from a Canadian firm and used it in a Siberian oil pipeline. The CIA then installed a 'logic bomb' (malware) into the system so that at a certain point the system would break, leading to what Thomas C. Reed described in 2004 as 'the most monumental non-nuclear explosion and fire ever seen from space' (Zetter, 2014:199). The story is often repeated but most commentators suggest there's little proof: as Rid says, 'The available evidence on the event is so thin and questionable that it cannot be counted as a proven case of a successful logic bomb' (Rid, 2013:6). Others have defended Reed's government sources and the fact that in the 1980s the FBI, CIA and DoD did run a joint operation to sabotage software and hardware headed for the USSR after a leak of a USSR 'wish-list' of items as part of the 'Farewell Dossier' (codenamed after the Soviet agent who provided the information).

Either way, the Idaho test demonstrated the reality of the problem, pushing the issue of cyber defence and critical infrastructure protection to the fore again. Most CI, in the west, was in private hands, with companies resistant to regulation and to spending money on an abstract threat that reduced their profitability. Experts recognized, however, the increasing vulnerabilities caused by company's cost-cutting attempts, such as the installation of SCADA (Supervisory Control and Data Acquisition) systems. Whereas previously control systems were local, with on-site staff overseeing safety, the introduction of SCADA systems connected to the internet now allowed remote diagnostics. The benefit was being able to cut local staff and systems, but the connection to the internet introduced a new vulnerability to penetration. These control systems are now everywhere, underpinning our industries and infrastructure, and have also become more compatible, in increasingly running commercial software, and more discoverable, through online searches. Zetter points to industrial accidents, such as the August 2009 Sayano-Shushenskaya power plant disaster in Siberia that killed 75 people when a turbine broke, as evidence of 'the extent of damage a cyberattack *could* wreak' (Zetter, 2014:150).

There is evidence that such systems have already been hacked. In March–April 2000 a former worker sabotaged the pumps at a water treatment plant in Maroochy, Queensland, Australia, spilling over a million litres of raw sewage into local parks and rivers, 'in what is considered to be the first publicly reported case of an intentional control-systems hack' (Zetter, 2014:136). Vitek Boden, who was eventually jailed for two years, had worked for the company that installed the SCADA system and had the software to control it on his laptop and the knowledge of how to operate the radio-transmitting equipment to communicate with the system, allowing him to take control of the sewage pumps. After 9/11 US intelligence agencies gradually discovered the extent of terrorist interest in critical infrastructure. A 2002 CIA Directorate of Intelligence Memorandum acknowledged that al-Qaeda had 'far more interest' in cyberterrorism than previously believed (Zetter, 2014:141).

Following a warning about US vulnerabilities, Bush issued the National Security Presidential Directive NSPD-54 on 9 January 2008, which established the 'Comprehensive National Cybersecurity Initiative' (CNCI), designed to establish a front-line defence against intrusions and strengthen cybersecurity. Its focus on government and military systems meant, however, it didn't address the broader public CI vulnerabilities. Then on 24 October 2008 the NSA discovered a beacon inside the air-gapped US Central Command networks, a breach that 'was unprecedented in military and intelligence history' (Harris, 2014:147). Given the code name 'Operation Buckshot Yankee', the NSA established it was a Russian worm that had infiltrated the system through a Russian-made USB drive used in Afghanistan and managed to command it to stand down. Although it was later discovered that the worm was a variant of a three-year-old worm that was mostly harmless, the NSA's success in finding and dealing with it meant it was rewarded with the creation of a new unit, uniting all defensive and offensive capabilities.

First proposed in a memorandum on 23 June 2009, United States Cyber Command (USCYBERCOM) was activated on 21 May 2010 and became operational on 31 October, under the NSA Director General Alexander. USCYBERCOM would now be a 'sub-unified command' under STRATCOM (Strategic Command), which had run cyber-operations since 2002. The Army, Navy and Air Force would continue to have cyberwarfare units run by USCYBERCOM, although the protection of civilian infrastructure wasn't included in its remit, being the responsibility of the Department for Homeland Security (DHS). With their expertise in surveillance and CNE, the NSA was uniquely placed to take on this role. Its TAO unit were especially busy. According to the Snowden revelations, by 2013 the TAO had hacked into and implanted spying devices on at least 85,000 computer systems in 89 countries. TAO reportedly relies upon a stockpile of otherwise unknown 'zero day' vulnerabilities, either discovered by its own researchers, or bought on the cybersecurity 'grey market' from individual hacker-traders or from private companies such as Vupen. Harris describes the NSA as 'the single largest procurer of zero day exploits', claiming the NSA has a budget of 'more than $25m' to buy them (Harris, 2014:94; 100). Critics argue that by stockpiling zero days rather than allowing them to be patched, the NSA contributes towards civilian vulnerabilities.

Europe was also acting on cybersecurity. The European Network and Information Security Agency (ENISA) had been set up in 2004 and in May 2008 NATO established the Cooperative Cyber Defence Centre of Excellence (CCDCE) in Tallinn, Estonia. In March 2010, the UK set up the Office of Cybersecurity and Information Assurance (OCSIA) and that October's Strategic Defence and Security Review identified cyberse-curity as a 'Tier 1' threat, awarding OCSIA £650m at a time of widespread cuts. On 31 May 2011 the UK placed cyberwar on the same footing as conventional war, asserting the same rules, norms and logic applied to the domain and established a cyberwar operations centre at Government Communications Headquarters (GCHQ). In April 2013, it set up the Cybersecurity Information Sharing Partnership (CISP) for government and industry to work together on protection, and on 31 March 2014 the government launched the Computer Emergency Response Team (CERT-UK) to coordinate defence. In October 2016, the government created the National Cyber Security Centre (NCSC), which absorbed most existing cyber initiatives including CERT-UK.

In America, the new President, Obama, had emphasized cybersecurity during his campaign and on taking office launched a 60-day review that led to the *Cyberspace Policy Review* published 29 May 2009. Although Obama oversaw the creation of USCYBERCOM, once again no real plan emerged to protect civilian, private CI, due to the complexity of the problem and resistance to regulation. Obama's most significant decision was to reauthorize a secret, Bush-era project codenamed 'Operation Olympic Games': an attempt to sabotage Iran's nuclear programme. In August 2005 Iran defied IAEA inspections by resuming uranium enrichment at its Natanz plant. The USA wanted to dissuade Israel from an airstrike and couldn't itself intervene in another Muslim country after Afghanistan and Iraq, so the USCYBERCOM/NSA proposed a cyberattack. They had begun preparations in late 2005 and Bush authorized it in 2006 as a joint operation with Israel's Unit 8200, with the CIA overseeing the weapon's delivery.

They began by developing espionage malware to gather intelligence for the weapon. The NSA's TAO created 'Flame', a multipurpose spyware worm with 650,000 lines of code that could steal files, monitor keystrokes and screens, turn on the microphone and record conversations, and use Bluetooth to steal data from smartphones within 20 metres. Using 80 C+C servers it infected about 1,000 computers from at least December 2007. A second espionage tool, created by Unit 8200 and named 'Duqu' by researchers, was a RAT (a remote access Trojan, sometimes called a remote administration tool). It acted as a backdoor to computers, allowing other modules such as keyloggers to be downloaded to discover information about the Siemen's Programmable Logic Controllers (PLCs) used in Natanz and to steal security certificates.

The main cyberweapon, named 'Stuxnet' by later researchers, went through several phases of development and deployment. The attack code was tested in the USA at a covert facility created in 2005 at the Oak Ridge National Laboratory on identical centrifuges intercepted on the way to Libya, with other tests conducted at Dimona, Israel. An initial version of Stuxnet ('0.5') was introduced, perhaps as early as November 2007. Obama reauthorized the programme and greenlighted the more aggressive version that became Stuxnet that was unleashed on 22 June 2009, with further updated waves on 23 March 2010 and from 14 April 2010.

Stuxnet was a sophisticated worm, using an unprecedented four zero days. It worked by infecting systems and checking their configurations. Unless it found exactly what it was looking for it would do nothing other than look for other systems to infect. Its targets were highly specific Siemens industrial PLCs being used in a highly specific configuration at one single facility: Natanz. Natanz's systems were air-gapped hence the worm was launched at four key Iranian companies connected to the plant, infecting it through an infected USB or laptop. Once inside, it acted as a rootkit, lurking on the system, intercepting commands to the PLCs to change their operation and disabling digital automated alarms. First it spent 13 days on the system recording normal operations, then, at certain intervals, it began its sabotage, forcing the centrifuges to spin beyond their tolerance levels until they broke, whilst sending the false, recorded data back to fool the operators. The code was designed to follow a complex pattern to further confuse Iranian scientists.

By early 2010 around 2,000 out of 8,700 centrifuges had been damaged beyond repair and Iran's small supply of uranium had been depleted. Stuxnet, however, had spread too far, with over 100,000 infections in over 100 countries, being discovered by security researchers in July 2010. Researchers pieced together that Stuxnet was a state-created cyberweapon to attack Iran and went public in August. Within days Iran had severed all connections to Stuxnet's C+C servers. By then another 1,000 of the remaining 5,000 centrifuges had been rendered inoperative. A *New York Times* article on 15 January 2011 revealed Stuxnet was a joint US-Israel operation.

Stuxnet succeeded in setting back Iran's enrichment programme, although opinions vary as to how damaging it was. Zetter says that any effect it had wore off quickly, with the Iranians showing 'a remarkable ability to recover from any damages and delays that Stuxnet and other factors had meted out' (Zetter, 2014:361). It could have been designed to be more destructive and faster-acting, but that would have had a greater risk of repercussions. Zetter argues it's slow and stealthy attack 'made it harder to achieve more extensive results but also made it harder for Iran to make a case for striking back' (Zetter, 2014:365). Stuxnet did succeed in staving-off a military strike and delayed Iran's programme, leaving the door open for historical negotiations that began in 2013, but its real significance lay elsewhere. First, it was the first known state-created cyberweapon. Second, it was operational proof that a cyberweapon could have a real-world destructive effect. As such, Zetter says, 'It wasn't an evolution in malware but a revolution' (Zetter, 2014:125). Third, it was the first state attack on another state's critical infrastructure. As such it set a dangerous precedent, accelerating research into cyberweapons, promoting them as an acceptable option and provoking an Iranian counter-reaction.

Iran was hit again in April 2012 when its National Oil Company was struck by a hard-drive-wiping virus it called 'Wiper'. Researchers found links to Duqu and Stuxnet, and suggested this was a separate, later Israeli attack. By then Iran had created its own cyberwarfare unit to study Stuxnet. In August 2012, it launched the Shamoon virus at the US-Saudi oil company Saudi Aramco, wiping 30,000 hard-drives and planting the image of a burning US flag on every computer. Then in September, the 'Izz ad-Din al-Qassam cyber-fighters' – assumed to be a cover for the Iranian government – took six major US banks offline in a DDoS attack 'unprecedented in its scale and sophistication', being 'several times larger than what Russia had directed at computers in Estonia in 2007' (Harris, 2014:193). Iran was also responsible for a hacker attack on the Las Vegas Sands Corporation in February 2014 that wiped data, downloaded sensitive data and defaced the company website. The attack, destroying 20,000 computers (costing at least $40m to replace) was a direct response to the pro-Israeli, part-owner Sheldon Adelson's public suggestion to use nuclear bombs on Iran. Iranian activities have continued. In July 2017, Iranian hackers were discovered to have set up a female 'honeypot' to contact international scientists on LinkedIn, Facebook and WhatsApp, to recruit them for Iran or to infect their computers with the PupyRAT malware.

US activities have also continued. More US-created cyberweapons have emerged, including the banking Trojan 'Gauss' discovered in August 2012, designed to track Hezbollah money-laundering through Lebanon's banks, and 'Regin', an espionage tool discovered in November 2014 that Symantec called 'one of the most sophisticated pieces of malicious software ever seen' (Cellan-Jones, 2014). The USA has also become more open about its cyber-policy. The May 2011 US government report 'International Strategy for Cyberspace' suggested online attacks could constitute war, and in December 2011 Congress authorized the Pentagon to wage cyberwar, giving it the power to launch 'offensive strikes' as part of the 2012 funding bill. In October 2012 Obama signed PPD-20, 'US Cyber Operations Policy', which institutionalized cyberattacks as an integral tool of military operations and it was later revealed that Obama had ordered a list of potential international targets to be drawn up. In August 2017 President Trump elevated USCY-BERCOM to the same status as STRATCOM and the eight other unified commands. In early 2011, the USA had recognized cyberspace as the 'fifth domain' of war (after land, sea, air and space) and in June 2016 NATO followed suit, confirming that a cyberattack on one of its members would trigger an article 5 collective military retaliation.

China and North Korea

China is also emerging as an important cyberpower. The 1991 Gulf War was a wake-up call for the Chinese. Having built their defences upon overwhelming conventional forces, they were shocked by the USA's technical capacities. Military leaders referred to the Gulf War as 'the great transformation' (Clarke and Knake, 2010:49) and began to downsize their military and invest in new technology. Through the 1990s they closely followed US debates about the RMA and Information Warfare (IW), integrating them directly into the PLA. In the early 2000s, following a new military doctrine of 'Information Confrontation', Departments of 'Information Security Research' were set up in universities to train a new generation of cyber-experts and hackers and China actively protected its own cyber-space and explored how to integrate cyberattacks into its military operations. In November 2004 China published a White Paper on National Defence which declared: 'The PLA, aiming at building an informationalized force and winning an information war, deepens its reforms, dedicates itself to innovation, improves its quality and actively pushes forward the RMA with Chinese characteristics with informationalisation at its core' (Chinese Government, 2004). This Chinese concept of IW includes both physical attacks and electronic warfare against an enemy's communication, information and intelligence systems, and military deception, operational secrecy and psychological warfare.

After both the 1999 NATO bombing of the Chinese embassy in Belgrade on 7 May 1999 and the 1 April 2000 'Hainan Island Incident' (an air collision between the USA and China that killed a Chinese pilot), the Chinese hacked US military and political websites and mobilized thousands of net users to send emails and viruses. The DDoS attacks and web defacements were more symbolic than destructive, signalling China's intention to take cyberspace seriously as a domain of conflict and activity. This became clear in the following years when a number of significant global Chinese incursions were identified. These included systematic, large-scale attacks, first identified late 2004 and dating back to 2003, codenamed 'Titan Rain' by the USA. Here, Chinese military hackers infiltrated DoD and defence contractor systems as well as the Pentagon's unclassified networks, hitting hundreds of computers and stealing sensitive military data. Another operation, codenamed 'Ghostnet' by researchers, was discovered in March 2009. Chinese hackers had infiltrated governments and organizations at least 103 countries, taking over 1,200 computers in numerous embassies and spearphishing with a RAT (a remote access Trojan malware). A top target were offices relating to NGOs involved with the Tibetan issue.

In April 2009, Chinese hackers were reported to have infiltrated the US power grid, leaving behind software that could be disrupt services or damage equipment. That December, Google discovered Chinese hackers had gained access to Gmail accounts, including those used by human rights activists. They eventually uncovered evidence 'of one of the most extensive and far-reaching campaigns of cyber-espionage in US history' (Harris, 2014:172). Dubbed 'Aurora', it included the penetration of over 20 companies, including Symantec, Yahoo, Adobe, Northrop Grumman and Juniper Networks. Google went public in a blogpost on 12 January 2010, accusing the Chinese government of attacking their infrastructure and afterwards worked with the NSA to monitor and track intrusions into their systems.

There may be a range of Chinese hackers, including patriotic groups and hackers hired by private companies for intellectual property (IP) theft, but most commentators see the Chinese state as implicitly supporting or as responsible for the systematic activity. China's hacking had earned it the description of an 'Advanced Persistent Threat' (APT), but the USA was hesitant in publicly challenging China for diplomatic reasons. However, in a ground-breaking 18 February 2013 report, the US security firm Mandiant exposed and named the Chinese threat, identifying the PLA cyberwarfare Unit 61398, located in Pudong, Shanghai, as responsible for systematic hacking activities, including stealing data from 142 organizations since 2006.

Most Chinese hacking that we know about has been for espionage, rather than destruction. Although this suggests it belongs on the lower end of the military scale and doesn't count as 'cyberwar', the scale of the infiltration makes it an extremely serious phenomenon. Much of the espionage is for economic benefit. This has included hacking computers in Europe in December 2010 for advantages in G20 trade negotiations and in March 2011 prior to a Brussels summit, whilst in August 2011 McAfee discovered 'the biggest series' of cyberattacks to date, with the infiltration of 72 organizations, companies, governments and the UN (Finkle, 2011). This economic espionage has become so persistent that the NSA's General Alexander described the theft of US intellectual property

(IP) as 'the greatest transfer of wealth in history' (BBC, 2015). Other espionage is for political purposes, such as the infiltration of computers of the Australian PM and two senior ministers in March 20112, where hackers from the Chinese intelligence services accessed several thousand emails. The other target for China is valuable military data.

In 2009, it was revealed that Chinese hackers had stolen terabytes of data about the design and electronics systems of the $300bn Joint Fighter programme, the F-35 Lightning II, from the computer systems of private contractor companies. Unconcerned at retribution, the hackers took little care to hide their tracks. The design directly influenced China's stealth fighter, the Shenyang J-31, as well as the Chengdu J-20 fighter jet. The 'Byzantine Hades' attacks (as they were codenamed) also stole data on the B-2 stealth bomber, the F-22 jet, space-based lasers, missile navigation and tracking systems, and nuclear submarine and anti-air missile designs. A leaked DoD Powerpoint listed at least 30,000 associated hacking incidents, including over 500 significant intrusions into DoD systems with over 1,600 DoD computers penetrated and 600,000 user accounts compromised: 'The presentation makes the point of equating the amount of data extracted (50 terabytes) to be equal to five Libraries of Congress' (Gady, 2015). In July 2011 the Pentagon admitted another attack had occurred, stealing 24,000 files from the defence industry computer network, whilst in March 2012 the USA said Chinese hackers had gained 'full functional control' over key NASA computers in 2011, taking over the Jet Propulsion Laboratory computers and accounts, and gaining 'full system access' (BBC, 2012).

The USA has begun to react to this. On 14 May 2014 a US Grand Jury charged five Chinese members of PLA Unit 61398 with cyberespionage against US corporations, in a landmark judgement trying to bring members of a foreign government to trial. The US position was exposed as hypocritical, however, as the Snowden leaks proved the USA was engaging in exactly the same economic and political espionage. In March 2014, it was revealed that the NSA's TAO had breached China's servers to spy on the government and Chinese companies in 'Operation Shotgiant', and that the USA had been hacking into the Chinese telecommunications company (and competitor) Huawei since 2009, with one NSA document admitting, 'We currently have good access and so much data that we don't know what to do with it' (Der Spiegel, 2014). On 25 September 2015, at a conference at the White House, President Obama and Chinese President Xi Jinping agreed a cyber economic-espionage truce, saying they would not 'knowingly support' these practices and would abide by 'norms of behaviour in cyberspace' (BBC, 2015). The truce didn't, however, rule out the continuing political and military hacking.

North Korea is also known to have advanced cyberwar capacities. In July 2009, just prior to 'Cyber Storm', a US cyberwar exercise, North Korea activated a 40,000-computer botnet, sending a huge DDoS attack against US and South Korean government websites and international companies. From 4 to 9 July US websites were hit with up to 1m requests a second, taking down the web servers for the Treasury, Secret Service, Department of Homeland Security, Federal trade Commission and Department of Transport and also hitting the NASDAQ, New York Mercantile, New York Stock Exchange and *Washington Post*. A second wave on the 9th saw 30–60,000 computers launching attacks on South Korean sites, and a final wave on the 10 saw 166,000 computers in 74 countries flooding South Korean systems. The South Korean National Intelligence Service (NIS) claimed a North Korean hacker team, Unit 110, was responsible, punishing South Korea for agreeing to take part in Cyber Storm.

Opportunities for response are limited as North Korea has such a limited internet that it is 'almost as cut off from the virtual world as it is from the real one' (Clarke and Knake,

2010:27). Despite this, Clarke identifies four cyberwarfare units in North Korea and suggests another 600–1,000 cyberwarriors may be based in China, taking advantage of its superior networks. Attacks on South Korea are now common, including cyberattacks in October 2010, March 2011 and March 2013. On 25 June 2013, on the anniversary of the start of the 1950–53 Korean War, there were coordinated attacks on the website of the presidential office as well as on other official and media sites.

The most unusual incident, however, began on 24 November 2014 when a hacker group called the 'Guardians of Peace' leaked confidential data from Sony Pictures Entertainment (of Sony Corps). Having had access for up to a year, it's claimed that they destroyed 3,000 computers and 800 servers and stole over 100 terabytes of data, embarrassing Sony with the leak of emails, star salaries, copies of unreleased films and employee social security numbers. FireEye (who'd bought Mandiant) and the FBI identified the group as 'Dark-Seoul' who worked for North Korea from across Asia. This was information warfare: Sony was targeted because of its forthcoming film *The Interview*, which, according to North Korea, in including an attempt to assassinate Kim Jong Un, constituted 'sponsoring terrorism' and 'an act of war' (Beaumont-Thomas, 2014). Sony cancelled the film's release following threats of violence on 17 December, but reversed that decision after criticism by Obama on the 19th. On 23 December Sony released *The Interview* on YouTube as a pay-per-view, with the film making $15m between 24 and 27 December.

Although US security experts queried the North Korean attribution, in January 2015, the *New York Times* revealed why the US government was so certain North Korea was responsible: the NSA had penetrated their systems since 2010 and hence had an inside-view of the state's activities. It wouldn't be the last time North Korea targeted popular culture. In October 2017, its hackers hit a British TV company, Mammoth Screen, in response to their proposed Channel 4 North Korean drama about a kidnapped UK scientist, 'Opposite Number'.

On 27 October 2017, the UK security minister Ben Wallace said the government believed North Korea was responsible for the global 'WannaCry' attack the previous May. The ransomware affected 30,000 computers in 150 countries, locking their contents and demanding Bitcoin payment to release them. In the UK, it crippled 48 NHS Trust computing systems running older software. Commentators attributed the malware to Unit 180 of North Korea's Reconnaissance General Bureau (RGB), their primary intelligence and clandestine operations agency, whose members often travel abroad for better connections and to cover their traces, and suggested that the intention was either anti-western disruption or to make money. The latter isn't so far-fetched, as North Korea has also been implicated in the February 2016 Bangladesh Bank heist when the Federal Reserve Bank of New York received SWIFT instructions to transfer $951m from a Bangladesh Bank account held there to fake accounts around the world. The New York bank stopped many transactions but $81m was lost when it was transferred to the Philippines.

The precise attribution is complex, however, as other security experts attribute both the WannaCry and Bangladesh Bank attacks to the 'Lazarus' cybercrime group, who they say is a North Korean, state-backed entity. There are claims that it was also behind many of the anti-South Korean DDoS attacks and the Sony Pictures attack (as both the 'Guardians of Peace' and 'DarkSeoul') as well as bank heists in Poland, the Philippines, Vietnam, Ecuador, Mexico and Taiwan. Ultimately, however, little is known about the composition of the group, or their actual relations with the North Korean authorities. Few, however, think Lazarus could operate without state support or a close relationship with the intelligence services, especially Unit 180 of which it is probably part. North Korea's aggressive cyber-operations continue. In May 2017, a panel of UN experts investigating violations of international sanctions on North Korea were hit by a sustained cyberattack, and in June 2017 the UK parliament was hit by a systematic attack attributed to North Korea, trying to access MP and staff email accounts.

Evaluating 'cyberwar'

As we have seen, 'cyberwar' is now an established and accepted concept, being employed by governments, enshrined in military policy, discussed by commentators and journalists, and referenced in popular culture. Richard Clarke, for example, is clear, arguing 'cyberwar is real', as advanced nations possess the ability to 'devastate a modern nation' at the speed of light; 'cyberwar is global', able to employ computer systems from around the world to strike against CI deep inside an enemy's country; and 'cyberwar has begun', with nations currently engaged in ongoing operations that blur the distinction of peace and war (Clarke and Knake, 2010:30–1). As I suggested in the chapter's introduction, however, definitions and claims of cyberwar are not so straightforward as the term doesn't easily fit into existing conflict categories.

First, there are problems with the term 'cyberattack', which is used for a wide-range of phenomena, from DDoS, to virus-laced emails, penetration for espionage or theft, web-defacements and cyberweaponry with real-world impact. However, does a successful DDoS that crashes a server really constitute an 'attack'? Perhaps if enough servers across a country are taken down we might accept that it does, but where is that threshold? Similarly, is sending spearphishing emails an 'attack'? If the emails aren't opened or are dealt with, is it still an 'attack'? How many, therefore, need to be sent and opened and to have an effect for it to constitute an 'attack'? Even penetration is complex, as access may only be gained to certain levels or accounts and we can ask again at what level of seriousness would this constitute an 'attack'?

Espionage has been an activity every government has engaged in and, although taken seriously, with repercussions for state relationships, it has rarely been considered an explicit 'attack'. But espionage today is 'easier, cheaper, more successful and has fewer consequences than traditional espionage', making it 'fundamentally and qualitatively different from what has gone before' (Clarke and Knake, 2010:232). At a certain point, therefore, should it be considered an 'attack', and if so, at what threshold? Even hacking to damage systems is ambiguous. Would implanting dormant malware constitute an 'attack', or only a 'potential attack'? Are web-defacements an 'attack'? Is there a point when their volume and content might be considered aggressive enough? Although cyberweaponry causing real-world damage represents the most obvious example of an 'attack', again we could ask what level of physical damage, and to what targets, would we consider serious enough to necessitate a state-level response?

Second, there is a temporal issue with cyberattacks. Unlike traditional, kinetic attacks, which have a defined temporal existence, usually with clear starting and end points, 'cyberattacks' can happen near-continuously, becoming so ubiquitous as to be normalized. If, therefore, they have been happening for so long, it is difficult to claim a particular start date and a certain point when they become unacceptable and require a response.

Third, there remains an 'attribution problem' with cyberattacks. Although digital forensics are improving, there are circumstances when attribution is easier (if, for example, you are inside an enemy's systems), and many attacks have blindingly obvious political origins, the problem of proof nevertheless remains. It is broadly accepted that even if you can trace an attack back to a specific address it could still be being used as a proxy by another system. Plus, you can't prove who was actually sitting at the computer, their affiliation or their motivation, with the range of possible actors today and the ambiguity of their relationship to the state blurring simple identifications. All of this adds up to 'plausible deniability' for states engaging in cyber-operations.

Even if attribution is secure, the fourth problem is how to respond to cyberattacks. What level of scale, temporality, access, espionage or electronic or physical damage is sufficient to trigger a state response? The March 2013 Tallinn manual on cyberwar described Stuxnet as 'an act of force', most likely against international law, although it was unsure whether it constituted an 'armed attack' that would justify an armed reaction by Iran (Zetter, 2013). Hence Rid's claim that, to date, 'there is no cyber attack, not even the over-cited Stuxnet, which unequivocally represents an act of war on its own' (Rid, 2013:166). Should an attack ever hit that threshold, then the next problem is how should a state respond. Should it respond in kind, using CNA, or, at some point, would a physical, military response be appropriate? Arguably, causing real-world damage would appear to justify a real-world kinetic response, but it didn't with Stuxnet and there is no agreement as to the level of damage required or what would constitute a commensurate military reply.

Researchers point out that a key problem complicating 'cyberwar' is that offence is always stronger than defence. The January 2013, US Defense Science Board taskforce report, 'Resilient Military Systems and the Advanced Cyber Threat', for example, concluded that there was no reliable defence against a resourceful, dedicated cyber-attacker, returning us to Ware's 1967 conclusion that networked systems are inherently insecure. One consequence may be that cyberattacks simply have to be accepted, leading to a higher threshold of response. Indeed, given the scale of global 'attacks' today, this could be the reality in which we are already living.

Another conclusion from this is that more emphasis needs to be put on deterrence, but this brings us to the fifth problem. Unlike conventional weaponry, cyber-capacities are secret: states don't advertise their ownership, don't admit their capabilities and don't hold parades of them for the international community. Their value is lost when they are exposed or used, hence cyberweapons will only be employed when a state believes their loss is worth it. As such, beyond the hypothetical threat of wondering what a state *might* do, cyberweapons have no real deterrent value. As Clarke admits, for all the USA's cyber-capacities, 'Other nations are so undeterred that they are regularly hacking into our networks', hence 'deterrence theory plays no significant role in stopping cyberwar today' (Clarke and Knake, 2010:195).

The secrecy, ease of concealment and problem of even measuring cyber-strengths also means that cyberwar is unregulable. No arms controls treaties are possible or enforceable and there are currently few international bodies or agreements attempting to prohibit the sale of spyware, hacking tools or zero-days. The Wassenaar Arrangement, a 41-nation organization trying to control 'dual use' technologies, has attempted to regulate the sale of cyberweapons, but it has had a limited effect and was not designed to curb their use by governments. In August 2017, it was revealed that 13 years of UN negotiations aimed at restricting cyberwar had collapsed in June following disagreements over the right to self-defence, with Russia, China and Cuba especially rejecting the idea of punitive attacks by states claiming to be victims of cyberwar.

Another issue around cyberwar is whether we should be worried at all. Compared to recent small-scale terrorist attacks in Europe involving trucks, vans and knives, cyberwar doesn't appear that dangerous. Thomas Rid's *Cyberwar Will Not Take Place* (2013) provides the best expression of this view. As he explains, 'Cyberwar has never happened in the past, it does not occur in the present, and it is highly unlikely that it will disturb our future. Instead the opposite is taking place: a computer-enabled assault on violence itself' (Rid, 2013:xiv). What we are seeing, he argues, is a reduction in conflict: sabotage, espionage and subversion aren't new, but what characterizes their contemporary cyber-forms is that they occur with less violence and with less chance of causing war. Hence, he suggests, 'Cyberattacks help to diminish rather than accentuate political violence' (Rid, 2013:xiv) as most attacks are non-violent or only violent non-directly, As Rid says: 'No cyber offense has ever caused the loss of human life. No cyber offense has ever injured a person. No cyber attack has ever seriously damaged a building' (Rid, 2013:166).

Rid accepts the psychological impact of cyber-weapons and agrees that they have the potential to cause loss of life, but argues that they exist on a spectrum with low-level weapons, being more like paintballs, with serious, high-level weapons being rarer, highly crafted and very specific in their targets. Ultimately, for Rid, the concept of 'cyberwar', 'has more metaphorical than descriptive value', like the 'war on obesity' (Rid, 2013:9). Security researcher Cris Thomas offers support for Rid here. His 'Cyber Squirrel 1' project was set up to counteract 'the ludicrousness of cyberwar claims by people at high levels in government and industry', demonstrating that, to date, the real threat to CI isn't cyberweaponry but animals. From 2013 to 2017, he says, squirrels, birds, rats and snakes were responsible for 1,700 power cuts affecting 5 million people. In 2013, for example, jellyfish managed to shut down a Swedish power station by clogging pipes carrying cold water to the turbines (BBC, 2017).

These are important points, but they risk underplaying the threat of cyberwar, which is increasingly demonstrating a real-world impact. Since Stuxnet there have been other examples of physical damage from cyber-intrusions. In January 2015 Germany's Federal Office for Information Security revealed hackers had infiltrated a steel mill through a spearphishing attack on their business network, working their way into the production networks to access systems controlling plant equipment. Their actions meant the plant was 'unable to shut down a blast furnace in a regulated manner' resulting in 'massive damage to the system' (Zetter, 2015).

In another incident on 23 December 2015, hackers using the BlackEnergy Trojan gained access to at least three regional power authorities in Ukraine, taking over the control systems to disconnect electrical substations, causing a power failure that left 250,000 homes without electricity, including in Kiev. The first proven blackout caused by a cyberattack, Symantec blamed it, and an attack on a Ukrainian media company, upon a group called Sandworm. Ukraine's intelligence agency, the SBU, blamed state-backed Russian hackers, following their ongoing military conflict over Crimea and eastern Ukraine. There was another outage on 17 December 2016, involving the Industroyer malware (sometimes called 'Crash Override'), which, like Stuxnet, was created specifically to target industrial controllers, being able to take control of them or to damage them, rendering them unbootable. Commentators consider the 2016 attacks as a Russian test for malware able to damage electrical power stations as well as other CI.

The west's vulnerability is also being increasingly recognized. The USA has publicly warned about the extent of (probably Chinese) hacker infiltration of the US energy system, power stations and networks, and in July 2017 GCHQ's NCSC warned that state-based hackers had compromised control systems in the UK in the energy sectors as well as engineering, industrial control and water sector companies. On 6 September 2017, Symantec warned that a hacking group called 'Dragonfly' was targeting the European and US energy sectors. Operating since 2011, they had gone dark in 2014 having been discovered placing backdoors in European and US power plant industrial control systems, before returning in 2015 (as 'Dragonfly 2.0') and penetrating facilities in the USA, Turkey and Switzerland. Symantec were unable to identify those behind the attacks, but warned that 'the group now potentially has the ability to sabotage or gain control of these systems should it decide to do so' (Hern, 2017a). On 22 September 2017, Ian Levy, a Technical Director at GCHQ's NCSC warned, 'sometime in the next few years we're going to have our first category one cyber-incident' (Hern, 2017b).

Ultimately, cyberwar is important because of this risk to civilian life. Whereas international laws of war prohibit deliberately targeting civilian sectors, these are precisely what cyberwar now threatens. The scale of what can be targeted is also significant. The USA recognizes sixteen CI sectors: chemical; commercial facilities; communications; critical manufacturing; dams; the defence industrial base; emergency services; financial services; food and agriculture; government facilities; health care and public health; information technology; nuclear reactors, materials and waste; transportation; and water and wastewater systems. Successful attacks on any of these – or, in a worst-case scenario, on many simultaneously – could cause chaos.

Every increase in networking increases our vulnerability. Mark Devost warned of this in his 1996 paper, 'Information Terrorism: Can You Trust Your Toaster?', and the contemporary development of 'the internet of things' – of 'smart', networked, household objects – realizes his fears of a world of objects that could be hacked and controlled. White-hat hackers have already demonstrated that 'smart' thermostats can be locked with ransomware, raising the possibility of hackers having full control of domestic heating systems, and potentially able to damage them as they can power-plant generators. But every networked object isn't just a potential target, it's also a potential weapon. In January 2014 hackers warned they had been able to take control of a 'smart' refrigerator and use it to send 750,000 virus-laced emails. The same risks apply to the self-driving cars in which many governments and companies are currently investing. Technology experts have already raised the possibility of hackers taking control of these to carry out terrorist attacks. Tomorrow's cyberwar attacks, therefore, may come from our own fridges, heating systems, home-assistants and cars.

What if, however, the military are also as vulnerable to attack? The most remarkable vision of the possibilities of cyberwar was published two years before *WarGames* in Vernor Vinge's 1981 story 'True Names' (Vinge, 2001). Here, working for the government, two hackers, 'Mr Slippery' and 'Erythrina' chase another entity, 'the Mailman', through cyberspace, with each side grabbing computational power and resources to use against each other in an escalating digital war. In a remarkable section of the novel, they enter the military systems, employing military forces against each other in the real world, intercepting helicopter payloads, taking over military satellite lasers and detonating warheads in their silos. This is clearly science fiction, but so too for many commentators was the very idea of real-world cyber-damage before Stuxnet, and even after Stuxnet this threat is still only grudgingly accepted. As in 1983, however, the future may owe more to science fiction than we think.

Ambient cyberwar

On 1 June 2017, Guillaume Poupard, the Director General of the National Cybersecurity Agency of France (ANSSI), warned the world was heading towards a 'permanent war' in cyberspace, with nations getting closer to 'a state of war that could be more complicated

... than those we've known till now' (Dearden, 2017). He was only partially right, as it isn't certain the concept of war even applies anymore here. International law recognizes only international armed conflicts (wars between states) and non-international armed conflicts (or civil wars). Non-state actors have rarely been considered capable of 'war', although the USA has argued that al-Qaeda and Islamic State qualify and hence are in 'armed conflict' with the USA. As we have seen, digital technologies are transforming traditional concepts of war and this is especially true with cyberwar which is today conducted by a wide range of non-state actors, including terrorist groups, hacking groups of varying affiliations and levels of state sponsorship, and even state-supported and employed criminal groups.

Hence the Russian DDoS attacks against Estonia, Georgia and Kyrgyzstan were sometimes credited to the Russian Business Network (RBN), a powerful cybercrime organization operating since at least 2007 with claimed links to the Russian government and intelligence services. Commentators suggest the group's botnets may have either been purchased or simply directed by the Russian state to launch the attacks. By 2010 the hacktivist collective Anonymous was also engaging in politically motivated cyberattacks such as 'Operation Payback', the DDoS defence of Wikileaks against the US government and US companies. Another group, emerging in April 2011, was the Syrian Electronic Army: state-backed, pro-Assad hackers, who engaged in a range of disruptive hacks and vandalism against the west. Their most important action was to hack into the Associated Press Twitter account on 23 April 2013, posting, at 1.08pm: 'Breaking: Two Explosions in the White House and Barack Obama is injured'. Within three minutes the DOW nose-dived 150 points, wiping $136bn in equity from US companies.

Islamist hackers have also emerged. Following the French magazine murders, Anonymous launched #OpCharlieHebdo on 10 January 2015 against Islamist websites. In response, Islamist groups such as Anonghost launched their own digital jihad against France, with Islamists hacking 19,000 French websites within days, posting black IS flags and messages such as 'Free Palestine. Death to France. Death to Charlie' (Akbar, 2015). What followed was a bewildering tit-for-tat. On the 12 January IS hackers accessed US CENTCOM's Twitter feed, renaming it 'Cyber-Caliphate' with an underline saying, 'I love you ISIS' and posting sensitive, hacked military information. In February 2015, Anonymous launched #OpISIS against Islamist websites and social media accounts, whilst in August a pro-IS Twitter account posted a spreadsheet with information about hundreds of US and UK government and military personnel. The USA began militarily targeting IS's Cyber-Caliphate, killing its leader, UK citizen Junaid Hussein, in a drone strike on 16 September 2015. Its activities continued, however, with the Cyber Caliphate hacking 54,000 Twitter accounts in November 2015, posting account details online along with the phone numbers of the CIA and FBI heads, as revenge for Hussein.

Not every Islamist hack was so successful. On 2 January 2015, the 'Arab Security team' hacked into a Bristol bus website, having mistaken 'TravelWest' for a far more significant part of the USA and Europe's transportation critical infrastructure than it was. On 26 January, another Islamist hacker succeeded in hacking into and defacing the website of non-league Chatham FC, having mistaken it for a Premier League club. The website had to be shut down prior to their match with Tilbury FC, which attracted a crowd of 63. Western governments, nevertheless, took the threat seriously. On 17 November 2015, UK Chancellor George Osborne announced increased cyber-investment, warned IS was trying to develop the ability to launch CI attacks, and revealed the UK had developed 'offensive cyber capability' and would employ it against IS (Reuters, 2015). On 16 April 2016, President Obama admitted the USA was already conducting cyber-operations against IS.

But cyberwar capabilities today extend beyond both states and organized groups, as *everyone* now has the potential to become a cyberwarrior. The ongoing Web 2.0 revolution has unleashed a hitherto repressed desire to *participate*, removing the barriers to creation and activity and allowing the sharing of tools, knowledge and skills. The internet also allows these dispersed individuals to collaborate in larger projects, in a process that Jeff Howe calls 'crowd-sourcing' (Howe, 2008). An example of this was seen in March 2009 when the US border patrol crowd-sourced its border protection by allowing anyone to register and monitor webcams along its 1,254-mile border. An Australian pub signed up, enabling its drinkers to email in alerts from the other side of the world when they saw suspicious activity.

The same crowd-sourcing has emerged in cyberwar. In August 2008, Evgeny Morozov published his article, 'How I became a soldier in the Georgia-Russia cyberwar'. Investigating 'how much damage someone like me … could inflict on Georgia's infrastructure, acting entirely on my own and using only a laptop and an internet connection', Morozov describes how he logged on to stopgeorgia.ru, a possibly state-backed, Russian website offering a free DOSHTTP download – a legal DDoS stress-tester used by security professionals – alongside a list of target Georgian URLs, with a real-time indication of which sites remained up. All he had to do was type in the URL of his target and press the 'start flood' button to begin his own, individual DDoS cyberattack against another state's online infrastructure. 'In less than an hour, I had become an internet soldier', Morozov said, noting how many patriotic Russians would be attracted to this mode of military participation (Morozov, 2008).

Soon after, in the 2008–09 Israel-Gaza War, a pro-Israeli group set up the 'Help Israel Win' webpage, which included a 'Patriot' Trojan tool that could be downloaded onto your computer, allowing the group to join it into a botnet used to launch DDoS attacks on Palestinian targets. Here, user participation required little more than handing over control of your computer. The December 2010 defence of Wikileaks by Anonymous demonstrated the same crowd-sourced cyberwar, with Anons downloading and using the 'LOIC' (Low-Orbit Ion Cannon) to engage in DDoS attacks against Visa, Paypal and Amazon. Ultimately, the operation's success owed more to the participation of key Anons with botnets than to this individual activity, but the fact that so many rushed to participate is significant. The US and UK authorities certainly thought so, as they traced and arrested numerous Anons who had used the LOIC without hiding their IP addresses.

DDoS attacks are unsophisticated and at the lower end of the cyberwar scale, but they are highly participatory and easy to launch, with Youtube hosting thousands of 'how-to' videos for aspiring cyberwarriors. They are also powerful. Patrick Worms, a Georgian government advisor argues, a DDoS 'is so easy to set up, it's so easy to activate and it is so potentially effective that any country is at potential risk of serious disruption from these tools' (Thibodeau, 2009). Malware is also easy to acquire and employ and, given that cybercriminals were quick to copy the design principles and coding techniques of Stuxnet, we can expect future cyber-weapons to become available online. Hence, just as professional journalism has been challenged or supplemented by 'citizen journalism', so, with cyberwar, we are already seeing the emergence of a 'citizen militarism'. Following a century in which civilians increasingly became the victims of industrial warfare, the twenty-first century is seeing the increasingly complex situation where local and global citizens are also becoming mediatory and even military participants – in reporting on conflicts and engaging in online informational war and even direct cyberwarfare.

Although a considerable literature emerged after 9/11 on 'radicalization', to explain how individuals became involved in terrorism, a more important trend may be *normalization*. The internet has transformed many activities, including 'friendship', personal communications, and sexual activity and expression, making what would once have been considered abnormal into common, everyday behaviours, and it has had a similar effect on war, normalizing our participation. As I argue in the next chapter, this is a new mode of 'participative war' in which individuals in conflict zones and around the world contribute to and engage in informational conflict in pursuit of their own political allegiances, including posting micro-propaganda for their preferred causes and posting abuse of or attacking their enemies. This individual, civilian engagement blurs the categories of military participation and, when this engagement includes actual cyberattacks, may have important implications in the future for the norms and laws of warfare, affecting questions of civilian targeting and their responsibility for war-crimes. Indeed, just as individuals can now participate in cyberwar, so too might they become the targets for warfare. In August 2009, a series of DDoS attacks took down Facebook, Google, Twitter and other Web 2.0 platforms, all, it was suggested, in order to attack one person: a Georgian blogger and Russian critic called 'Cyxymu'. In the future, we may see much more aggressive attacks on individuals and their homes, whether using cyber or physical weaponry.

Finally, to complicate things further, consider the contemporary Russian information war against the west. Russia has long embraced the concept of information war. In 1998, Sergei P. Rastorguev, a Russian military analyst, published *Philosophy of Information Warfare*, in which he argued that one of the most important weapons was disinformation, which allowed nations to be weakened from within. It was an idea applied within Russia itself by Putin's aide Vladislav Surkov, a 'political technologist' who, from 1999 to 2011, as Pomerantsev says, directed Russian society like a 'great reality show' (Pomerantsev, 2015:78, 81). It was Surkov's tactics that were employed against Crimea and the Ukraine in 2014, where Russian media immersed every political event in a maelstrom of competing claims and lies in order to confuse all discourse, with the mass dissemination of fake news allowing the control and redefinition of reality itself. Surkov's 2014 novel *Without Sky* had explained some of this, describing a 'non-linear warfare' – a state of war where instead of two sides, there were multiple sides, all with shifting allegiances, in a war of 'all against all' (Pomerantsev, 2015:275).

The year before, in February 2013, in an influential article in the Russian journal *Military-Industrial Courier* entitled 'The Value of Science is in the Foresight', Valery Gerasimov, the chief of the general staff of the Russian military, had set out a vision of precisely this 'hybrid' or 'non-linear' warfare where 'the lines between war and peace are blurred'. As he argued there: 'The very "rules of war" have changed. The role of non-military means of achieving political and strategic goals has grown, and, in many cases, they have exceeded the power of force of weapons in their effectiveness. … All this is supplemented by military means of a concealed character, including carrying out actions of informational conflict'. Ultimately, 'Long distance contactless actions against the enemy are becoming the main means of achieving combat and operational goals' (Gerasimov, 2013:24). These ideas help explain the Russian informational campaign against the west, active since at least March 2016.

This campaign has several, interlinked elements. The first is political hacking. Hacking groups such as 'Fancy Bear' (named APT28 by the west) and 'Cozy Bear' (APT29), with possible links to GRU, the Russian Main Intelligence Directorate, have hacked western political parties and campaigns including the Clinton campaign and Democratic National Committee and the Macron French presidential campaign among others. Emails and other material has then been leaked through Wikileaks or Pastebin in order to damage particular candidates and promote others (such as Trump and Marine Le Pen) whose authoritarian, nationalist and nativist politics are preferred by Russia. A second, related element has been the explicit funding of, or political support for, particular parties, as seen again in personal campaign-connections between Trump and Russia, Russian loans for Le Pen's National Front and personal talks with Putin, and Russian connections with the German far-right AFD (Alternative for Germany) party.

A third element has been the use of social media, including Twitter, Facebook, YouTube and Instagram. 'Troll Farms' such as the pro-Kremlin, Internet Research Agency (IRA) in St Petersburg hired hundreds of workers to produce online stories, videos, photos, memes, comments and contributions promoting Russian interests. Through these farms and other operatives, 'fake news' is created and disseminated, with many fake stories spread using expertly spoofed legitimate news websites; fake profiles ('sock-puppets') are created, posting about controversial and divisive issues such as race, immigration and Islam; Twitter bots automatically produce or retweet stories and comments; Facebook ads are bought (the IRA spent $100,000 on 3,000 Facebook ads in the two years before May 2017); and Russian international news outlets such as RT (Russia Today) and Sputnik pick up on and report these manufactured stories and controversies.

All this could just be considered a traditional propaganda or electoral-interference campaign (of the sort the USA has often indulged in) rather than 'information warfare'. However, the scale of the campaign, its ability to reach deep inside the everyday media platforms and informational experiences of the population, the way these messages were reported by mainstream media and reposted by both political supporters and by other, ordinary civilians, all give this campaign a diffuse, ongoing, real-time, efficacy – one dissolved through the everyday life of the polity. Russia's campaign is best understood as an attack on the *social and political* critical infrastructure of the west, being designed to manipulate the political discourse; to exacerbate and widen existing divisions; to polarize political beliefs; to sow disorder in order to weaken politics and create real internal conflict; and to spread fake stories and create contradictory realities to undermine political debate and faith in the democratic political process and in the concept of democracy itself. What Russia wants most is a weakened Europe, a USA in its own image, internally divided democracies, authoritarian nativist western nations accepting of Russia's own politics, the discrediting of democracy itself and a relativistic international order that would jettison discussion of human rights in favour of great powers and their sphere of influence. Military weapons will never achieve this, but informational weapons might.

The chapter began by discussing the definition of cyberwar and its limitations. Given the range of cyber acts and attacks, their temporality, the range of actors and motivations, the possibilities of individual participation, the range of impacts and effects of informational and cyberwar, and the way cyber-operations dissolve throughout the polity and through all social and media life, we can see that all attempts to define cyberwar or relate it to traditional concepts of conflict are problematic. Depending upon your definition, you could claim that cyberwar *isn't* happening and has never happened, as states haven't yet gone to war with cyberweapons, and you could claim that cyberwar isn't even a real form of warfare, being, at best, an adjunct to physical, military operations, representing an unknown or even limited potential threat. However, as Paul Virilio has noted, 'We are always one war behind' (Virilio, 2002:35) and it may be that traditional conceptions of conflict and its operation are now anachronistic. In this view, cyberwar is something already happening, and indeed *happening all the time* as our ambient everyday political and military reality. Here we are all active, mobilized participants in an emergent, ideational, informational conflict without end, all subjected to and ourselves firing our own, tiny, Gramscian hegemonic bullets in a real-time battle for our political beliefs and causes. Marshall McLuhan saw this coming: after World War I's railway war, World War II's radio war, and Vietnam's TV war, he suggested, 'World War III is a guerrilla information war with no division between military and civilian participation' (McLuhan, 1970:66). *WarGames*, therefore, got it wrong, as it was too optimistic. It wasn't the genius, teen-hacker who we had to worry about in the future: it was all of us.

Key reading

The best general overviews of the origin and development of cyberwar and the issues surrounding it are found in Kaplan (2016) and Harris (2014), although Healey (2013) and Shakarian et al (2013) are also worth looking at. Beyond this, Zetter (2014) provides an excellent, extremely detailed exploration of Stuxnet, but also includes broader discussions of cyberwar and cybersecurity, making it an important text. Clarke and Knake (2010) is worth reading in coming from a US government counter-terrorism advisor, but critics have pulled apart some of its more alarmist claims. Rid (2013) sounds an important note of caution over the concept and dangers of cyberwar, although his actual argument is very nuanced and still includes a recognition of the damage cyberattacks might do. The most detailed discussions of the cyberattacks mentioned in the chapter are found online, although this requires considerable skill in searching and source evaluation as the information is often contradictory, confused or uncertain. Many attacks were only under-stood later, hence contemporary commentary and attributions may be contradicted by later evidence and interpretations. When you read reports of cyberattacks, it is worth remem-bering that, outside of governments, there is little certainty as to what has taken place and even the best security commentators and organizations remain limited in what they can definitely establish. As a final point, bear in mind that most of the texts here have a US bias, explaining the history and development of cyberwar from the US perspective. I've followed this in the chapter in order to summarize the agreed-upon information and timeline, but there are other, as yet untold, histories of cyberwar that could be explored and written.

References

Akbar, J. (2015) 'Death to France. Death to Charlie': Pro-ISIS Hackers Launched 'Unprecedented' Wave of Cyber-Attacks on 19,000 French Websites', *Daily Mail*, 15 January, http://www.dailymail. co.uk/news/article-2912280/Death-France-Death-Charlie-Pro-ISIS-hackers-launched-unprece dented-wave-cyber-attacks-19-000-French-websites.html.

BBC. (2012) 'Hackers had "Full Functional Control" of NASA Computers', *BBC News*, 8 March, http://www.bbc.co.uk/news/technology-17231695.

BBC. (2015) 'US and China Agree Cybercrime Truce', *BBC News*, 25 September, http://www.bbc.co. uk/news/world-asia-china-34360934.

BBC. (2017) 'Squirrel "Threat" to Critical Infrastructure', *BBC News*, 17 January, http://www.bbc.co. uk/news/technology-38650436.

Beaumont-Thomas, B. (2014) 'North Korea Complains to UN about Seth Rogan Comedy The Interview', *The Guardian*, 10 July, https://www.theguardian.com/film/2014/jul/10/north-korea-un-the-interview-seth-rogen-james-franco.

Cellan-Jones, R. (2014) 'Regin, New Computer Spyware, Discovered by SYMANTEC', *BBC News*, 23 November, http://www.bbc.co.uk/news/technology-30171614.

Chinese Government. (2004) 'China's National Defence in 2004', *Chinadaily*, http://www.chinadaily.com.cn/english/doc/2004-12/28/content_403913_4.htm.

Clarke, R. A. and Knake, R. A. (2010) *Cyberwar*, New York: Ecco Press.

Dearden, L. (2017) 'World Heading Towards "Permanent Cyber War", France Warns', *The Independent*, 1 June, http://www.independent.co.uk/news/world/europe/cyber-war-world-warning-france-criminals-extremists-russia-countries-guillaume-poupard-anssi-a7767886.html.

Der Spiegel. (2014) 'NSA Spied on Chinese Government and Networking Firm', *Spiegel Online*, 22 March, http://www.spiegel.de/international/world/nsa-spied-on-chinese-government-and-networking-firm-huawei-a-960199.html.

Finkle, J. (2011) 'Biggest Ever Series of Cyberattacks Uncovered, U.N. hit', Reuters, 3 August, https://www.reuters.com/article/cyberattacks/biggest-ever-series-of-cyber-attacks-uncovered-un-hit-idUSN1E76R26720110803.

Gady, F.S. (2015) 'New Snowden Documents Reveal Chinese Behind f-35 Hack', *The Diplomat*, 27 January, https://thediplomat.com/2015/01/new-snowden-documents-reveal-chinese-behind-f-35-hack/.

Gerasimov, V. (2013) 'The Value of Science is in the Foresight', *Military Review*, Jan–Feb, pp. 23–29, http://usacac.army.mil/CAC2/MilitaryReview/Archives/English/MilitaryReview_20160228_art008.pdf.

Gibson, W. (1995) *Neuromancer*. London: Harper Collins.

Graham, M. (2008) 'Welcome to Cyberwar Country, USA', *Wired*, 2 November, https://www.wired.com/2008/02/cyber-command/.

Harris, S. (2014) *@War: The Rise of Cyberwarfare*. London: Headline Books.

Healey, J. (ed.) (2013) *A Fierce Domain: Conflict in Cyberspace 1986–2012*. Vienna: Atlantic Council, Cyber Conflict Studies Association.

Hern, A. (2017a) 'Hackers Attacking US and European Energy Firms could Sabotage Power Grids', *The Guardian*, 6 September, https://www.theguardian.com/technology/2017/sep/06/hackers-attacking-power-grids-in-us-and-europe-have-potential-to-sabotage.

Hern, A. (2017b) 'Major Cyber-Attack will Happen Soon, Warns UK's Security Boss', *The Guardian*, 22 September, https://www.theguardian.com/technology/2017/sep/22/major-cyber-attack-happen-soon-warns-uks-online-security-boss.

Howe, J. (2008) *Crowdsourcing*. London: Random House Books.

Kaplan, F. (2016) *Dark Territory: The Secret History of Cyber War*. New York: Simon & Schuster.

McLuhan, M. (1970) *Culture is Our Business*, New York: McGraw Hill.

Morozov, E. (2008) 'An Army of Ones and Zeros: How I Became a Soldier in the Georgia-Russia Cyberwar', *Slate*, 14 August, http://www.slate.com/articles/technology/technology/2008/08/an_army_of_ones_and_zeroes.html.

Pomerantsev, P. (2015) *Nothing is True and Everything is Possible*. London: Faber and Faber.

Reuters.. (2015) 'Britain to Develop Offensive Cyber Capability, Osborne Warns', *Reuters*, 17 November, https://www.reuters.com/article/us-britain-security-cybersecurity-osborn/britain-to-develop-offensive-cyber-capability-osborne-says-idUSKCN0T619D20151117.

Rid, T. (2013) *Cyberwar Will Not Take Place*. London: C. Hurst & Co.

Shakarian, P., Shakarian, J. and Ruef, A. (2013) *Introduction to Cyberwarfare*. London: Elsevier.

Thibodeau, T. (2009) 'Russia's Cyber-Blockade of Georgia Worked. Could it Happen Here?', *Computerworld*, 27 April, https://www.computerworld.com/article/2524062/government-it/russia-s-cyber-blockade-of-georgia-worked–could-it-happen-here-.html.

Vinge, V. (2001) *True Names and the Opening of the Cyberspace Frontier*, New York: Tom Doherty Associates LLC.

Virilio, P. (2002) *Ground Zero*. London: Verso.

Ware, W. H. (1967) 'Security and Privacy in Computer Systems', Spring Joint Computer Conference, Atlantic City, 17–19 April, Santa Monica, California: RAND Corporation, https://www.rand.org/content/dam/rand/pubs/papers/2005/P3544.pdf.

Ware, W. H. (1979) 'Security Controls for Computer Systems: Report of Defense Science Board Task Force on Computer Security – RAND Report R-609-1', Published for the Office of the Secretary of Defense, Santa Monica, California: RAND Corporation, https://www.rand.org/pubs/reports/R609-1/index2.html.

Zetter, K. (2013) 'Legal Experts: Stuxnet Attack on Iran was illegal "Act of Force"', *Wired*, 25 March, https://www.wired.com/2013/03/stuxnet-act-of-force/

Zetter, K. (2014) *Countdown to Zero Day: Stuxnet and the Launch of the World's First Digital Weapon*, New York: Crown Books.

Zetter, K. (2015) 'A Cyberattack has Caused Confirmed Physical Damage for the Second Time Ever', *Wired*, 8 January, https://www.wired.com/2015/01/german-steel-mill-hack-destruction/.

10 #ParticipativeWar

Social media in Gaza and Syria

Full spectrum access

In the opening chapter we saw how, in the 1991 Gulf War, the USA created a model of media management that enabled it to overcome the 'Vietnam syndrome' and fight a war in front of the world's media, with domestic and international support. The following chapters traced how it redeployed this model in Kosovo in 1999, Afghanistan in 2001 and Iraq in 2003. Subsequent chapters demonstrated how this model began to break down with developments in digital technologies. The internet had already allowed ordinary people's voices to be heard during the Kosovo War, whilst the Iraq War saw the rise of soldier and civilian blogs and, soon after, the Abu Ghraib torture scandal was exposed when the perpetrators' own photographs leaked to the global media. From 2004 to 2005 a 'Web 2.0' revolution, encompassing improved, multi-media mobile technologies, improvements in connectivity and free, participatory, online platforms, enabled anyone to become a media producer and distributor. As we saw in chapter 6, this led to the end of military informational control as it struggled to manage the media activities of their own personnel.

This wasn't what the military had expected: it had foreseen instead a future of complete control. We saw in chapter 3 how the US military had re-theorized warfare in the 1990s to include developments in information technology and new concepts of 'information warfare'. One of the most important papers of the era was *Joint Vision 2010 – America's Military: Preparing for Tomorrow*, published by the Joint Chiefs of Staff in July 1996. This emphasized the need for 'full spectrum dominance', encompassing the control of both the informational/electronic and the physical battlefield. The 13 November 2001 bombing of Al-Jazeera's Kabul office during the Afghanistan War and 8 April 2003 bombing of their Baghdad television station and Abu Dhabi television during the Iraq War should be understood in this context. For the USA, 'full spectrum dominance' meant the full control of the electromagnetic broadcasting spectrum and the interdiction of any attempts to produce and distribute information outside of their sanctioned outlets.

Very simply, Web 2.0 blew this policy apart. Instead of a future of reduced information and tightly controlled, on-message broadcasters, the explosion of popular online platforms for 'user-generated content', cheap internet connections and public Wi-Fi, and multi-media phones (and eventually smartphones) combined with a remarkable cultural revolution in the use of digital technologies and the way we lived our life through them with ourselves at the centre of our own 'me-dia' worlds, to create a perfect storm that swept aside government messages, military control and even the centrality and dominance of broadcast-era media themselves. It meant there was now no way to stop the creation and sharing of information from and about the battlefield. Now, not only are governments, militaries and their soldiers producing, but so too are those within the conflict zone, whether terrorists, government or opposition militias and civilians, as well as anyone else around the world with anything to say, something to add, or a need to share and get involved by adding their voice or opinion.

Information now pours onto, from, about and around the battlefield. Within a few years we have moved from the US concept of 'full spectrum dominance' to the new military reality of *full spectrum access*. This is a new mode of participative warfare, where everyone can experience and take part in conflict. The result, as we saw in the last chapter, is a diffusion of that 'information warfare' the military theorized and laid claim to throughout the population and even across the globe. Every interested person, of any age, experience, expertise, and qualification, can now fire their own hegemonic bullets in a fractal, digital infowar aimed at exposing their situation or promoting their preferred political interpretation. In this chapter I want to take two examples – the 2014 Gaza War and the 2011–present Syrian Civil War – to explore this new mode of global participation in conflict.

Box 10.1 The 2014 Gaza War

The 2014 Gaza War was a small event in a much broader Arab-Israeli conflict. The basis of this conflict is a contest between the two groups over one geographical territory, the land between the Mediterranean Sea and the Jordan river. Jewish claims are based on biblical promises and the region being the site of the iron-age Jewish kingdoms of Israel and Judah. Palestinian claims are based on their continuous residence and, until 1948, their clear demographic majority. The conflict's origins lie in the late nineteenth-century Zionist movement, which wanted a national homeland for Jews and advocated immigration to Palestine, a territory within the Ottoman empire. Successive waves of Jewish immigration increased the Jewish population by 1914–59,000, compared to 657,000 Muslim and 81,000 Christian Arabs. After fomenting an Arab revolt in World War I against Turkish rule, the British reneged on promises of Arab independence, whilst their 1917 'Balfour Declaration' supported a Jewish homeland in Palestine. After the Ottoman empire's defeat, Britain formally took over the 'Palestinian mandate' in 1923, ruling the territory until 1948 – a period that saw rising Jewish immigration and increased Arab-Jewish violence. In response to European anti-Semitism and the Holocaust, Jewish immigration increased during and following World War II and Jewish pressure grew for a national home. The UK gave up its mandate and on 27 November 1947 the UN voted to create two states with majority populations. The Arabs rejected this, arguing no amount of immigration gave the right to claim a national state. Fighting soon broke out between Arabs and Jews and by the time Jewish leaders declared the establishment of Israel, on 14 May 1948, the better-organized Jewish forces controlled more territory than the UN had planned. The next day, neighbouring Arab nations invaded. The subsequent Arab-Israel War ended with an Israeli victory. The armistice, signed on 10 March 1949, left Israel in control of 77 percent of the former Palestinian mandate territory with the Arab population split between Israel, the West Bank (under Jordanian control) and Gaza (under Egyptian control). Israeli regional control increased after their victory over Syria, Egypt and Jordan in the 'Six Day War', from 5 to 10 June 1967. They seized the Golan Heights from Syria, the Egyptian Sinai Peninsula, the West Bank and Gaza, holding the 'occupied territories' in defiance of UN Resolution 242, adopted 22 November 1967, which called for the withdrawal of Israeli forces from land seized in the war. Despite the November 1947 UN vote, the Palestinian territories have never become a Palestinian nation, being thwarted by post-1948 events.

The Gaza Strip is a sea-facing Palestinian territory 141 square miles in size with a 2015 population of 1.85 million. It was occupied by Israel from 1967 until 1994 when, after the Oslo Accords, it mostly came under the control of the Palestinian Authority (PA). In 2005 Israel repatriated Israeli settlers from the strip and unilaterally withdrew their forces. Despite this, Israel remains the de facto occupying power due to the security barrier begun in 1994, its control of entry points and the movement of people and goods, its control over Gaza's airspace and coastal waters, and its punitive military strikes and incursions. The harsh conditions aided the rise of the Islamist group Hamas, who won legislative elections on 25 January 2006. Following a US-backed failed Fatah coup attempt, Israel declared Gaza a 'hostile state' and enforced a comprehensive economic and political blockade on it. On 27 December 2008 Israel launched 'Operation Cast Lead' to stop rocket fire from Gaza and weapons smuggling. Lasting until 16 January 2009, the invasion led to the deaths of 13 Israelis and 1,166–1,417 Palestinians, including 926 civilians. Following mutual attacks, Israel launched 'Operation Pillar of Defence' from 14 to 21 November 2012, resulting in the deaths of two Israeli soldiers and up to 120 Palestinian combatants and 105 civilians. After the 2013–14 Israeli-Palestinian peace talks broke down, Fatah and Hamas were reconciled in a unity PA government on 23 April 2014. Israel refused to deal with the PA.

On 12 June 2014, the abduction and murder of three Israeli teenagers in the West Bank by independently acting Hamas members provoked the IDF's 'Operation Brother's Keeper' in the West Bank and soon after, on 8 July, 'Operation Protective Edge' against Gaza in order to stop rocket attacks and the 'terrorist' use of tunnels into Israel. An air campaign against Gaza was followed by a ground invasion lasting until Israel withdrew on 26 August. Sixty-six Israeli soldiers and six civilians were killed. The UN claims up to 2,251 Palestinians were killed, including 1,462 civilians (513 of them children). The operation led to huge numbers of internally displaced persons and caused massive damage to power generation and electrical supplies, water supplies, sewage pipes, schools and health facilities, and to economic infrastructure such as factories and farms. The UN calculates 7,000 homes for 10,000 families were razed, with another 89,000 homes seriously damaged.

Gaza online

The primary event preceding the 2014 Gaza War was the kidnapping of three Israeli teenagers from the Israeli West Bank settlement of Alon Shvut in Gush Etzion on 12 June 2014 by two independently acting Hamas members. Although the government had evidence the teens had already been killed, the Israeli Defence Force (IDF) launched

'Operation Brother's Keeper' in the West Bank which, in the following eleven days, arrested 350 Palestinians, killing five. The boys' bodies were found on the 30th and the suspects were killed in an IDF shootout on 23 September.

Whilst Palestinian deaths – such as the recent killing of two unarmed boys during a Nakba Day protest on 15 May – attracted little domestic or international attention, the Israeli kidnappings sparked a media storm. The Twitter hashtag #BringBackOurBoys was promoted in a photo at the UN by the Israeli Permanent Representative Ron Proser on the 17 June, leading to an Israeli 'propaganda blitz' as Israel's 'trained online propaganda brigades' flooded social media with the hashtag (Blumenthal, 2015:11). The hashtag was directly adapted from that used for the kidnapped Nigerian schoolgirls that April, and Prime Minister Netanyahu's wife Sara copied Michelle Obama's famous photo of support for the schoolgirls by similarly posing with a handwritten hashtag sign that was widely shared online. The social media campaign reverberated around the world, especially among Jewish communities.

With tensions running high, there were numerous revenge attacks on Palestinians reported, including the murder of 16-year-old Mohammed Abu Khdeir on 2 July. Social media stoked this hatred. One pro-Israeli Facebook page demanded the execution of one Palestinian prisoner for each hour the teens were missing, whilst another, 'The People of Israel Demand Revenge', gathered 35,000 members in days. Active duty IDF soldiers 'took to Facebook to demand revenge, posting photos of themselves with the weapons they said they were aching to use', whilst Ayelet Shaked, of the right-wing Jewish Home party, gathered thousands of 'likes' for her posts calling for the killing of Palestinian civilians and the destruction of their homes, 'otherwise more little snakes will be raised there' (Blumenthal, 2015:17).

The online Palestinian response was equally vociferous. Many Palestinians criticized the emphasis on Israeli child deaths, citing the Nakba Day killings as evidence of hypocrisy, and appropriating #BringBackOurBoys to draw attention to Palestinian suffering and deaths. Meanwhile, a cartoon on the Fatah Facebook page portrayed the teenagers as rats wearing Stars of David dangling on fish hooks with the caption 'masterstroke' (Soffer, 2014). Gazan Palestinians released a song on social networks mocking the kidnappings and calling for more abductions, whilst a Palestinian group posted a parody video of the kidnappings on YouTube, making fun of the Israelis and presenting the kidnapping as an act of the 'Abu Saqer el Khalili Brigades, the Kick Ass Branch' (Ho, 2014). Palestinian popular support for the kidnapping proliferated on social media, with the #ThreeShalits hashtag (in reference to a kidnapped IDF soldier Gilad Shalit) and celebratory three-finger photos being posted in Palestine and across the muslim world.

The IDF 'Operation Protective Edge', launched against Gaza on 8 July, was technically a response to Hamas rocket fire and 'terrorist' use of tunnels to enter Israel, but the assault was closely linked to the kidnappings and West Bank operation as part of a systematic attempt to clamp down on Hamas. From the beginning the war was accompanied by a government and military-led multi-platform, multi-device, information war by both the Israeli state and IDF and the Palestinian Authority and Hamas. Both groups immediately used Facebook, YouTube, Twitter and Instagram, producing a range of captioned photographs, cartoons and infographics for each platform, and both responded quickly to combat enemy claims with their own online rebuttals.

The IDF, for example, tweeted a cartoon captioned, 'Where do Gazans hide their weapons?', with images of houses, mosques, hospitals and schools, commenting 'Hamas commits a double crime by hiding weapons intended to kill civilians in mosques, schools, hospitals and homes'. The image was retweeted by Hamas' Izz ad-Din al-Qassam Brigades, with the claims crossed out as 'lies' and a new caption reading, 'Main goal of this graphic is justifying destroying residential houses in Gaza. Israel has the heavy weapons, F-16 & tanks and used it against Palestinian civilians'. Adding the comment, 'Israel tries to justify its crimes against Gaza residents', the tweet included the hashtags #Gazaunderattack, #PrayforGaza and #PalestineUnderAttack (Fowler, 2014).

Cartoons became an important informational weapon. The IDF produced a cartoon showing two houses with underground basements: the Israeli one contains a young girl sheltering in her bedroom in fear, whilst in the other, labelled 'Hamas', the basement is stuffed full of rockets to send. Meanwhile, Hamas tweeted a cartoon reversing these charges, showing an IDF soldier pointing to a tiny missile lodged uselessly in a wall and saying 'You see! We're just defending ourselves', as a torrent of giant Israeli missiles fly overhead, obliterating Gaza (Fowler, 2014). Governments and militaries, therefore, were major producers of propaganda graphics, crafting and pushing out simple messages designed to appeal online and to be sharable across platforms, attempting to conscript civilians as informational warriors in their campaigns and to weaponize virality itself.

Twitter became an important battleground. The IDF posted dozens of updates every day about the operation, live-blogging events on the ground and giving their view of the conflict. They posted updates on Hamas rocket fire (including a 'rocket-counter' listing the total fired since the operation began) and on the success of the 'Iron Dome' anti-missile shield. The IDF tweeted warnings about their military activities and even made personal phonecalls and sent text messages to people in Gaza telling them to leave areas they would be operating within and target sites. These messages were publicized by Israel as evidence of its concern for human life, but Gazans complained they had nowhere to go as the entire territory was bounded by Israel's security barrier. Israeli and IDF propaganda emphasized the terroristic nature of Hamas, making continued reference to their rocket fire, use of tunnels for weapons smuggling and armed attacks, their arms caches around Gaza and their claimed use of 'human shields' for their attacks.

Mobile apps were also weaponized. The IDF blog discussed their new smartphone app that invited readers to consider the question what if Hamas fired a rocket at your city? 'We created an application to help you put Hamas' threat in perspective. Type in your home town and it will show you just how far these rockets can reach'. The app showed an M-302 rocket with a range of 90–160km with its circumference imposed on a map of Israel, asking you to 'imagine' the reach of the rocket by superimposing the same red warning circle over a map of your country. The IDF created maps of the United States, Canada and UK for illustrative purposes and asked readers to 'share these with your friends and neighbors to show them what living under Hamas terror looks like' (Israel Defence Forces, 2014a).

Hamas was also active on Twitter, through its al-Qassam Brigades accounts. These included Arabic, Hebrew and English language accounts, highlighting the different, global constituencies for its propaganda. Its accounts carried updates on casualties and reports on its own rocket responses to IDF actions. Its feed defended the Brigade and highlighted the plight of Palestinian civilians and civilian deaths. Hamas's multi-lingual Twitter accounts and the IDF rocket mapping app both highlight the importance Israel and Hamas placed on international audiences. The physical battlefield was small, but the informational battlefield was global and real-time: for both sides, what the world experienced now and what it thought, *mattered*. The bulk of the IDF and Hamas Twitter followers lived abroad, with the feeds being closely followed by journalists who used them to produce mainstream media stories, but even retweets and sharing by ordinary people was important, in crowd-sourcing the informational war.

This led to a new phenomenon: hashtag wars. The ability to create a hashtag that was taken up and widely used was important. Pro-Palestinian hashtags such as #GazaUnder-Attack, #StopIsrael and #PrayforGaza all went viral, for example, beating equivalent Israeli hashtags: whereas #GazaUnderAttack was used over 4m times, #IsraelUnderFire only managed 170,000 uses. But Hamas was at a disadvantage on Twitter. As an American company, Twitter followed US policy which designates Hamas as a terrorist organization, hence Hamas faced the additional problem of its accounts being removed. Twitter was a valuable battlefield, but it wasn't necessarily trustworthy. A BBC investigation discovered that many of the images of Gazan bombing and explosions posted under hashtags such as #GazaUnderAttack were reused from the 2008–09 war, or were even taken from Iraq and Syria.

YouTube was another important propaganda outlet, with both sides publishing videos promoting their cause. On 2 July, the IDF YouTube channel posted a video called '15 Seconds: Not Enough Time', showing people running for shelter from Hamas rockets. Their 15-second warning was so little, the IDF argued, that even Usain Bolt, 'the world's fastest man', couldn't make it in time (Israel Defence Forces, 2014b). The Palestinians' most popular video was a catchy Hamas Arabic song called 'Shake Israel's Security' (or 'Attack, Do Terror Attacks'). First produced by the al-Qassam Brigades in 2012, it was given new, Hebrew lyrics and re-issued as part of the new propaganda war, designed to strike fear into Israelis.

Things didn't go as expected, however, as it became an immediate hit in Israel. Israelis embraced the badly pronounced pop song with its gibberish Hebrew, and immediately started adapting the lyrics or reposting it with new videos or in new versions. There were a capella versions, acoustic versions, a soulful piano rendition, videos of Israelis dancing in the streets to it, a Smurf version, a Lion King cartoon version with the soundtrack put into the mouths of hundreds of tunnel-digging prairie dogs, dancing Teletubbies versions and even a version by a parrot. This was real participative war: spontaneous, emergent, 'bottom-up' rapid responses to official propaganda weapons with an entire ecology of peer-produced parodies. YouTube kept removing the original video for promoting terrorism due to its exhortation to 'eliminate all the Zionists' and calls to 'do terror attacks', but it was the popular Israeli troll-warfare that did most damage to Hamas.

Probably the most powerful video intervention was a version of the then-popular Facebook 'ice-water bucket challenge' – a challenge originally designed to raise money for Amyotrophic lateral sclerosis (ALS) and going viral in June 2014. Jordanian comedian Mohammed Darwaza came up with the idea of the video 'rubble-bucket challenge', which was launched by Gazan student Maysam Yusef then popularized by Ayman al-Aloul, the editor-in-chief of *Arab Now Agency*, a Gazan newspaper. Lacking water, or the electricity to freeze it, the Palestinians raised awareness for their plight by pouring the destroyed remains of their own houses over their heads. A Facebook page was set up and people posted videos on Facebook, YouTube and Twitter under the hashtags #RemainsBucket, #RubbleBucketChallenge and #RemainsBucketChallenge. In his video, Aloul stands in front of a horrific, ruined vista of destroyed buildings as far as you can see, explaining the challenge and its meaning, before a colleague pours a bucket of the rubble over his head. 'We don't ask for material aid', he says, 'we ask for solidarity, especially from those who have followers and an audience' (Aloul, 2014). Yusef similarly emphasized the importance of the message:

> our campaign is more of a social media revolution, where people show their solidarity with Gaza and publicly reject the killing of civilians. We are trying to form a worldwide movement to pressure Israel to stop this genocidal act against Palestinians of Gaza.
>
> Toor (2014)

Another example of civilian-producers were those who gained fame on Twitter, such as Farah Baker, a 16-year-old living near Shifa hospital who gained 166,000 followers for her emotive tweets of teenage and family life during the war. Her live tweets of the Israeli bombing in which she expressed her fear she wouldn't last the night were especially powerful. 'I can't stop crying. I might die tonight', she posted during attacks on 28 July (Getlen, 2017). In one photo, she is seen holding up a sign with the simple message, 'I am Farah Baker, Gazan girl, 16 years old. Since I was born I have survived 3 wars and I think this is enough #SaveGaza' (Getlen, 2017). Her tweets had a huge impact, with their authenticity and the global fear for her safety impacting on the global perception of the conflict. *Foreign Policy* magazine named Baker as one of its 'global thinkers' of 2014.

Other propaganda, however, had a more uncertain origin, such as Israel's 'Hasbara' ('Explaining') rooms. These were facilities set up within institutions such as universities giving students the opportunity to produce pro-Israeli material and comments and to counter online Palestinian propaganda. The most famous was the 'war room' at the Interdisciplinary centre (IDC) Herzliya, set up by a student in a computer lab and staffed with 400 student volunteers working 8am–9pm. The IDC 'advocacy room' was first created for the 2012 'Operation Pillar of Defence' campaign when 1,600 students spread social media messages to 21 million people in 62 countries, across 31 languages. Reactivated in 2014, it posted under the handle and hashtag #IsraelUnderFire, gathering thousands of followers across numerous platforms and opening a dedicated website available in 13 languages with information, texts, videos, pictures and testimonials.

Critics, however, point to close links between these explaining rooms and the government. The Prime Minister's office, the Minister of Foreign Affairs, and the Defence Force considered the university's 2012 efforts so successful they sought to collaborate during subsequent military actions. One of the key 2014 organizers, Lidor Bar David, was later revealed to be a former IDF captain and in the 2014 campaign Prime Minister Netanyahu launched a 'diplomacy war room' from his own office, and also purchased promoted tweets for selected posts to increase their visibility. There have been claims the government paid students in these rooms, to create the impression of spontaneous, mass support, though Israeli sources point out the support itself was genuine.

This was a social media war, but it wasn't certain that each side was talking to the other. Reports showed a surge in Facebook 'unfriending', as people's politics polarized. 'Nearly 20% of Israeli Facebook users cut off "friends" during the Gaza war', as people blocked feeds or unfriended others to avoid receiving opposing views (Tobin, 2014). This lends support to claims social media acts as an 'echo chamber' or, as Pariser says, a 'filter bubble' (Pariser, 2011), bringing us – whether by choice or algorithmic decision – experiences, comments and ideas that reflect back our own. Gilad Lotan's analysis of Twitter during the war confirms this. Despite methodological issues (dealt with in follow-up research), his network graph showed few connections between pro-Israeli and pro-Palestinian Twitter users, with a clear polarization of reporting and links. As he commented, 'users make deliberate choices about what they choose to amplify'. In their choices, they aren't seeking out new information but choosing what they already believe, hence, 'messages passed along in one side of the graph will never reach the other' (Lotan, 2014).

The mainstream media reported the war as best they could, although they were hampered by a lack of access to Gaza. Accusations of bias were common, with the right and pro-Israeli commentators claiming a left-wing, anti-Israeli bias in the media's focus on the destruction of Gaza, although that was the reality of a war that pitched a small-scale militia against F-15s in residential areas of Gazan cities. Meanwhile, the left and pro-Palestinian commentators saw an implicit bias in the mainstream media's fear of criticizing the Israeli state and fear of being accused of 'anti-Semitism'. Israel's global political communications and military media management has traditionally been strong. It has been very successful at representing its actions within western media and a western culture that has often been sympathetic to its cause and it has imposed strong controls on conflict zones preventing independent or critical reporting. In the 2014 Gaza War Israel's mainstream media strategies continued to work as their spokespersons offered defences of the operation and countered critical claims. Online, however, it was a different matter.

Very simply, Israel easily won the 2014 Gazan War on the ground, but it decisively lost the global, online, 'participative Gazan War'. Probably the major reason for this was the sharing of images of destruction and death. Footage of Hamas or IDF fighters was rare, meaning that most of the images experienced globally were images of explosions, of ruined houses, of people sifting rubble for their loved ones, of living, dust-covered bodies being pulled from the wreckage and of lifeless and bloodied bodies being cradled by agonized rescuers and screaming family members. These images had a truth-value that cut through all official attempts at justification and spin. Media appearances by telegenic, experienced, calm and suited representatives of the Israeli state, whose well-crafted messages were subjected to only the lightest grilling by broadcasters, simply didn't work anymore when people's social media feeds were filled with such direct and visceral horror. The most horrific images were those of dead children. These images were treated cautiously by mainstream television news and newspapers, but once available they were widely shared online and it was difficult for *any* produced propaganda to compete with their power.

The most famous images of the impact of this conflict were captured by journalists on 16 July, when the Israeli Navy targeted shells directly on a small group of children playing on the beach. The photographs form a horrific sequence. Close-up shots show them running. Other shots from the front show them desperately sprinting towards the cameraman for cover and finally other photographs show the aftermath. In one, a mangled corpse of a child, limbs at horrific angles, lies face down in the sand; in another a distraught man stands over a body, hands held apart in disbelief at the unthinkable scene. A year later, the Israeli military exonerated itself over the killing of the four 9–11-year-olds, claiming the strike had targeted a 'compound', 'known to belong to Hamas's Naval Police and Naval Force (including Naval Commandos)' (Beaumont, 2015). Journalists who saw the killings, however, testified to the targeting of the children and said they saw only a small fisherman's hut nearby containing a few tools where the children had been playing hide-and-seek.

Israel's later self-exoneration was irrelevant: the images had already been seen around the world. Israel was aware of the growing problem. Prime Minister Netanyahu complained on CNN of Hamas deliberately using images of 'telegenically dead Palestinians', to which Benjamin Wallace-Wells in the *New York* magazine pithily replied, 'If Netanyahu is so bothered by how dead Palestinians look on television, then he should stop killing so many of them' (Flanagin, 2014). In the mainstream media, Israeli discussions of 'terror tunnels' went unchallenged by news presenters and the phrase was widely repeated on TV. Little was said, however, about how the tunnels had been dug to bring in basic supplies into what is commonly described as 'the world's largest outdoor prison' (Dawber, 2013), nor about the absurdity of comparing tunnels to the F-15 fighters currently being deployed against civilian targets. In the mainstream media, rules of impartiality meant a false equivalence was drawn between the two competing military, technological systems.

No discussion of tunnels, however, could eclipse the images of civilian casualties. Civilians had died in large numbers too in 2008–09 but the major difference now was the spread of multi-media smart-phones that enabled anyone in the conflict zone to capture the reality of the conflict and send it out to the world. Arguably this distorted the reality of the war, as casualties have no necessary relationship to whether a war is justified or not and many acts of war with high civilian casualties (such as the use of atomic weapons on Japan) have been defended as justifiable and necessary. The journalistic emphasis on human-interest stories and suffering may also miss the more complex underlying political reality. These are valid points, but what struck most commentators was the vast differential between the Israeli and Palestinian casualties, with the 66 IDF soldiers and six Israeli civilians contrasting with reports of up to 2,251 Palestinians killed, including 1,462 civilians (513 of them children). This wasn't simply a tragic outcome of aerial bombing: Blumenthal describes the IDF's 'open-fire' policy, their creation of 'free-fire zones' where 'soldiers could kill people regardless of their involvement in battle' and numerous incidents of civilians being deliberately killed, often at close range (Blumenthal, 2015:48;60). The scale of civilian deaths together with the massive destruction of housing, the critical civilian infrastructure and economy, and the resulting internal displacement and humanitarian crisis were all a measure of the superiority of Israeli forces. Online, however, they became a powerful sign of the illegitimacy of the operation. This may be a distortion of the war, but, alternatively, if we consider the reality of war as the military destruction of lives and bodies, then social media may have been the most truthful outlet.

There were other important symbols of this indifference tarnishing Israel's global reputation. On 9 July, the Danish journalist Allan Sorenson posted a photograph on Twitter showing 'Sderot cinema. Israelis bringing chairs to hilltop in sderot 2 watch latest from Gaza. Clapping when blasts are heard' (Mackey, 2014). Photographs are snapshots, and their ability to misrepresent reality is well known. We don't know what the people shown had themselves experienced. We know Sderot had been hit by rocket fire so there was certainly a deeper story to tell here, but the image of smiling, happy Israelis enjoying the spectacle of the Gazan bombing when contrasted online with images of children in coffins was devastating. Other online activities added to the feeling of Israeli pleasure in death. Two mobile phone game apps, 'Bomb Gaza' and 'Gaza Assault: Code Red' were developed and available on the Google Play store from 16 July until they were removed by the 14 August. This smart-phone, video-game war was part of the participatory information war, with 'casual gaming' representing a casual view of the Palestinian suffering.

In conclusion, Israel won the 2014 Gaza War and achieved a control over its representation in mainstream media outlets, successfully countering claims of excessive violence with reasoned explanations of its self-defence against Hamas terrorism. Online, however, the propaganda offensive failed. In this participatory war – a war that united people within Gaza with people from all over the world – Israeli actions were widely condemned and the images of the destruction of Gaza and bodies of its residents ensured Israel suffered an overwhelming informational defeat.

Box 10.2 The Syrian Civil War

The Syrian Civil War is an armed conflict in Syria between Ba'athist government forces and its regional and global supporters, and those who want to overthrow it and their supporters. In 2011, Syria had a population of 22.5 million people, the majority of whom were Sunni Muslims, though the country was run by an Alawite (Shia) minority. The Arab Socialist Ba'ath Party had ruled since a 1963 coup, with Hafez al-Assad serving as president from 1971 until his death in 2000 when he was succeeded by his son, Bashar al-Assad. The catalyst for the civil war was the 'Arab Spring', which saw the overthrow of Ben Ali in Tunisia in January 2011 and Mubarak in Egypt in February and spurred protests across the Arab world. In Syria, the effects of the 2006–11 drought, the failure of Assad's economic reforms and lack of opportunities for the young combined with unhappiness at police repression and a desire for increased political freedoms, leading to pro-democratic protests from 28 January 2011. Major demonstrations began from 15 March in Damascus and Aleppo, but it was in Daraa, following the arrest and police torture of 23 children for anti-regime graffiti, that the revolt came to life. Significant protests began there on the 18th, leading to a regime backlash killing at least 37 on the 23. Protests spread through the country and Assad used the security forces to put them down. By April the army was being used against demonstrators and the protests became an armed rebellion, leading to a civil war.

The rebel, Free Syrian Army (FSA) was formed on 29 July 2011 and the Syrian National Council was formed on 23 August to organize the anti-Assad forces, but the opposition was split into too many local, competing militia groups. Over the following years the secular, reformist moderates would be eclipsed by the rise of Islamist and jihadi groups and a bewildering on-the-ground war emerged with multiple groups forming and reforming into temporary or more long-lasting alliances or new organizations. The more radical Islamist groups such as Jabhat al-Nusra, formed in January 2012 as a Syrian branch of Islamic State in Iraq (ISI) (and hence also of al-Qaeda), often cooperated with pro-democratic rebels, though at other times they fought both government forces and other opposition groups. The most extreme jihadis were ISI who, in April 2013, split from al-Nusra and al-Qaeda and began their own rival Syrian operations as the Islamic State of Iraq and Syria (ISIS). Headquartered in Raqqa from January 2014, ISIS became the major opposition military force, relentlessly pursuing its dream of a Caliphate. The war devastated Syria. All sides were accused of war-crimes, including murder, torture, rape, disappearances, massacres and the indiscriminate use of weapons. Islamic State carried out atrocities on captive populations and the Assad regime used barrel bombs and chemical weapons on civilians. The 21 August 2013 chemical attack on Ghouta, Damascus, for example, killed up to 1,729 people. By December 2017 the Syrian Observatory for Human Rights estimated up to 481,612 people had been killed. By March 2016 the UN OCHR estimated 13.1 million Syrians required humanitarian assistance, with 6 million internally displaced and over 5 million becoming refugees.

The Syrian Civil War became a complex, multi-layered conflict, fought on many levels. (1) The war began as a political conflict between Assad's authoritarianism and those wanting democratic reform. (2) As it developed it also became a theological conflict between those wanting a secular state and those promoting Islamist Sharia. (3) As such, the conflict was divided between native Syrians fighting for their own country and the thousands of foreign fighters who arrived to fight for global jihad. (4) The war was also a local, sectarian conflict between the ruling Shia minority and the majority Sunni population. (5) As a result, the war became a regional, proxy conflict between the local, pro-Assad Shia powers, Iran, Lebanon (and Hezbollah) and Iraq, and the competing regional Sunni powers, Saudi Arabia, Qatar and Turkey. (6) This division was complicated by the success of the stateless, northern Kurds, whose People's Protection Units (YPG) proved a powerful rebel force, becoming the main component of the anti-Assad Kurdish-Arabic Syrian Democratic Forces (SDF), simultaneously fighting as part of the revolution and as a force pursuing its own state. (7) The rebel Kurds were supported by the West, though opposed by the otherwise pro-rebel Turkey, who consider the Kurds a terrorist force. In early 20018 Turkey launched a separate series of attacks on the Kurds, splitting the war again. (8) The war also drew in other regional actors, with Israel pursuing its own agenda and conflict, with ongoing airstrikes against Assad's forces and facilities and against Iranian military assets. (9) The war can also be seen as a key battleground in the decades-long, state-level theological conflict between Shia Iran and Sunni Saudi Arabia for the soul of Islam itself, a war simultaneously fought out in Yemen. (10) The Syrian Civil War also became a global proxy conflict between Russia and the west, with Russia intervening on behalf of its Cold War ally, Syria, first with weapons, then, from 30 September 2015, with troops and airstrikes. The Russian aid proved decisive, enabling Assad to push back opposition gains and secure his position. After Afghanistan and Iraq, the USA was unwilling to intervene elsewhere and even Assad crossing Obama's 'red line' of using chemical weapons didn't lead to action. The USA tried to train and arm 'moderate rebels', but the rise of Islamist groups made it difficult to identify these and the USA feared its weapons would end in jihadi hands, hence the failure of its policy. Assad used chemical weapons again, firing Sarin gas at Khan Shaykhun on 4 April 2017, killing over 100 and injuring over 550, prompting President Trump to launch the first US strikes on the Assad regime, firing 59 Tomahawk missiles against the Syrian airbase responsible. (11) Finally, the Syrian Civil War was the background to what the west saw as a broader conflict – an existential battle between itself and Islamic State and Salafi-Jihadism. The USA and its allies began airstrikes on ISIS (by then known as Islamic State) in Syria from September 2014, but competed for airspace with Russian airstrikes against both Islamic State and the rebels the USA wanted to support. Arguably it was this battle against Islamic State, rather than the Civil War itself, which the west saw as most important. By early 2018 the Syrian Civil War was continuing, though Assad was in an improved position and rebel groups were on the verge of defeating Islamic State.

Syria's global civil war

The Syrian Civil War has been described as 'the most socially mediated conflict in history' (BBC, 2014). Inspired by the Arab Spring revolutions in Tunisia and Egypt, it began very simply, with calls for political reform and protests against Assad's repressive regime. Lacking access to state media, the protestors adopted Web 2.0 platforms such as Facebook, Twitter and YouTube as their own global media system. There was a sufficient audience for this. In 2011, 4.6 million Syrians (22.5% of the population) had access to the internet, up from 1.4 million in 2006. Mobile phone use was also reasonably high – according to the CIA World Factbook there were 87 mobile phones per 100 of the population by 2014. When the protests started, therefore, the Syrians began sharing what was happening in order to communicate with others in the country and encourage their participation, to document the abuses of the Assad regime and to influence international opinion against it.

Armed with mobile phones, protestors took videos and uploaded them to YouTube. The footage was raw, amateur and angry, all of which added to its evidentiary nature. Some of the most powerful early footage was of the tortured body and funeral of 13-year-old Hamza Ali al-Khateeb who had been detained in a Daraa protest on 29 April 2011 and killed by the security forces. His body was returned to his family on 25 May, showing signs of torture including severe bruising, whip-marks and burns consistent with electro-shock devices. It had three gunshot wounds and his genitals had been severed. The family would once have been helpless, but now they distributed images and video of the body on YouTube, which was broadcast on Al-Jazeera. State television claimed the injuries were faked, but Ali al-Khateeb became a rallying point for protestors, with a Facebook page, 'We Are All Hamza Ali Khateeb' (named after the Egyptian page, 'We Are All Khaled Said'), gathering 105,000 followers by the end of May.

Copying Mubarak's policy in Egypt (and following advice from Tehran), the government tried to silence the protestors with a communications shutdown, including restricting and arresting foreign journalists and removing the 3G telephone network and some satellite phones. The regime also deliberately targeted those taking photos or videos, killing them in the streets or torturing them in prison. Activists found ways to work around the blocks and access the internet (including using neighbouring countries' networks) and to film without being seen. As Ghrer says, 'the activists came to utilize small cameras hidden in shirt buttons, pens, ties and glasses, for instance, and installed them in cars, shop doors and activist's clothes so as to be able to tape the policemen's violations close up' (Ghrer, 2013).

As the situation deteriorated, fewer journalists were able or willing to access the battle-field and so the social media productions of the protestors became the main source of information. A huge volume of videos were uploaded and re-broadcast in the west and on Al-Jazeera and Al-Aribaya. The regime tried to discredit these by producing and uploading fake videos and sending them to news organizations in order to expose them afterwards. The protestors countered this by holding up signs showing the date and place or that day's local newspapers to prove their veracity. The protestors also countered the government's system of citizen spies by setting up 'Awainiyya' pages on Facebook to expose the 'watchers'. The names, addresses and photos of suspected spies were posted online, although they were also used for the settling of scores and often led to vigilante action.

As the protests turned into an armed insurrection, YouTube found new uses. The Free Syrian Army (FSA) announced their formation in a YouTube video on 29 July 2011 and other emerging militias followed suit, recording their own videos to publicize themselves, gain an informational advantage, and encourage regime defection and recruitment. Unable to impose themselves on the battlefield, the smaller opposition groups used social media as a force multiplier, to appear more significant and successful than they were to attract more recruits and – hopefully – international support. Each group advertised its regime recruits and displayed its insignia to promote its name and encourage donations through their linked Facebook pages. The videos also allowed each group to display their forces and weapons, whilst Twitter accounts gave real-time updates on their operations. Following the March 2011 multi-state NATO intervention in Libya, the Syrian opposition fully expected western support and used YouTube and social media to position themselves as important groups and potential recipients of help.

Social media also served as a training aid for the non-professional rebel armies. In July 2012, it was reported rebel fighters were sharing tips on Facebook and YouTube. The channel 'FSAHelp' included a high quality, Arabic 15-minute HD video explaining the basics of assault rifles, with others teaching how to shoot from a prone position, how to creep up on an enemy and hand-to-hand combat. FSAHelp also ran a Facebook help page, including photos about turning off ringtones and videos demonstrating how to use anti-tank missiles. YouTube was also used to spread disinformation. There were claims the FSA doctored videos to highlight their successes, advertise regime desertions and expose opposition atrocities. Captured regime forces were forced to recite FSA claims on camera.

One of the most important uses of social media was for donations. Hajaj al-Ajmi, a Kuwait sheikh, used his 347,000 Twitter followers to call for funds for the militias. The *Washington Post* described him as a 'money machine' for the rebels and an eastern Syrian group renamed itself the Hajaj al-Ajmi Pilgrims in honour of his donations (O'Neil, 2013). In 2012, the Popular Commission to Support the Syrian People, a Persian Gulf Group of online organizers, was thanked by rebel factions for donations ranging from $10,000 to $600,000. The total sum of online donations ran into millions of dollars. Messaging services such as WhatsApp were also employed in the war. As it didn't require a lot of data, WhatsApp was widely used to coordinate rebel groups and to organize help. The Syrian Civil defence Volunteers ('The White Helmets') used WhatsApp to hold meetings with colleagues, to contact areas under siege or inaccessible, and to organize more help where it was needed, for example at Khan Shaykhun after the chemical attack.

Twitter proved one of the most popular platforms, being used to globally highlight the plight of the Syrian people. Building on the earlier #SaveKessab campaign by Armenian-Americans (Kessab was a north-west Syrian village with a large Armenian population), the #SaveAleppo or #Save_Aleppo hashtag campaign was launched on 4 April 2014 to publicize civilian suffering in the midst of rebel-government fighting. The hashtags were used 120,000 times by the 9th, with 35,000 'likes' on the linked Facebook page. A green #SaveAleppo profile photo also spread online for use on Facebook and Twitter, whilst the Facebook page carried more images and photographs of the situation in the city.

The Syrian opposition was also internationally based, producing material for global consumption. The National Coalition of Syrian Revolutionary and Opposition Forces was formed in exile in Doha, Qatar, in November 2012, becoming the executive body of the Syrian opposition (although not accepted by elements such as the SDF and many Islamist groups). Its central media offices were based in Istanbul, producing Arabic press releases and social media content, whilst English-language releases and content were produced by Syrian groups in North America and Europe. Crilley discusses the importance of the Coalition's English-language Facebook page, which was set up in November 2012, although not actively used until March 2013. Over the next two years it published 1,174 posts, consisting of links to the official Coalition website, links to the Coalition's Twitter account, press releases and statements, as well as photos, videos and infographics. Visual media was especially important online, with an evolution in their propaganda over from 2013 to 2015 as their focus moved from Assad's use of chemical weapons, to the humanitarian crisis, to Assad's war-crimes, to the threat of ISIS and Assad, to the plight of the refugees and back to the issue of war-crimes. Crilley concludes the Coalition 'used visual media on Facebook pages to project a visuality of suffering in order to gain the support of global audiences' (Crilley, 2017:152).

Social media didn't just report the war, it was also employed tactically, as part of military operations. In September 2012, for example, a rebel attack in northwest Syria on the town of Harem was posted on Facebook. The Local Coordination Committee posted a Google Maps image on Facebook marking government troop positions around the town. The resulting battle was live-blogged on the newsfeed with posts that included calls for help from a pinned down rebel unit and, a day later, a detailed description of the unit's escape from harm. Like the videos of the protests, footage of combat proliferated, being posted by each group. Each, however, represented only a tiny slice of reality, being implicitly politically positioned, without any independent means of verification. Rather than adding to the public knowledge of the war, they only added to the impossibility of establishing a coherent understanding.

Not every use of social media was successful. In May 2013, a video emerged showing a Syrian rebel leader taking a bite from the heart of a dead soldier. US-based Human Rights Watch identified him as Abu Sakkar, a well-known insurgent leader from Homs, and described his action as a 'war-crime'. In the video, Sakkar, the leader of the independent Omar al-Farouq Brigade (an FSA offshoot), is heard to say, 'I swear to God we will eat your hearts and your livers, you soldiers of Bashar the dog', and insults the Alawite ruling minority (BBC, 2013). The video attracted international condemnation and had a global impact. When pro-Assad, Russian President Vladimir Putin visited London for the G8 Summit in June 2013, he embarrassed PM David Cameron by criticizing the west's support for the rebels: 'One does not really need to support the people who not only kill their enemies, but open up their bodies, eat their intestines in front of the public and cameras. Are these the people you want to support?' (Anishchuck, 2013).

Other embarrassments were more minor. In 2014 a video released by Zahran Alloush, the head of the jihadi rebel group Jaysh al-Islam, showing him addressing his forces attracted widespread online ridicule when social media users spotted the 'Hello Kitty' notebook he was using on the table in front of him. Other videos also found international fame, such as the September 2017 footage of three IS fighters who filmed themselves being blown up by a Syrian tank, with one fighter's headcam capturing the moment a shell hit their vehicle. They had been filming for propaganda purposes, but their footage was discovered and used against them. Some footage was deliberately faked. In November 2014 video went viral of a Syrian boy braving sniper fire to rescue a young girl, although it later emerged it had been filmed with child actors in Morocco by Norwegian filmmakers 'to see if the film would get attention and spur debate, first and foremost about children and war', and 'to see how the media would respond to such a video' (Dearden, 2014).

Within a few years the Syrian opposition had developed its own media outlets. The Syrian National Council (SNC), National Coordination Committee for Democratic Change (NCC) and Local Coordination Committees of Syria (LCCs) all operated their own networks of professionally designed websites and social media accounts. Two of the main opposition outlets were Ugarit News and Omawi New Live, which reported online on YouTube, Facebook and Twitter, with channels in Arabic and English, posting footage of security force attacks and atrocities and rebel operations. Despite their theological disdain for the modern world and image making, the jihadi Islamist groups were also active multi-media producers. Following the precedent of other groups, Jabhat al-Nusra announced its formation on YouTube on 23 January 2012 and produced propaganda videos via its Al-Manara Al-Bayda Foundation for Media Production that were posted on YouTube as well as jihadi websites and chatrooms. It later developed its own website and linked online outlets for its productions. ISIS, or Islamic State, went further, developing an entire, globally effective media ecology to promote its cause that will be considered in detail in the next chapter.

The Syrian embrace of social media caused problems for the western, Web 2.0 companies. Facebook had never envisaged itself as a news and reporting platform and at the start of the conflict took down many opposition pages. Dozens of pages of non-aligned NGOs and opposition citizen journalism pages were removed due to graphic imagery and calls for violence that violated Facebook's terms of use. Many LCCs, for example, produced Facebook pages to highlight violence against civilians but fell afoul of these rules. Others were taken down for odd reasons: the Daraa al-Mahata LCC Facebook page had 42,000 'likes' but was removed in October 2013 for posting an old photo of a man they said had been killed by the Syrian Army. Rebels and activists believed the government and Assad supporters were gaming Facebook's community standards and reporting system. The pro-Assad Syrian Electronic Army (SEA) even boasted of these tactics, saying, 'We continue our reporting attacks' against LCCs (Pizzi, 2014). The SEA was also suspected of taking down the Facebook page of the London-based NGO, the Syrian Network for Human Rights in October 2013. Rebels complained Facebook acted without warning and with little room for appeal.

In other ways, the west tried to help the opposition media. In May 2016, it was revealed that, under its 'Conflict and Stability Fund', the UK government was paying £2.4m to private contractors to boost the 'moderate armed opposition' (MAO) in Syria. After the government failed to get parliamentary support for military action in Syria in August 2013 it gave a contract to Regester Larkin, whose head set up the company Innovative Communications & Strategies (InCoStrat) which took over the contract in November 2014. Hired by the Foreign Office and overseen by the MoD, the Istanbul-based contractors were tasked to run an anti-ISIS 'MAO Central Media Office' with 'media production capacity', and to train MAO spokespersons and media coach 'influential MAO officials' (Cobain, Ross, Evans, and Mahmood, 2016). Effectively the UK government was producing rebel propaganda for Syrian civilian and military audiences, plus targeted messages for particular audiences. The output encompassed photographs, videos, military reports, radio broadcasts, print products and social media posts, all branded with the logos of fighting groups and circulated within the Arabic broadcast media and online without acknowledgement of the UK's role.

Western hopes for an effective moderate opposition, however, were naive. The opposition had splintered into too many militias and become dominated by Islamist groups. Even keeping track of the opposition was difficult, as groups formed and reformed into endlessly shifting short-term or long-term alliances and new organizations. In-fighting was common, especially against the jihadi Islamists who were themselves tacitly supported by Assad to further undermine the opposition. Russia trolled the west's confused foreign policy, with the Russian Embassy, UAE tweeting on 5 October 2015, 'Find the right #Syrian rebel to arm – non-trivial task to start your day with', alongside a cartoon showing cut-outs of a bazooka and of 8 identical fighters labelled the Al-Nusra Front, Islamic Front, Free Syrian Army, Army of Mujahideen, Anjad Al-Sham Islamic Union, Ansar Al-Deen Islamic Front and so on (Bartlett, 2015).

Amidst this confusion, accurate information about Syria was difficult to obtain. Foreign and independent journalism was barred by the regime and, as in Iraq, the civil war was highly dangerous. By October 2017, the Committee to Protect Journalists counted 103 reporters, filmmakers and editors killed in Syria since March 2011, most by government bombing. Most of the opposition welcomed reporting as aiding its cause, although Islamic State didn't, deliberately targeting and executing journalists. Home-grown citizen-journalist groups emerged to fill the reporting vacuum. The most famous was Raqaa is Being Slaughtered Silently (RBSS), a non-partisan, independent news organization established in April 2014, when Islamic State began to seize Raqaa. It began with 17 undercover journalists who used false names and encrypted channels to post stories about IS atrocities to its news page. By October 2017 its Facebook page had over 668,000 followers and its Twitter account had over 80,000 followers. IS installed public security cameras, controlled mobile phone ownership and prohibited public photography to avoid scrutiny. It found and executed journalists such as 30-year-old Ruqia Hassan, who was working in Raqaa under the name Nissan Ibrahim, who was executed as a spy by IS in September 2015, and RBSS co-founder Naji al-Jerf who was murdered by them in Turkey on 27 December 2015.

Individual citizen-journalists also operated in Syria, such as Malek Blacktoviche, working in Aleppo, who reported rebel activities and worked for foreign news organizations such as the USA's NBC. The independence of these citizen-journalists, however, has been questioned. Reporters Without Borders spokesperson Alexandra El Khazen commented, 'Citizen journalists in Syria do not have the same access to information. Each has [their] own reality. They depend on the local group who controls the area they work in'. Working in isolation, 'you don't always have other information that contradicts or balances yours' (Baraniuk, 2016). Many citizen-journalists were anti-Assad, although others were more deliberately neutral. The Syrian Red Crescent, for example, manages a set of Facebook pages and Instagram and Twitter feeds to keep local and international observers informed.

Individuals have also become important conduits of information, in sharing the reality of life in the warzone. On 24 September 2016, 7-year-old Bana al-Abed began tweeting, with her mother, from the account @alabedbana. She picked up 4,000 followers in a week and by October 2017 had over 363,000 followers. Her pinned tweet reads 'My name is Bana, I'm 7 years old. I am talking to the world now live from East #Aleppo. This is my last moment to either live or die – Bana' (Alabed, 2017). Tweeting from a residential area suffering from government and Russian airstrikes, including barrel bombs, phosphorus bombs and cluster bombs, she posted photos and videos of herself playing with her family ('Drawing with the brothers before the planes come'), images of the nearby carnage, including destroyed buildings and her neighbours trapped in the houses ('All of them under the rubble now. Is my house next?'), as well as pleas for help. Under the hashtag #MassacreinAleppo she said, 'Children like me are dying here in Aleppo please stop the bombing'. In October 2017, her book *Dear World*, telling her life story was published.

There were other attempts by the Syrians to alert the world to their plight. After the July 2016 release of the AR app Pokemon Go, reports emerged of teens in Syria playing the game in rebel-held Douma, downloading it when a connection was available via the proxy Viber, with players risking their lives climbing round rubble to 'collect' Pokemon.

Activists also tried to raise awareness by printing out photos of Pokemon and photographing them in the rubble of destroyed houses to post online. Perhaps the most powerful use of the game was the images produced by the Turkey-based Revolutionary Forces of Syria (RFS) Media Office, which published photos on Facebook and Twitter of Syrian children holding up cards with pictures of Pokemon on, asking to be collected instead: 'I am here, come save me' (Graham-Harrison, 2016).

The Syrian government controlled the state-run mass media but it was also active online, recognizing the importance of the global informational war. The Syrian Arab News Agency (SANA) had its own website and produced material online. The Assad regime tried to push the idea that this was an illegitimate foreign, proxy war by the USA, Saudi Arabia and Qatar, trying to undermine Syria through sectarianism and Islamic militancy. Assad wasn't overly worried by the civil war. In July 2013, he found time to set up the account 'Syrianpresidency' on Instagram, posting photos showing himself and his wife Asma smiling and greeting civilians, ignoring the reality of a conflict where estimates of the dead already surpassed 100,000 people. Government PR efforts have often been clumsy. On 25 June 2015, the SANA English Twitter account posted, 'Now that #Summer is upon us, snap us your moments of Summer in #Syria using the hashtag #SummerinSyria'. Syrian users trolled the government by hijacking the hashtag, sharing photos of injured civilians, devastated cities, bomb sites, explosions and refugee camps. The US Syrian Embassy account joined in, posting images of destruction, commenting, 'Assad regime barrel bombs Al Bayan hospital in #Aleppo, kills 5 nurses + 1 doctor', under the #SummerinSyria hashtag' (Ghani, 2015).

The Assad regime was helped by a cyberwar campaign run by the patriotic hacking group, the Syrian Electronic Army (SEA). Emerging on Facebook in April 2011, it announced its existence and aims on it website in May 2011 when they also set up a Twitter account and YouTube channel. Although most commentators believed it had state links and backing, its webpage presented its members as ordinary citizens: 'We're all Syrian youths who each have our specialized computer skills, such as hacking and graphic design. Our mission is to defend our proud and beloved country Syria against a bloody media war that has been waged against her. The controlled media of certain countries continues to publish lies and fabricated news about Syria' (Harding and Arthur, 2013). Opposition activists have claimed the SEA was bankrolled by Assad's billionaire cousin Rami Makhlouf, with the group moving in 2012 from Damascus to a secret base in Dubai, operating out of one of Makhlouf's shadowy companies there. Members, it is claimed, were given food and accommodation and paid $500–1000 for high-profile hacks against western targets.

The SEA launched defacement attacks against Syrian opposition Facebook pages and websites, posting pro-regime messages and images of the Syrian flag or Bashar al-Assad. It also made software available on its Facebook page to encourage DDoS attacks and recommended supporters post pro-government messages on popular Facebook pages, such as Oprah Winfrey's, 'as a way to reach out to, and influence the American public opinion' (Strickland, 2011). The SEA's most famous attacks, however, were aimed at western websites, highlighting again the global informational battle the civil war had become. Its first hack was of Harvard University servers, replacing the homepage with a photo of Assad and a banner reading 'Syrian Electronic Army were here'. It followed this with DDoS or web hijacking and defacement attacks against Al-Jazeera, Sky News Arabia, Human Rights Watch, CBS News, FIFA, Associated Press, *The Guardian*, the *Financial Times*, Thomson Reuters, Skype, the *New York Times*, the *Huffington Post* and the US Marines Corps.

Its most spectacular attack was hijacking the Associated Press Twitter account on 23 April 2013 to post 'Breaking: Two Explosions in the White House and Barack Obama is Injured', which caused the DOW to nose-dive 150 points in three minutes, wiping $136.5bn in equity off US companies (Foster, 2013). Its tweet, 'Ops! @AP owned by Syrian Electronic Army!' confirmed its responsibility. Its activities continued through to early 2015 with attacks on Skype, Xbox, Microsoft, *The Sun, The Sunday Times, The Independent*, Ebay, Paypal, Facebook, Reuters and *Le Monde*. On 27 November 2014, it attacked Gigya's comment system, posting messages to hundreds of websites, saying 'You've been hacked by the Syrian Electronic Army'.

The SEA's stated aim was to punish western news organizations seen as critical of Syria's regime and push its alternative narrative that it was terrorists who were destroying the country. As its targets, however, included *The Guardian*'s GuardianBooks' and BBC Weather's Twitter accounts and Italian online shops and tourism guides then the hackers were either after soft targets or popular sites that might gain viewers, or they misunderstood what they were attacking. In May 2011, for example, as part of an apparent attack on the British government for its Syrian policy, the SEA hijacked the websites of Bournemouth and Poole Borough Council and Royal Leamington Spa Town Council, mistaking them for more important parts of the UK government structure than they were. The message they posted on the Royal Leamington Spa website – 'We, the Syrian, don't harm anybody, but if you dare to interfere in the Syrian Internal Affairs we are able to put appropriate limits. STOP to interfere in the Syrian Internal Affairs. Leave us alone' – didn't really specify what the council itself had done wrong (Noman, 2011). The SEA's Facebook page even crowd-sourced its hacks, running opinion polls on what kind of websites people wanted to see attacked, with US and then British sites coming top.

Another cyberwar operation was exposed by US cybersecurity firm FireEye in February 2015 who found from late 2013 to early 2014 hackers had posed as beautiful women in a 'honeytrap' for Syrian rebels, contacting them on Skype and sending selfies loaded with malware allowing data to be stolen from their laptops and phones, as well as creating fake Facebook pages with malware for those clicking on links. The stolen data included operational plans for a rebel military attack on Khirbet Ghazaleh in Southern Daraa. FireEye traced the servers used to outside Syria but couldn't say who the hackers worked for.

The opposition scored their own cyberwar victory early on in March 2012, when hackers, aided by a government mole, managed to access and leak over 3,000 emails from the personal accounts of Bashar al-Assad and his wife Asma. Assad's emails included poems from his wife to him and flirtatious photos from women, whilst his wife boasted of her power over him, declaring in an email 'I am the real dictator'. Ultimately email records that showed Assad swapping entertaining internet links on his iPad and downloading music from iTunes and his wife ordering designer goods and furnishings proved embarrassing for the regime. Asma was seen to be discussing shoes costing over $5,000 and ordering candlesticks, tables and chandeliers from Paris for over $10,000 whilst Homs suffered under a government siege that killed hundreds of people.

As part of its support for Assad, Russia extended its ongoing global information war to Syria, carrying out what Solon calls 'an extraordinary disinformation campaign' against the 'White Helmets', the humanitarian volunteer organization who save civilians from the aftermath of airstrikes and bombs (Solon, 2017). The White Helmets' activities have won international recognition and the footage they film on handheld and helmet cameras has exposed the war-crimes of the Assad regime, hence the Russian desire to undermine and discredit them. Since September 2015, Russian state media and a network of supportive alternative news networks and individuals have tried to create a counter-narrative to the mainstream account, with claims of 'staged' attacks and fake rescues with recycled victims and accusations that the White Helmets are part of Islamic State. The White Helmets' participation in the video 'mannequin challenge' that went viral in November 2016, for example, was repurposed by their critics and used out of context as evidence of their staging of rescues. Solon argues that 'The Russian strategy has been very successful at shaping the online conversation about the White Helmets. By gaming the social media algorithms with a flood of content, boosted by bots, sock-puppet accounts and a network of agitators, propagandists are able to create a "manufactured consensus" that gives legitimacy to fringe views'.

In all these ways, a local civil war became a global, multi-layered informational conflict. But it also became global in a more physical sense with the refugee crisis. The use of airstrikes, barrel bombs, phosphorus bombs and chemical weapons against residential areas and the destruction of homes and infrastructure led to huge numbers fleeing the country. The refugees faced continual danger. There were stories of Turkish border guards shooting people fleeing Syria and in August 2017 video emerged of border guards beating and abusing refugees. With poor conditions in the Jordanian camps, many chose to save up to reach Europe through the Balkan route, which brought further risk of death. In 2015 alone, 3,200 Syrians were killed crossing the ocean. The refugees led to a Europe-wide debate, although not about western foreign policy but instead about 'immigration', the cultural threat to Europe and fears the refugees were terrorists.

Right-wing mainstream media pushed these ideas, but they were given free reign online with the spreading of anti-refugee memes on right-wing Facebook pages such as the English Defence League (EDL), South East Alliance and Pegida. These included photos erroneously claiming to prove individual refugees were previously IS fighters, images of large, muscled men with the tagline 'refugees?', and bizarre claims that these men were actually body-builders. One such image had the captions, 'Please help feed and house this poor, defenceless refugee', 'I heard we can get free steroids in England!', and 'Don't be racist and let me in'. That image, along with numerous others, was debunked as actually being taken in 2013 in Christmas Island, off the coast of Australia. Other memes tried to 'dehumanize refugees in Europe by claiming they are all cowardly men who have left behind women and children in war-zones' (Kleinfeld, 2015). The EDL, for example, shared one image contrasting British soldiers captioned 'Go to war-zone. Leave women and children in safe country', with all-male, smiling refugees captioned, 'Go to safe country. Leave women and children in war-zone'. Other images posted included fake photos of people fleeing ships in huge numbers, a photo of refugees captioned 'ISIS. Coming to your town very soon' that actually showed Syrians heading into Iraq, and photo memes claiming Syrians were being given places in plush hotels. Syria was a global informational war fought for numerous political agendas.

A few images briefly brought the reality of the Syrian civilian suffering to life for the west. One was the death of three-year-old Aylun Kurdi, which made headlines on 2 September 2015. That day Aylun and his family boarded a small inflatable boat that sank a few minutes after leaving Bodrum, Turkey. A series of images by the Turkish photo-journalist Nilüfer Demir became famous, including one showing Aylun dead on the beach and another showing his body being carried from the shoreline. Posted by journalists on Twitter, the images went viral, with 53,000 retweets per-hour. Placed on the front page of every newspaper and widely shared and commented on, the images appeared on 20 million screens worldwide in just 12 hours. They seemed to herald a shift in attitudes in the west towards Syrians, with researchers finding 'a transformation in the language around what was happening in Europe, with the use of the word "refugee" outstripping "migrant"' (Press Association, 2015). The moment soon passed, however, and little was done for the refugees. Immigration remained an important issue in European elections in 2016 and a factor in the June 2016 UK 'Brexit' vote.

Another important image came in the video shot by Aleppo journalist and member of the Aleppo Media Centre, Mustafa al-Sarout on 17 August 2016. His images of 5-year-old Omran Daqneesh sitting in an orange ambulance chair in shock, with his hair and face covered in a paste of grey dust and matted blood, having been pulled from the rubble after a government airstrike in Qaterji in Aleppo, went viral and made global news. The images were a visceral reminder of children's suffering in the war and even made CNN's anchor Kate Bolduan cry on air, although the boy's pro-government family later complained that rebel groups were exploiting the images for their propaganda value. As with Aylun Kurdi, Omran Daqneesh's moment passed and the plight of civilians continues in the ongoing war.

Conclusion

Gaza and Syria represent just two examples of contemporary participative war. Today, every conflict is, to a greater or lesser extent, hypermediated through a range of outlets by governments and the military, by mainstream news producers and by groups, and individuals, using every technological system, platform, service and device available to them. Now anyone around the world can take part in these conflicts, adding to the informational war and sharing images, posts, video, memes and propaganda. This represents an ecological transformation in how conflicts are fought, mediated and experienced, making every warzone a global battlefield. What is less certain is what this adds up to.

The Gazans, for example, won the online participative war in 2014 but this had little effect on the ground. They remained under Israeli control, their infrastructure had been decimated, there had been a huge loss of life and the international powers had done nothing to change any of this. On 6 December 2017, President Trump controversially recognized Jerusalem as the capital of Israel, effectively legitimating Israel's right to land not granted under the original UN plan. In the following days Israel responded to Palestinian protests with more airstrikes on Gaza. The Gazan online 'victory' was real, but Gaza as an issue was soon replaced in the fast-moving and fickle world of social media. In Syria, too, social media haven't necessarily had the effect we might have expected. The sheer volume of information pouring from the country's battlefields and the lack of context, explanation and verification have produced a bewildering and fractal disintegration of knowledge. Whilst the mainstream media have often been criticized for their partial, biased or constructed coverage and representation of warfare, it isn't necessarily the case that the new, bottom-up coverage automatically solves these problems and brings us closer to the 'truth' of war. These are issues we will return to in the Conclusion.

Key reading

There are few analyses yet of the 2014 Gaza War so Blumenthal (2015) is essential reading, though his book is intended as a critique of the Israeli military operation. Background reading about the broader Arab-Israeli conflict can be found in Bunton (2013), Gelvin (2014), Harms and Ferry (2017) and Sherbok and El-Alami (2015), whilst Bregman (2014) and Filiu (2014) provide detailed discussions of the occupied territories and Gaza. More information about the use of social media in the war can be found online. The examples I've discussed are easily discoverable by following the references or Googling them. There are many books about the Syrian Civil War, though few say much about the role of social media and digital technologies. I'd recommend Abboud (2016) as the best student introduction to Syria, with Glass (2015), Hokayem (2013), Lesch (2013), Lister (2015), and Phillips (2016) as further reading. Again, the footnotes and Google provide the best means of research into the role of social media. This chapter hasn't discussed issues around Islamic State in detail as that is the subject for the next chapter. The texts I recommend there on Islamic State are also extremely useful for understanding the Syrian Civil War, with most of the books containing chapters on how it began and how IS became involved. Just published as I write this, Patrikarakos (2017) is an essential overview of what I call 'participative war' and should be the first call for anyone interested in what I cover in this chapter as well as chapters 6 and 11.

References

Abboud, S. N. (2016) *Syria*, Cambridge: Polity Press.

Alabed, B. (2017) '@Alabedbana', Pinned Tweet, *Twitter*, https://twitter.com/AlabedBana/status/808599057794998272.

Aloul, A. (2014) 'Ayman al-Aloul – Bucket of Rubble', *YouTube*, 23 August, https://www.youtube.com/watch?time_continue=2&v=duSBig1-VUA.

Anischuck, A. (2013) 'Putin Warns West Not to Arm Organ-Eating Syrian Rebels', *Reuters*, 16 June, https://uk.reuters.com/article/uk-syria-crisis-putin/putin-warns-west-not-to-arm-organ-eating-syrian-rebels-idUKBRE95F0AF20130616.

Baraniuk, C. (2016) 'Citizen Journalism is Playing a Crucial Role in Aleppo – But it Comes at a Cost', *Wired*, 2 November, http://www.wired.co.uk/article/syrian-citizen-journalists.

Bartlett, E. (2015) 'The Russian Embassy is Making Jokes about the Civil War in Syria on Twitter', Indy100, *The Independent*, https://www.indy100.com/article/the-russian-embassy-is-making-jokes-about-the-civil-war-in-syria-on-twitter–byoad1CmDg.

BBC. (2013) 'Outrage at Syrian Rebel Shown "Eating Soldier's Heart"', *BBC News*, 14 May, http://www.bbc.co.uk/news/world-middle-east-22519770.

BBC. (2014) 'Syria: Report Shows How Foreign Fighters Use Social Media', *BBC News*, 15 April, http://www.bbc.co.uk/news/world-27023952.

Beaumont, P. (2015) 'Israel Exonerates Itself over GAZA Beach Killings of Four Children Last Year', *The Guardian*, 11 June, https://www.theguardian.com/world/2015/jun/11/israel-clears-military-gaza-beach-children.

Blumenthal, M. (2015) *The 51 Day War: Resistance and Ruin in Gaza*, London: Verso Books.

Bregman, A. (2014) *Cursed Victory. A History of Israel and the Occupied Territories*, London: Penguin.

Bunton, M. (2013) *The Palestinian-Israeli Conflict: A Very Short Introduction*, Oxford: Oxford University Press.

Cobain, I, Ross, A., Evans, R., and Mahmood, M. (2016) 'How Britain Funds the "Propaganda War" Against Isis in Syria', *The Guardian*, 3 May, https://www.theguardian.com/world/2016/may/03/how-britain-funds-the-propaganda-war-against-isis-in-syria.

Crilley, R. (2017) 'Seeing Syria: The Visual Politics of the National Coalition of Syrian Revolution and Opposition Forces on Facebook', *Middle East Journal of Culture and Communication*, 10, pp.133–58.

Dawber, A. (2013) 'Tales From Gaza: What is Life Really Like in "The World's Largest Outdoor Prison"', *The Independent*, 12 April, http://www.independent.co.uk/news/world/middle-east/tales-from-gaza-what-is-life-really-like-in-the-worlds-largest-outdoor-prison-8567611.html.

Dearden, L. (2014) 'Footage of Syrian Boy "Braving Sniper Fire" to Rescue Girl was Faked by Norwegian Filmmakers', *The Independent*, 14 November, http://www.independent.co.uk/news/world/middle-east/footage-of-syrian-boy-braving-sniper-fire-to-rescue-girl-was-faked-by-norwegian-filmmakers-9862600.html.

Filiu, J. P. (2014) *Gaza: A History*, London: C. Hurst & Co. Ltd.

Flanagin, J. (2014) 'War and Media in the Gaza Strip', *The New York Times*, 22 July, https://op-talk.blogs.nytimes.com/2014/07/22/war-and-media-in-the-gaza-strip/.

Fowler, S. (2014) 'Hamas and Israel Step Up Cyber Battle for Hearts and Minds', *BBC News*, 15 July, http://www.bbc.co.uk/news/world-middle-east-28292908.

Foster, P. (2013) '"Bogus" AP Tweet about Explosion at the White House Wipes Billions off US Markets', *The Telegraph*, http://www.telegraph.co.uk/finance/markets/10013768/Bogus-AP-tweet-about-explosion-at-the-White-House-wipes-billions-off-US-markets.html.

Gelvin, J. L. (2014) *The Israel-Palestine Conflict*, New York: Cambridge University Press.

Getlen, L. (2017) 'How wars are Now Won and Lost on Social Media', *New York Post*, 2 December, https://nypost.com/2017/12/02/how-wars-are-now-won-and-lost-on-social-media/.

Ghani, A. (2015) 'Syrians Respond to 'Summer in Syria' Tweet with Photographic Realities of War', *The Guardian*, 25 June, https://www.theguardian.com/world/2015/jun/25/summer-in-syria-tweet-photo-realities-war-hashtag.

Ghrer, H. (2013) 'Social Media and the Syrian Revolution', *Westminster Papers*, 9(2), pp. 113–22.

Glass, C. (2015) *Syria Burning*, New York: OR Books.

Graham-Harrison, E. (2016) 'Syrian Campaigners use Pokemon Go to Ask World to Save War Children', *The Guardian*, 21 July, https://www.theguardian.com/world/2016/jul/21/campaign-pokemon-go-craze-attention-syrian-conflict.

Harding, L. and Arthur, C. (2013) 'Syrian Electronic Army: Assad's cyber- warriors', *The Guardian*, 30 April, https://www.theguardian.com/technology/2013/apr/29/hacking-guardian-syria-background.

Harms, G. and Ferry, T. M. (2017) *The Palestine-Israel Conflict*, London: Pluto Press.

Ho, S. (2014) 'Palestinians Mock Kidnapping in Video', *The Times of Israel*, 23 June, http://www.timesofisrael.com/palestinians-parody-kidnapping-in-video/.

Hokayem, E. (2013) *Syria's Uprising*, London: Routledge.

Israel Defence Forces. (2014a) 'What if Terrorists Could Shoot This Rocket at Your Country?', *Israel defence Forces*, 9 July, https://www.idfblog.com/2014/07/09/what-if-terrorists-could-shoot-this-rocket-at-your-country/.

Israel Defence Forces. (2014b) '15 Seconds: Not Enough Time', *YouTube*, 2 July, https://www.youtube.com/watch?v=qs1R5RAC02c.

Kleinfeld, P. (2015) 'Calling Bullshit on the Anti-Refugee Memes Flooding the Internet', *Vice*, 10 September, https://www.vice.com/en_uk/article/zngwz9/kleinfeld-refugee-memes-debunking-846.

Lesch, D. (2013) *Syria: The Fall of the House of Assad*, New Haven, Conn: Yale University Press.

Lister, C. (2015) *The Syrian Jihad*, London: C. Hurst & Co.

Lotan, G. (2014) 'Israel, Gaza, War & Data', *Medium*, 4 August, https://medium.com/i-data/israel-gaza-war-data-a54969aeb23e.

Mackey, R. (2014) 'Israelis Watch Bombs Drop on Gaza From Front-Row Seats', *The New York Times*, 14 July, https://www.nytimes.com/2014/07/15/world/middleeast/israelis-watch-bombs-drop-on-gaza-from-front-row-seats.html.

Noman, H. (2011) 'The Emergence of Open and Organized Pro-Government Cyber attacks in the Middle East: The Case of the Syrian Electronic Army', *OpenNet Initiative*, 30 May, https://opennet.net/emergence-open-and-organized-pro-government-cyber-attacks-middle-east-case-syrian-electronic-army.

O'Neil, P. H. (2013) 'Why the Syrian Uprising is the First Social Media War', *The Daily Dot*, 18 September, https://www.dailydot.com/layer8/syria-civil-social-media-war-youtube/.

Pariser, E. (2011) *The Filter Bubble*, London: Penguin Books.

Patrikarakos, D. (2017) *War in 140 Characters*, New York: Basic Books.

Phillips, C. (2016) *The Battle for Syria*, London: Yale University Press.

Pizzi, M. (2014) 'The Syrian Opposition is Disappearing from Facebook', *The Atlantic*, 4 February, https://www.theatlantic.com/international/archive/2014/02/the-syrian-opposition-is-disappearing-from-facebook/283562/.

Press Association. (2015) 'Alan Kurdi Image Appeared on 20m Screens in just 12 Hours', *The Guardian*, 15 December, https://www.theguardian.com/media/2015/dec/15/alan-kurdi-image-appeared-on-20m-screens-in-just-12-hours.

Sherbok, D.C. and El-Alami, D. (2015) *The Palestine-Israeli Conflict*, London: Oneworld Publications.

Soffer, A. (2014) 'Sick: Fatah Posts Cartoon Mocking Kidnapped Teens as "Rats"', *Arutz Sheva*, 15 June, http://www.israelnationalnews.com/News/News.aspx/181726.

Solon, O. (2017) 'How Syria's White Helmets Became Victims of an Online Propaganda Machine', *The Guardian*, 18 December, https://www.theguardian.com/world/2017/dec/18/syria-white-helmets-conspiracy-theories.

Strickland, E. (2011) 'The Syrian War is Raging on Facebook', *IEEE Spectrum*, 18 May, https://spectrum.ieee.org/tech-talk/telecom/internet/the-syrian-war-is-raging-on-facebook.

Tobin, A. (2014) 'Gazan War Politics Drove 1 in 6 Israelis to "Unfriend"', *The Times of Israel*, 31 October, http://www.timesofisrael.com/gaza-war-politics-drove-1-in-6-israelis-to-unfriend/.

Toor, A. (2014) 'Rubble Bucket Challenge Aims to Raise Awareness about Gaza', *The Verge*, 26 August, https://www.theverge.com/2014/8/26/6068763/rubble-bucket-challenge-aims-to-raise-awareness-about-gaza.

11 Viral war

Islamic State's digital terror

A global brand

In the last chapter, we explored the rise of 'participative war', a new mode where networked technologies and online public platforms allow anyone within or outside of a conflict zone to participate in informational war, to tell their story, expose events, offer support and contribute towards or expose propaganda. This chapter explores this participation in more detail through a case study of one of the most important military groups operating in recent years in Syria and Iraq, Islamic State (IS). It examines how Islamic State employed digital technologies to establish an online presence that helped them eclipse al-Qaeda as the leading force in Salafi-Jihadism, as well as spread their ideology, recruit new members, and cause terror and shock at the heart of western societies. It considers how, from its base in the chaos of war-torn Syria and Iraq, a small, ideologically extreme, apocalypse-inspired, cult group not only achieved success on the local battlefields, tearing up the Sykes-Picot line, but also built a powerful, global terrorist brand, advertised its atrocities to the world, and weaponized virality to digitally supercharge jihadism, mobilize the hidden Muslim resentment at the 'War on Terror', appeal to young people from around the world and present itself as an existential threat to the entire west.

The rise of Islamic State

In Chapter 7, I traced the aftermath of the Iraq invasion from 2003 to 2010. We saw there that the Bush administration had no plans for post-war Iraq and had underestimated the reconstruction needed. The CPA's plans to transform Iraq into a western-style democracy, with a neo-liberal market economy were naïve, even without the deteriorating security situation. Post-Saddam Iraq was chaotic, with widespread looting, and whilst network-centric warfare had easily defeated the Iraqi military, a small, mobile military force simply couldn't control a country. Saddam's plan had always been for a guerilla war anyway, hence the USA immediately faced an organized Baathist insurgency that was soon joined by disillusioned Sunnis and complicated by the arrival of foreign jihadis coming to fight the 'far enemy', the USA.

One of the most successful jihadi groups was Jama'at al-Tawhid Wal-Jihad (The Party of Monotheism and Jihad), formed in 1999 by the Jordanian Abu Musab al-Zarqawi (born Ahmad Fadeel al-Nazal al-Khalayleh). He had met Osama bin Laden in Afghanistan, but differed in his visceral hatred of Shia Muslims and desire to take territory for a caliphate.

After fleeing Afghanistan in 2001, al-Zarqawi came to Iraq by 2003 and began a terror campaign with the Baghdad bombings of the Jordanian embassy on 7 August and the UN headquarters on the 19th. He was inspired by Islamist theorist, Abu Bakr Naji's 2004 online book, *The Management of Savagery*, which argued for 'the power of vexation and exhaustion', recommending constant violence to deplete enemy forces and morale, render a region ungovernable due to the 'chaos and savagery', and polarize Muslims, forcing the Sunnis to choose the jihadis (Hosken, 2015:46.). Al-Zarqawi also showed an understanding of the role of the media. On 11 May 2004 he released an online video of the murder of US hostage Nick Berg, in response to the Abu Ghraib revelations. The black uniforms, Guantanamo Bay-style orange clothing for the victim, accusatory statement and decapitation would set the template for later Islamic State videos.

Al-Zarqawi made his operational base in Anbar province, surviving the second battle of Fallujah in November to December 2004 and becoming the USA's most wanted insurgent. His success won Bin Laden's support and on 17 October 2004 he paid 'bayat' to Bin Laden and changed his group's name to Tanzim Qaidat al-Jihad fi Bilad al-Rafidayn, or al-Qaeda in Iraq (AQI). Bin Laden criticized his killing of Shia Muslims, advocating attacks purely against the USA, but the Shia were always al-Zarqawi's main target and his 22 February 2006 bombing of the Shia shrine, the al-Askari mosque in Samarra, finally provoked a full-scale sectarian war. Over 1,300 Sunni bodies were found around Baghdad in the next few days and the following years saw an explosion of murders, kidnappings and torture, attacks by Shia militia and the Iranian-influenced, Shia-dominated Iraqi government (including Interior Ministry police death squads) as well as by Sunni insurgent groups, and an ethnic cleansing of neighbourhoods. Al-Zarqawi had got the chaos he wanted.

AQI merged with other groups into the Mujahideen Shura Council (MSC), on 15 January 2006, although al-Zarqawi was killed in an airstrike on 7 June 2006. On 13 October, the MSC, under Abu Ayyub al-Masri, declared the creation of the Islamic State in Iraq (ISI), with Abu Hamza al-Baghdadi as emir. ISI's predominantly foreign fighters, violence, harsh laws and move into the smuggling trade led to clashes, however, with Sunni tribes. The tribes created the Anbar Salvation Council in response and by January 2007 they were in armed revolt against the jihadis in a movement known as 'the awakening' (Sahwa). The USA took its chance, creating a Sunni tribal police organization, 'The Sons of Iraq', and implementing a 'surge' of 21,000 troops from June 2007, which, along with NSA surveillance and cyberwar operations (see chapter 9), combined to nearly defeat ISI by 2008. The Shia-led Iraqi government, however, destroyed the opportunity due to their anti-Sunni hostility. They disbanded the Sons of Iraq, arrested its leaders and repressed Sunni tribesmen, protestors and politicians, driving Iraqi's Sunnis back to the radical Islamists.

On 18 April 2010, al-Masri and al-Baghdadi were killed by US rockets in a safe house. Under the influence of former Baathist officers, ISI reformed around Iraqi fighters, announcing Abu Bakr al-Baghdadi (born Ibrahim Awad Ibrahim al-Badri) as its new leader on 16 May 2010. This new phase of ISI had been organized through the USA's own prison system in Iraq, with prisons such as Camp Bucca helping the jihadis to network and organize. When the Iraqi government took over the prison system in 2009 it released many jihadis, allowing them to reorganize (ironically, retaining members of the 'awakening' in prison due to their cooperation with the USA). The US withdrawal from Iraq was officially completed on 8 December 2011, leaving the country controlled by a Shia government under the dictatorship of Nouri al-Maliki.

Whilst ISI regrouped in Iraq, it opened another jihadi front in Syria. Syria's president Assad had helped jihadis enter Iraq to destabilize the US occupation and when a civil war developed in Syria from March 2011 he had aided the rise of jihadi Islamists to split the opposition and reduce the possibility of western intervention on the rebels' side. In 2012 al-Baghdadi sent a delegation to Syria to set up the ISI/AQ affiliate, Jabhat al-Nusra, under the Syrian militant Abu Mohammad al-Julani. The group announced its formation on 23 January 2012. By 2013, al-Baghdadi feared it might eclipse ISI so on 8 April he announced al-Nusra and ISI were uniting as the Islamic State of Iraq and the Levant (ISIL), otherwise known as the Islamic State of Iraq and Syria (ISIS). Al-Julani and the head of al-Qaeda, al-Zawahiri, rejected this. In response ISIS rebuilt its Iraqi army through jailbreaks and foreign fighters and launched its own Syrian campaign against al-Nusra as well as the other rebel groups and the Assad regime. ISIS became the most powerful militia in Syria, taking Raqqa on 13 January 2014 as its new capital. On 16 January, al-Qaeda expelled ISIS.

Over the following months, ISIS launched an astonishingly successful military offensive, wiping out the Sykes-Picot line that separated Syria and Iraq, taking Fallujah by 14 January and Mosul by 10 June. The Iraqi army fled in disarray, leaving ISIS to carry out a series of massacres, including the Sinjar massacre in August 2014, where thousands of Yazidis were executed. On 29 June 2014, an audio recording was released declaring the foundation of a new caliphate named 'Islamic State' (IS) under al-Baghdadi, and calling upon the world's Muslims to 'obey' him. On 5 July 2014, a video was uploaded to YouTube showing the secretive al-Baghdadi appearing in public at Mosul's Grand al-Nuri mosque the previous day, proclaiming himself Caliph of a state that forcibly included over 6 million people. The threat of Islamic State was significant enough for the USA to begin airstrikes against it in Iraq and Syria. Outflanked and looking increasingly obsolete, al-Qaeda attempted to delegitimize IS through Islamist scholars, but IS's on-the-ground success galvanized Islamism, eclipsing the older movement. By November new IS affiliates were swearing loyalty in Egypt, Yemen, Pakistan and Libya. Islamic State and its brand of Salafi-Jihadism appeared unstoppable.

The rise of IS was a disaster for the western powers, representing the opposite of what their 'War on Terror' had wanted. As Cockburn drily notes,

> Whatever they intended by their invasion of Iraq in 2003 and their efforts to unseat Assad in Syria since 2011, it was not to see the creation of a jihadi state spanning northern Iraq and Syria, run by a movement a hundred times bigger and much better organized than the Al-Qaeda of Osama Bin Laden.
>
> Cockburn (2015:38.)

IS represented a new phase in Islamist militancy, moving from al-Qaeda's model of western-directed, spectacular terrorism to one of local, sectarian conflict, with the aim of gaining and holding territory and drawing in new fighters to expand it. Al-Qaeda betrayed a western, Marxist-Leninist influence in its belief in a revolutionary vanguard as a 'base' or 'foundation' ('al-qaeda') and in its always-futural ideal state. Whilst al-Qaeda, therefore, never expected to take power, and assumed instead it would have to remain select and secretive, IS rejected this defeatism and cowardice, asserting itself publicly and globally and forcing the caliphate into existence as an act of political and military will. IS, therefore, was simultaneously more grounded than al-Qaeda, in its desire to take land and establish political structures, and more religious, in its belief in realizing the divine kingdom *now* and in the idea that this presaged an apocalyptic conflict in the Middle East. Hence its dual nature as both a local insurgent force fighting to establish a state and as a global, terrorist brand, attempting to provoke that final war with the west.

Box 11.1 Salafi-Jihadism

'Radical Islamism', or 'political Islam', is perhaps better described as Salafi-Jihadism. This is an extremist Sunni Muslim theological position, political philosophy and mode of activism. It is a minority position within Salafism and Islamic thought. One way to understand Salafi-Jihadism is through 'Wahhabism', a fundamentalist, puritanical, literalist variant of Islam, founded by the eighteenth-century Sunni scholar Muhammed ibn Abd al-Wahhab (1703–1792), who lived in the Najd area of what would become Saudi Arabia. Al-Wahhab was influenced by the fourteenth-century scholar Taqi al-Din ibn Taymiyyah (1263–1328), who believed, following a famous hadith, that only the first three generations of Muslims followed real Islam ('al-salaf al-salih' – 'the pious predecessors' – hence 'Salafism'), with his contemporaries being marked by a moral decay. Taymiyyah urged Muslims to revert to the pure origins of the faith found in the Quran and the Sunnah (the verbally transmitted record of the teachings, deeds and saying of the prophet Muhammad), established the principles of 'takfir' (the denunciation of apostates), and urged a holy jihad (a fight or struggle) against the 'kafir' (infidels, or non-believers) to create an Islamic State ruled by a Caliph following the rules of the Quran. Al-Wahhab revived these ideas, promoting a rigid view of 'tawhid' – the doctrine that only Allah is to be worshipped – that tolerated no difference. In 1744 al-Wahhab united with the tribal leader Muhammad Ibn Saud (1710–1765), who saw Wahhabism as a means to seize political power. The two embarked on a brutal campaign for domination and by 1803 the Saud family controlled most of the area, including Mecca and Medina. The Ottoman empire reestablished control by 1818, but following the end of the empire after World War I, the Saud family succeeded in uniting the region by force, with Abdul-Rahman al-Saud proclaiming the kingdom of Saudi Arabia in 1932. The kingdom was marked by the continuing alliance of the Saud family, who held power, and the al-Wahhab-descended theocracy, although the oil-based wealth of the royal family and their pro-western policies and modern lifestyles caused tension with the state religion.

1979 proved a key year for Saudi Wahhabism. The Shia Iranian revolution in February established a rival for the leadership of Islam, whilst the November occupation of the Grand Mosque in Mecca by radical Wahhabists calling for the end of the corrupt and decadent royal family exposed an internal threat. The December 1979 Soviet invasion of Afghanistan provided a solution. Many Arabs joined the Afghan mujahideen resistance, drawn by exhortations such as Abdullah Azzam's claim that a defensive jihad (understood now as an armed struggle) to protect invaded Muslim lands was the duty of every Muslim regardless of where they were. By funding and supporting this jihad, Saudi Arabia was able to globally export radical Wahhabist elements, ensuring its own domestic security. It also escalated its global promotion of Wahhabism, 'to reinforce their own religious credentials at home whilst increasing their influence overseas, allowing them to reassert their claim to both religious and political leadership in the Islamic world' (Burke, 2015:43.). It used oil revenues to fund Mosques, Islamic Centres, schools, colleges, universities, madrassas, teacher training, and literature around the world, to push its rigorous, intolerant and conservative vision of Islam, leading to the Wahhabization of the Sunni faith. As an ideological ally against Communism, Saudi Arabia's actions were ignored by the west. The end of the Cold War and rise of Salafi-Jihadism through the 1990s, however, led to increasing criticism of the Saudi-funded programmes.

Following the Soviet withdrawal from Afghanistan in December 1989, and influenced by Azzam's call for a global defence of Muslim lands, many jihadis looked for their next cause, taking part in the Algerian Civil war (1991–2002), the Bosnian War (1992–1995), the Tajikistani civil war (1992–1997) or the first and second Chechen wars (1994–1996; 1999–2005). One of the most influential Saudi jihadis was Osama bin Laden. With his associates, he created al-Qaeda in August 1988 as a vanguard Islamist movement, funding and promoting Salafi-Jihadism. After a failed Islamist project in Sudan, Bin Laden returned to Afghanistan in 1996, finding the country held by the Taliban, a Salafist Islamist movement originated in Saudi-funded Afghan-refugee madrassas in Pakistan (which taught a Deobandi-based Islamism close to Wahhabism). The Taliban and al-Qaeda shared a Salafist purity, a desire for Sharia law, a violent, jihadi zeal and intolerance of the kafir. Where they differed was the Taliban was a parochial group, whose only interest was in ruling Afghanistan, whereas Bin Laden and al-Qaeda had a vision of fighting the global enemies of their cause. In this Bin Laden was influenced by the anti-westernism of the key Egyptian Sunni Islamist thinker, Sayyid Qutb. Qutb's 1964 book *Milestones* described the state of contemporary 'jahiliyyah' ('ignorance') in the world, characterized by the rule of humans by humans. Hence his denunciation especially of the evils of western democracy and his call for the duty of returning to a pure Islamic society based on God's laws not man's. Bin Laden shared Qutb's anti-westernism, hence his hatred of the 'far enemy', his claim the west was at war with Islam and his signing of al-Zawahiri's 23 February 1998 fatwa, 'World Islamic Front for Jihad Against Jews and Crusaders', which ruled it was 'an individual duty for every Muslim' to kill Americans and their allies 'in any country in which it is possible to do it'.

Eventually, Islamic State would challenge al-Qaeda as the leading Salafi-Jihadist organization. Its roots lay in the organizations and ideals of Abu Musab al-Zarqawi, whose mentor was the Jordanian-Palestinian Salafist, Abu Muhammad al-Maqdisi. Al-Zarqawi shared the same Wahhabist Islamism as Bin Laden, but differed in his extreme takfirist hatred of non-Sunni Muslims and desire to take land for a Caliphate now. For all their differences, movements such as the Taliban, al-Qaeda, Islamic State, Boko Harem and al-Shabaab are linked by their common Wahhabist principles. As Moubayed writes, 'Spreading the faith by the sword, killing infidels and purifying the Islamic world from foreign ideas and lifestyles is the crux of Wahhabism and forms the cornerstone of jihadi thought and doctrine. It is the ideological blueprint for all the Sunni jihadi movements that have dominated world affairs over the last generation' (Moubayed, 2015:10.). They share a strong monotheism, applying tawhid as a weapon against all diversity, as well as a belief in the harshest Sharia law, the subjugation of women, the destruction of anything haram and the waging of permanent war against, and justified killing of, all non-believers and apostates. Violent jihadism is a duty, and self-sacrifice and martyrdom are symbols of the highest faith and commitment.

Islamic State's media operations

Islamic State's interest in media wasn't a new phenomenon. Islamist ideas had previously been spread by pamphlets, audio cassettes and, later, videos. Bin Laden, especially, understood the value of propaganda and tried to use film makers, press conferences and press releases for publicity. He took advantage of the development of local-language satellite channels such as Al-Jazeera, to reach a wider audience through video messages, but complained of editorial control over what was shown. Hence Burke's suggestion that 9/11 was a desperate attempt 'to grab the attention of the planet's mass-media' (Burke, 2015:54.).

Al-Qaeda had turned to the internet for communications, using email lists to disseminate information from 1995, using encrypted communications to plan its 1998 bombings and launching its first website in 2000. Its 2001 two-hour video *The State of the Ummah* was a memorable attempt at propaganda, showing al-Qaeda fighters in training, but al-Qaeda took time to recover following its Afghan defeat. By 2003 'Azzam the American' (Adam Gadahn) was beginning to modernize al-Qaeda's media production and produce new videos but there was a gradual decline in al-Qaeda's media output and its effectiveness in the years leading up to Bin Laden's death.

It was al-Zarqawi, in post-invasion Iraq, who realized the potential of the internet. He established a 'media department' to issue press releases and speeches, employed a dedicated web master, Abu Maysara al-Iraqi, and, beginning with 2004's *Heroes of Fallujah*, which showed an IED attack on a US armored personnel carrier, posted videos of insurgent activity online. In June 2004, he posted an hour-long compilation, *Wings of Victory*, and in 2005 published a 46-minute war footage video, *All Religion Will be for Allah*, on his website in multiple formats for easy viewing. In 2005 his media team set up the online magazine *Zurwat al-Sanam* ('Tip of the Camel's Hump') and began regular internet news broadcasts. Al-Zarqawi's most famous video was 'Abu Musab al-Zarqawi Slaughters an American', showing the execution of Nick Berg, which was posted on the al-Ansar Web forum on 11 May 2004 and subsequently downloaded millions of times. No broadcast media organization would ever have aired it: as Seib notes, 'the web provided a way for Zarqawi to avoid traditional media filters and deliver whatever images he wanted to a vast audience' (Seib and Janbek, 2011:35.). Dozens more people were subsequently beheaded by al-Zarqawi's group, with many being videotaped.

Al-Zarqawi's successor organization, Islamic State, continued and expanded these media strategies, becoming known for its sophisticated use of media. This seemed at odds with its Salafist, aniconic, anti-western and anti-modern world-view, but, as Burke argues, Islamic militancy is not a historically regressive movement but rather is 'fundamentally, profoundly contemporary, a product of the same global interaction of politics, economics, culture, technology and social organisation that affects us all' (Burke, 2015:11–12.). Violent extremists often anticipate trends, he argues, being 'ahead of the curve not behind it', and IS proves this point. Indeed, Atwan even refers to IS as a 'digital caliphate', arguing that 'without digital technology it is highly unlikely that Islamic State would ever have come into existence, let alone been able to survive and expand' (Atwan, 2015:1.).

Media were central to Islamic State. Indeed, Stern and Berger describe IS as 'a publicity whore' (Stern and Berger, 2015:244.). It had set up the al-Furqan Institute for Media Production in Iraq in November 2006 to produce ISI propaganda and expanded upon this by creating the al-Itisam Media Foundation, a film production unit, in Syria in March 2013 and the Syrian-based al-Hyat Media Centre in May 2014. Its media was specifically targeted at particular audiences for particular purposes. First, media was used for global publicity, targeting extremists in order to challenge al-Qaeda as the leading Islamist group and present the caliphate as the realization of the movement's goals. Second, it also aimed this publicity at the global, civilian Ummah, to politicize them, engage their support and gather recruits for its state. Third, media were used as part of a global, informational war, to terrorize the west and other enemies of the caliphate. It aimed to spread terror, destabilize the political and social order, and provoke the apocalyptic conflict it desired.

Fourth, media were used for local terror, to create a sense of divine unstoppability that would sap the enemy's will to resist and make them flee. Cockburn calls IS 'experts in fear' (Cockburn, 2015:xiv), arguing its videos of its fighters machine gunning kneeling Shia Iraqi army soldiers played a major part in terrifying and demoralizing the Iraqi army at the time of its capture of Tikrit and Mosul. Fifth, media were used for local propaganda, to convince civilians throughout Iraq and Syria not to be afraid of life under Islamic State. Images of markets, of abundance, and of happy, playing children were designed to reassure the civilian population, normalize life in the Caliphate and promote it as an existing and successful state. Sixth, the media were used against the state's own captive citizens who, cut off from external sources of information, with mobile phones and internet use controlled, were fed IS propaganda about their life in the Caliphate. This included the use of 'media points', or mobile kiosks run by the provincial media offices, which distributed indoctrination materials to the residents of newly conquered areas on USB drives or SIM cards.

Islamic State understood the power of branding, building its identity with simple, clear and emotively appealing messages and imagery aimed at key demographics, achieving a contemporaneity and cool that contrasted with more traditional Islamist communications. IS had established its media reputation with al-Furqan's *The Clanging of the Swords* video series. Produced from June 2012 to May 2104, these had grown in professional quality, with part four being viewed millions of times. However, the establishment of the al-Hyat Media centre proved to be the real turning point.

Al-Hyat produced multilingual videos shot in HD, demonstrating a greater professional quality and aesthetic. Their first video, the 13-minute *There is No Life Without Jihad*, released on 19 June 2014, immediately made global headlines as it depicted several western recruits, including three from Britain, sitting before the black flag, cradling weapons and talking directly to the camera about why others should join them. As one of the Britains, Abu Bara' al-Hindi (Abdul Raqib Amin, from Aberdeen) says, 'To all my brothers living in the west, I know how you feel [from] when I used to live there. In the heart, you feel depressed. The cure for the depression is jihad … all my brothers, come to jihad and feel the honour we are feeling, feel the happiness we are feeling' (Siddique, 2014.). With this one video, the psychological and geographical distance between the UK and Syria imploded.

One of the biggest al-Hyat productions was *Flames of War*, a 55-minute, professionally edited film released in September 2014, showing IS's seizure of the Syrian Army's 17 Division base near Raqqa including footage of the attack, captive Syrians digging their own graves and the executions. As Svirsky says,

> The film utilizes romantic imagery carefully crafted to appeal to dissatisfied and alienated young men, replete with explosions, tanks and self-described mujahedeen winning battles. Anti-American rhetoric provides the voice-over to stop motion and slow-motion action sequences. The use of special effects such as bullet-time is interspersed with newsreel footage.
>
> Svirsky (2014)

Rose explains how the film mythologizes IS fighters:

> It's a good example of what seems to be the Al Hayat style. Virtually every frame has been treated. The colour is so saturated, the combatants appear to glow with light. Explosions are lingered over in super slow motion. There are effects giving the feel of TV footage or old photographs. Transitions between clips are sheets of flame and blinding flashes. Graphics fly across the screen. Sonorous, auto-tuned chanting and cacophonous gunfire reverberate on the soundtrack. The Isis regime might have outlawed music, singing, smoking and drinking alcohol, but it clearly embraces Final Cut Pro.
>
> Rose (2014)

This remains, Rose says, a propaganda film: there's no footage here of IS's treatment of its civilian population or life under Islamic State 'and there's no telling how much of it is actually staged'.

As well as videos, IS produced the glossy, high quality, western-style online magazine, *Dabiq* (named after a town in Northern Syria where they believed the final battle with the infidels would occur). Published by a small team of European journalists, with the Syrian, US-educated IT specialist Ahmed Abousamra as its Chief Editor, it was published in a downloadable PDF form in Arabic, English, German and French, to target an international audience, with 15 issues appearing from 5 July 2014 to 31 July 2016. Other magazines were more specifically targeted, including the French-language online magazine *Dar al-Islam*, the Russian-language *Istok* and the Turkish-language *Konstantiniyye*. These magazines ran through to 2016 when they were replaced by *Rumiyah*, a new multilingual magazine first appearing on 5 September. IS also ran the Amaq News Agency and the al-Hyat YouTube channel, which broadcast reports from the captive British journalist John Cantlie, such as his September 2014 series *Lend Me Your Ears*, which described life in IS.

In addition to longer videos, al-Hyat also produced the 'Mujatweets', a series of shorter, Twitter-friendly videos showcasing ordinary soldiers talking about their experiences and encouraging others to join them. They were well-produced, using wandering, hand-held cam shots, point-of-view shots and close-ups to depict ordinary life in IS and create a personal intimacy with the fighters. The contrast with al-Qaeda couldn't be greater: whereas Bin Laden's videos mostly focused on himself, as the leader, comprising static shots of himself reading long theological and political statements, IS produced more dynamic, emotionally appealing, personal videos, focused on young recruits and their motivations. The Mujatweets demonstrated a western aesthetic, coopting the tropes of reality TV in the focus on the individual and their 'journey', creating a powerful identification from afar. The videos made joining Islamic State appear desirable, suggesting to young people that it could easily be *them* appearing before the camera.

This western aesthetic permeated IS's media productions, reflecting both the European background of many of the media staff and a deliberate decision to appeal to western youth. A September 2014 recruitment video, for example, used screen-grabbed scenes from *Grand Theft Auto 5* with messages telling viewers by joining IS they could act out these video-game fantasies killing people and destroying vehicles: 'Your games which are producing from you, we do the same actions in the battelfields!!' (Thornhill, 2014.). Video-game aesthetics appeared in another video, in March 2015, this time showing IS's Tigris River branch with HD cameras attached to their gun barrels to film point-blank executions in the style of point-of-view (POV), first-person-shooter (FPS) games like

Doom. As Rose comments, 'ISIS is in competition with western news channels, Hollywood movies, reality shows, even music video, and it has adopted their vocabulary' (Rose, 2014.).

IS placed a great value on those with technical and media skills, giving them higher wages and better accommodation. One Syrian interviewed for a Raqqa media centre said, 'They offered me $1,500 a month [five times the average Syrian salary], plus a car, a house and all the cameras I needed. I remembered looking around the office. It was amazing the equipment they had in there' (Thornhill, 2014). Media workers, however, were tightly controlled. Decisions were made from above as to what to film and what messages to create and film makers were then picked up in cars, driven to filming sites, then left in video editing centres to produce the final product. IS revenues from oil sales, kidnaps, extortion and robbery paid for the centres, equipment and wages. IS did allow decentralized media production in each 'Wilyat' (Libya, Algeria, Sinai, Yemen, Afghanistan, Nigeria and the North Caucasus), with each province having a media centre to produce local propaganda.

The IS media centres kept up a significant volume of production. The October 2015 Quilliam Foundation report, 'Documenting the Virtual Caliphate' found IS were producing 38 items a day, including 20-minute videos, full-length documentaries, photo essays, audio clips and pamphlets across a range of languages. Once created, IS crowd-sourced their distribution, relying on supporters around the world to share it and maintain it if it was taken down. IS were notable for their embrace of the most up-to-date digital technologies. They continually looked for new technologies, services and platforms to support their communications, employing them more radically and effectively than the authorities that opposed them.

One of the first platforms IS exploited was JustPaste.It, a site set up by a Polish student, Mariusz Żurawek, that copied Pastebin but allowed the posting of photographs and videos as well. The site was free, didn't require registration, was easy to use, worked for right-to-left languages and loaded quickly on mobiles without requiring a strong connection. Users posted material and then received a link that could be spread easily around social media. IS used it to compile photo sets of their massacres (such as the Sinjar massacre of Yazidis in August 2014) with links being widely posted on Twitter. IS also proved highly adaptable. When, in late 2014, JustPaste.It began deleting IS content following UK police requests, IS turned to the website Diaspora. This was a decentralized network, being composed of nodes, or 'pods', run by individuals or institutions, each of which operates a copy of the Diaspora software. As the developers boasted: 'Diaspora* isn't housed in one place, and it's not controlled by any one entity (including us)' (Diaspora*, 2011.). 'Joindiaspora.com', a pod run by Maxwell Salzberg, became IS's main hub until it was discovered and taken down for violating the group's zero tolerance for hate speech and violence. Other open networks such as Frendica, Quitter and even VKontakte were also used by IS until their accounts were shut down.

One IS solution had been to create its own app, 'Dawn' (short for 'Dawn of Glad Tidings'). Launched in April 2014 and available through the Google Play store, it was a news portal for information about Iraq and Syria that also included permission for it to auto-tweet links, hashtags and photos to the user's Twitter account, with sufficient randomness to avoid spam detectors. 'Dawn' was downloaded 5,000–10,000 times – tweeting 40,000 times in one day when Mosul was captured – before Twitter shut it down in June. A 2015 Brookings Study showed IS supporters responded to the loss of 'Dawn' by moving to clusters of Twitter bots, organized so that if one 'family' of bots was shut down, the others could continue to tweet (Berger and Morgan, 2015.).

With its ease of posting and its public nature and global reach, Twitter remained IS's primary focus. It had opened its first 'official' account in October 2013 under the al-Itisam name, quickly and quietly gaining 24,000 followers before it was suspended in February 2014. After that, official accounts, IS-organized individual accounts and supporter accounts kept up a Twitter barrage, engineering and playing the site's rules to keep IS material in the public eye. Key figures in the media departments acted as 'nodes', being responsible for distributing material and directing supporters. Attaining a celebrity status for their privileged position, they warned supporters to expect new material and to have back-up accounts ready, then tweeted links to JustPaste.It or the Arabic equivalents, Nashar.me and Manbar.me.

Foreign fighters were also important conduits, as figures attracting a lot of attention, and there's evidence IS used them deliberately. Klausen's study of 59 Islamic State foreign fighters showed that from January to March 2014, these accounts posted 154,120 tweets, with an average of 85 images and 91 videos each, to a network of 29,000 followers in a stream of controlled communications. Klausen concludes, 'What appeared to be a bottom-up movement to share jihadist videos was, it turned out, the product of a tightly controlled outreach effort by jihadi organizers' (Klausen, 2015:17). The Brookings study, tracking IS Twitter activity from September to December 2014, also found evidence of central organization. It estimated that at least 46,000 accounts were active at this time, and possibly up to 90,000, concluding that 'a highly engaged core' of 'mujahideen' (consisting of between 500–2,000 'hyperactive users' sending at least 50 tweets a day) were responsible for most of the production (Berger and Morgan, 2015:2–3; 7; 29.).

This was a difficult system to police. It had never occurred to the authorities that these platforms would be used for a global, terrorist propaganda campaign as terrorists usually hid, hence they were left playing 'whack-a-mole' with pro-IS accounts. Suspensions became a badge of honour for supporters. IS promised divine rewards for those who fought for it and that included those taking the battle to the enemy online: even *virtual jihadism* had its rewards. On 26 September 2015, the success of Twitter suspensions led IS's official media outlets to move to the encrypted messaging service Telegram, which had recently introduced a 'channel' function, allowing broadcasting to unlimited users. By August 2016 the official Telegram accounts had been repeatedly suspended so IS operatives set up mirror channels called the Nashir News Agency, copying the official outlets. By June 2017 there were about 130 active mirror channels. Telegram proved relatively ineffective at combating IS use of the app and it eventually surpassed Twitter as the main avenue of propaganda, also being used to plan terrorist attacks (such as the 19 December 2016 Berlin Christmas Market truck attack) and distribute propaganda about terrorist successes.

Not all jihadi media activity was so serious. In June 2014, the #CatsofJihad hashtag spread in Syria and Iraq, with jihadis posting pictures of jihadi cats next to IS flags, or guns and munitions. The Twitter account 'Islamic State of Cats' (@ISILCats) began posting on 25 June, tweeting dozens of photos of jihadis playing with cats, now referred to as 'mewjahids' or the 'mewjahideen'. The tweets demonstrated again the modernity of the jihadi recruits, and (if they were authorized) they were a recognition that jokes and memes could play a role in increasing IS's youth appeal. IS's interest in cats later waned, however, with the group issuing a fatwa in October 2016 against cat breeding in Mosul as opposed to its 'vision, ideology and beliefs' (Winter, 2016.).

Humour was evident in other tweets too. When the Iraqi Army fled Mosul on 10 June 2014 it left behind a huge amount of US supplies, including over 2,300 Humvees. On 18 June, IS

Twitter accounts shared a photoshopped image of Michelle Obama holding up a placard reading '#Bringback our Humvee' (referencing her #Bringbackourgirls response to Boko Harem's seizure of 276 female students in Nigeria in April). Another online IS-craze was the jihadi-selfie, with young fighters posting images of themselves posing with consumer goods or military hardware. As Diab notes, 'The jihadist selfie is helping transform the Spartan and puritanical image of holy war circa 1980s mujahideen in Afghanistan to make it resemble a mix between a lad's teen movie and an 18+ shoot'em-up video game' (Diab, 2015). Some questioned how spontaneous this was, however, seeing it as a deliberate IS strategy to increase its youth appeal.

IS was continually caught between the need for secrecy and the desire for publicity. It recommended strong operational security for its supporters, including VPNs, TOR and encrypted messaging services such as bitmessage.ch and Chatsecure, and wanted to be able to operate on services such as Telegram and Twitter without its accounts being noticed and suspended. At the same time, however, it wanted to employ these platforms for propaganda and recruitment, which meant it needed to operate in public, with maximum publicity. This openness was what marked IS out as an organization. Historically, most political movements have tried to hide their atrocities (the Nazis, for example, went to great lengths to destroy evidence of the death camps), in contrast to terrorist groups who want publicity for their acts. As Stern and Berger argue, IS is best understood as 'a hybrid terrorist and insurgent organisation' (Stern and Berger, 2015:11). As both a local, political movement and a global, terrorist movement, IS combined the two tactics, carrying out atrocities as part of its local military campaign and globally publicizing them in order to create terror.

But this openness wasn't just intended as terroristic. As befits the age of social media, selfies and 'TMI', IS documented its killing sprees for its own extremist pleasure and to globally promote its brand, gain followers and likes from around the world and increase its ranks. In contrast to al-Qaeda, whose web use prioritized secrecy (with hidden sites and hierarchical forums with layers of access based on trust), IS adopted an open and participatory approach, echoing that of Anonymous. Whereas hacking groups of the 1980s to 90s, such as the Legion of Doom, were secretive, controlled access and hid their identities, Anonymous were public, accessible and advertised an open membership for anyone to join in with whatever level of participation they wanted. At the lowest level, you could simply buy the mask or say you were Anonymous and you were. IS promoted a similar self-identificatory and participatory model, thereby increasing its appearance of support and the group's apparent threat level. What it wanted was participation, whether online or in the real world. By confidently seizing the public platforms of the internet, globally advertising itself, crowdsourcing membership and opening its arms to all Muslims to travel to join it, it turned a small, regional movement into what appeared to be a global, existential threat to the west, with a huge and lasting psychological impact upon the authorities and public.

The price of openness for lower-level members, however, was vulnerability to arrest. In response to online and real-world 'hate preachers', the UK had introduced the Terrorism Act 2006. S.1 of this had created a range of new offences including the 'encouragement of terrorism' and 'dissemination of terrorist publications', and these were now employed against IS supporters on social media. They were very broadly cast offences, with 'encouragement' being committed if anyone published a statement directly or indirectly encouraging anyone, anywhere in the world, or was 'reckless' as to whether anyone might be encouraged, whilst S.1(3) also made it a crime to glorify the commission or preparation of terrorist acts, regardless of whether anyone actually was encouraged to act. On 18

January 2013, Craig Slee of Preston was jailed for five years for posting videos on Facebook of al-Qaeda beheading captives and links to AQIM statements, and over the following years many were imprisoned for offences ranging from posting a single video online to systematic, multi-account Twitter campaigns.

IS's greatest media fame came from their beheading videos. Al-Zarqawi's 2004 video of Nick Berg's murder may have been inspired by Naji's recommendation in his *Management of Savagery* to combat the west's 'deceptive media halo' by copying the same techniques and using them back at them, and it is difficult not to see parallels between this violence and the post-9/11 rise of the 'torture porn' genre as epitomized by *Hostel* (2003) and *Saw* and its sequels (2004–09). As the US film maker Eugene Jarecki comments about the west, 'Not only are we a pace-setter in production values, we are also a pace-setter in murderous, amoral, profoundly disturbing content the world over… If we are watching [IS] come up to speed, it's to our own apparent obsession with gore and depravity' (Rose, 2014). IS returned to the beheading videos in August 2014, in response to US airstrikes from the 8. On the 19th it released a video entitled 'Message to America' showing US journalist James Foley being murdered, and this was followed by videos of, or about, the murder of Stephen Sotloff on 2 September, David Haines on the 13th, Alan Henning on 3 October, Peter Kassig on 16 November and Haruna Yukawa and Kenji Goto in January 2015.

Created by al-Furqan, the videos shared a simply, and powerful visual iconography, showing a black-clad figure (named 'Jihadi John' in the west), threatening the west then decapitating a Guantanamo Bay-referencing, orange-suited hostage against a stark, biblical background of sand dunes and blue skies. Like a reality TV serial, each video ended with a 'next-time' feature, showcasing the next hostage who would be killed. These videos became the basis for an entire video campaign of graphic hyperviolence, beginning with the murder of the downed Jordanian pilot, Muath al-Kasabeh on 3 January 2015. IS began by crowd-sourcing his death, calling on Twitter for recommendations of how he should be killed, with thousands offering suggestions. The final 22-minute al-Furqan video, *Healing the Believer's Chests*, distributed online on 3 February, showed the orange-suited pilot being burnt to death in a cage. Subsequent videos went further, including showing death by bazooka, homosexual men being thrown from rooftops and stoned, victims being driven over by a tank, prisoners in a cage lowered underwater with a waterproof cam capturing their deaths, mass executions of lined-up prisoners, often by children, the butchering of victims hung up on meat hooks in a slaughterhouse, the blowing-up of journalists with bombs in their cameras strapped round their necks, and youths tied to posts and cut in half with chainsaws.

Again, the comparison with al-Qaeda is important. Although IS threatened the west (in their 9 August 2014 Twitter warning showing the black IS flag on the White House railings, and their 19 August 2014 video of a blood-spattered stars-and-stripes with the message 'we will drown all of you in blood'), the reality was it was unable to mount any operations at that time, let alone enact anything on the scale of 9/11. Unable to produce terrorism, therefore, Islamic State instead created a new, powerful mode of *horrorism*, using hyperviolence to sicken global audiences. It discovered that, thanks to the internet, cheap, small-scale acts of *absolute body horror* could impact upon the global psyche almost as forcibly and traumatically as the downing of the World Trade Centre. Under the excuse of imposing sharia law, therefore, IS transformed its prisoners and captive civilian populations into snuff-victim extras in its film-factory, horror production line. Raqqa became a film set of severed heads and public executions carried out as much for a global as a local audience. The actual impact of these videos in the west may, however, be more complex. In a culture interested in horror gore and fascinated by extreme online material, many actively sought out IS's videos as entertainment, undercutting the horror IS intended.

Other videos showed IS destroying historical sites. Following the Taliban's destruction of Afghanistan's Bamiyan Buddhas in 2001, IS's Salafist hostility to idolatry was expressed in attacks on Nimrud, Hatra and Palmyra. There was some hypocrisy here as IS looted the sites for treasures to sell and globally distributed the resulting images of their hatred of 'idolatry', but acts like the blowing up of Palmyra's Temple of Bel on 30 August 2015, can best be understood as a deliberate act of cultural terrorism and an extension of their campaign to *horrify* the world. With the aid of the internet, a few, cheap explosives could have a powerful global impact, seizing the headlines for days.

Islamic State, therefore, systematically employed digital technologies to promote their ideology, state and activities. Theirs was a *digital blitzkrieg*: the speed of their military successes in 2014 was accompanied by the force-multiplier of a storm of tweets, photo sets, massacre videos, threats and celebrations that caught the west off-guard and created both a local and global panic. Through this, as the former State Department staffer Jared Cohen said, IS became 'the first terrorist group to hold both physical and digital territory' (Brooking and Singer, 2016). Thus, the information warfare theorized by the USA in the 1990s as a military strategy to control a battlefield had now been democratized and globalized. From 2014 onwards, Islamic State employed an ongoing, real-time, highly adaptable, multi-platform and crowd-sourced informational war that weaponized virality to reach deep within the heart of enemy territory, unsettle the authorities and threaten attacks. It also used it to recruit the west's own citizens as its army.

Islamic State recruitment

As a lasting and successful state needed citizens to people and defend it, one of the primary aims of IS's media strategy was recruitment. Given that IS's territorial ambitions were theoretically unlimited – in opposing the entire kafir world – it needed an ongoing supply of fighters to replace those killed in the permanent war it found itself in. By declaring a caliphate and demanding the global Ummah 'obey' and support the Caliph, al-Baghdadi, IS opened the way for a huge global influx of fighters. By 2017 the USA was estimating that, in total, up to 40,000 foreign fighters had travelled to Syria and Iraq, from up to 120 countries.

Echoing its crowd-sourced media strategies, IS relied on bottom-up peer-to-peer recruitment. Supporters would make contact by DM with a recruiter on Twitter or Facebook with the conversation then being transferred onto encrypted messaging apps and services such as Whatsapp, Kik and Surespot. The anonymous social media platform Ask.fm was also popular, and Skype was favoured as real-time conversations couldn't be monitored. Recruitment was gradual, through chats, messages and Q&As with key 'disseminators' and recruits who had already travelled to Syria. This peer-to-peer strategy proved to be a powerful way of attracting and inspiring youth in the west. It established a *personal* link with them, ensuring that before they had left they had already experienced a narrative of their journey, advice on preparations, what to pack and travel arrangements, and a romanticized image of their life in Islamic State. Although much of the academic literature has focused on individual 'radicalization', in reality the more important process was this *normalization*, in making the decision to join and travel appear easy and the life to come seem familiar.

IS's propaganda was designed to make joining it appealing. Much of the propaganda focused on the realized utopia of the caliphate where everyone was a 'brother' or 'sister' and where one could live an emotionally happy and fulfilled life under Sharia law. Images of everyday life, markets and shops, full dining tables and happy and playing children all built on this to present an idyllic life in IS. The other side to this was propaganda emphasizing the unhappiness of life among the kafir. As Abu Bara' al-Hindi says in *There is No Life Without Jihad*: 'To all my brothers living in the west, I know how you feel … In the heart, you feel depressed. The cure for the depression is jihad'. *Dabiq* magazine echoed this in 2015, turning the zero-hour contract, minimum wages of western austerity into an insult to Islamic youth:

> The modern-day slavery of employment, work-hours, wages, etc. is one that leaves the Muslim in constant feeling of subjugation to a kafir master. He does not live the might and honour that every Muslim should live and experience. Dedication of one's life towards employment, if the employer is a kafir, only leads to humiliation.
>
> Burke, (2015:210–11)

Jason Burke makes similar points about what was being offered to young British Asians:

> The conflict offered sexual opportunity, status and adventure – opportunities that could be seen, at least in the eyes of a particularly naïve and ignorant young person, as more inspiring than trying to scratch together fifty pounds for a night out in a run-down British port city, before another week's work of flipping burgers or studying for a, low-grade, low-utility degree.
>
> Burke (2015:211)

Jihadism, therefore, offered an empowerment and global level of meaning that was highly appealing to western Muslim youth caught within familial expectations, casual, everyday racism, and a depressed economy with few prospects of a good job or chance to get on the housing ladder.

With Islamic State, Islamism left behind the old-fashioned political activism of al-Qaeda, and connected to modern lives and youth in unprecedented ways. It was aided here by the rise of 'jihadi cool'. Over the previous decade, jihadism had rebranded itself online to appeal to youth, appropriating black, US hip-hop culture and music, using rap videos, clothing, style and fashion and consumer items and slang to create a counter-cultural phenomenon for western Muslims. The most famous example was the rap-song and video 'Dirty Kaffar', produced by the British Muslim rappers, Sheikh Terra and Soul Salah Crew in response to the Iraq War in 2004. The video has been downloaded and shared innumerable times as well as being remixed and made available on countless jihadi websites and it found a new audience as IS grew in influence.

Islamic State rode this wave. Its jihadis posting selfies online suddenly looked Hollywood-and-music-biz *cool*. Gangsterish, thuggish, dressed like ninjas and with cool weapons and nasheeds, these jihadi Che Guevaras became the poster boys for Islamism, creating an aspirational image for other males and helping bring in valuable female recruits, dreaming of their dishy future husbands. As the blogger 'Bint Emergent' explains: 'Salafi-Jihadism made being pious cool. It became cool to quote aya [verse] and study Quran. And CVE [Countering Violent Extremism] has absolutely no defense against this … I love jihadi cant – dem, bait, preeing, binty, akhi [brother] … It's like Belter dialect in the Expanse. And it borrows from all languages – because jihad draws from all races and ethnicities. The voice of youth counter-culture and revolution for an underclass. Like ghetto culture in the USA – the inexorable evolution of cool' (Cottee, 2015).

Westerners who had already made the journey became some of the most important recruitment tools. Thanks to Twitter they could easily be followed and messaged. Their lives in the warzone were visible and supporters could follow their personal 'journey'. One example was Mehdi Hassan, a 19-year-old from Portsmouth who travelled to Syria in October 2013 and tweeted as Abu Dujana RK (@abudujanaisis). He admitted to missing Coco Pops, sliced bread and *Coronation Street*, gave advice to those joining him to pack light, bring a spare phone charger and good trainers, gave advice on weapons, and also posted inspirational quotes. Alongside a photo of a machine gun he wrote, 'Check this badboy modified Glock out. Dawlah rock it hard'. On one occasion, he hosted a Facebook Q&A for friends at home. He was killed fighting in Kobani on 24 October 2014.

Another recruiter was 20-year-old Reyaad Khan from Cardiff, who travelled to Syria in November 2013 and died in an RAF drone strike on 21 August 2015. He was highly active on Twitter, posting photos of tables of food in response to US airstrikes and another showing himself in a supermarket, holding a jar of Nutella and saying, 'I was so terrified by the US airstrikes, I had to buy myself some Nutella to comfort my brittle heart'. The London amateur rapper Abdel-Majid Abdel Barry, who travelled to Syria in June 2014, also posted to his Twitter followers, including an image in August 2014 of himself in Raqqa holding up a severed head, commenting, 'Chillin with my other homie, or what's left of him'.

Islamic State also differed from earlier jihadism in actively recruiting women. One reason was to solve the problem of jihadi sex. Through Afghanistan and Bosnia there had been a recognition that young, foreign male jihadis, high on violence, ideology and drugs, needed a sexual release that wasn't available in strict, local Muslim communities. IS tried to solve this issue through sexual slavery, by exploiting the tradition of temporary marriages ('Nikah Mu'tah') to pass women around, and by encouraging female recruits who could marry jihadi fighters. The second reason was for these wives to help populate the caliphate. A state needed a population, hence women were needed to have children and raise the next generation of fighters, IS's 'lion cubs'. Hence IS propaganda was also aimed at women. The summer 2014 edition of *Dabiq* included a 'myth-busting' article, designed to appeal to women, illustrated with photos of handsome jihadis cuddling kittens (dogs are haram) and helping themselves from bowls full of chocolate bars, and images of girls in classrooms, whilst the IS Twitter feed posted photos of a local shop with a woman in a full niqab browsing a huge selection of handbags. In mid-2014 a Twitter account called 'Jihad Matchmaker' was set up for women seeking a match in Syria. Atwan suggests women made up about 10 percent of those travelling to join IS, estimating at least 2,500 recruits by 2015.

One important female recruiter was the Glasgow teenager Aqsa Mahmood, who travelled to Syria in November 2013, aged 19, and blogged on Tumblr under the name Umm Layth. Her 'Diary of a Muhajirah' offered advice to young western women thinking of travelling to Syria, including a packing guide and photos of Raqqa. Umm Layth also emphasized the material benefits of joining IS and marrying a jihadi, including 'kitchen appliances, from fridges, cookers, ovens, microwaves, milkshake machines etc., hoovers and cleaning products, fans, and most importantly a house with free electricity and water provided to you due to the Khilafah and no rent included' (Steele, 2015). She was also active on Twitter, with over 2,000 followers, promoting IS and calling for terrorist attacks in the west.

IS, however, had a highly traditional vision of women and their role. In late October 2014, for example, they established the 'Zora Foundation', an online Arabic media arm designed to offer advice for women and help them understand their contribution to jihad. It included a webpage, Facebook page and Twitter channel, offering videos, health tips, first aid advice and advice on acceptable 'feminine manual labour' such as sewing and creating online propaganda. It also included 'fast and easy recipes' such as dates, flour and butter: 'a quick recipe that can be served to the mujahedin with coffee or can be eaten at any time with water, especially during breaks in battles. They contain significant calories and will extend the power and strength of the mujahedin' (Fears, 2014). By the end of the month the Twitter account had 2,700 followers, with the Facebook page attracting 300 'likes'.

Despite this traditional vision, IS appeared empowering to young, western women in offering a break from the constraints of family life and influence of celebrity and consumer culture, at a time when austerity, high rents and prohibitive house prices significantly limited their opportunities at home. Women were not passive recruits to IS, but actively embraced its ideology and sought out the group and travel. IS gave them the thrill of illicit adventure, a new meaning in life, and a fast track to adulthood and independence, with a home, a husband, children and the exciting life of a jihadi bride. As the popular meme referencing the first wife of Mohammed put it: 'In a world full of Kardashians be a Khadijah'. Joining IS, therefore, was seen as a positive act, enabling women to leave the racism, islamophobia, Burka bans, feminism and gender norms of the west behind and make a new identity for themselves within their faith, living life as they wanted to. It was a narrative IS strongly encouraged.

There were some active female roles in Islamic State. Fanaticism was rewarded by being allowed to join the morality police (the all-women al-Khansaa Brigade or Umm al-Rayan Brigade), membership of which brought more freedoms and a salary, or serve as online IS 'cheerleaders' (Moubayed, 2015:182). These were women who managed IS-affiliated pages on Facebook, along with accounts on Twitter, Instagram, Ask.fm, Tumblr and Kik, taking part in online Q&As and posting material that was, Moubayed says, 'cutting-edge, trendy and well-planned'. They painted 'a rosy picture of family life under the Islamic State', promoting ideals of matrimony and the 'honour' of raising children, posting selfies and photos of themselves carrying babies, attending weddings, sewing clothes, eating ice cream and making Nutella pancakes, as well as 'pictures of their husbands playing snooker or sunbathing by large swimming pools, making it look like jihad is merely a "cool vacation"' (Moubayed, 2015:183).

The reality was very different. As Moubayed says, 'In contrast to the image they try to present, the women of ISIS live the average day-to-day life of a Syrian housewife: cooking, cleaning and taking care of children, away from the glitz and glamour posted on social media' (Moubayed, 2015:184). The housing stock was dilapidated, the infra-structure was in poor condition, with electricity and water supplies affected, and the quality of life was far below that in the west. Jihadi brides were under their husbands' control and often had to remarry as their husbands were killed in fighting. There was also competition between the wives and jealousy at the sex slaves the husbands took, and women had to watch their children being indoctrinated into the murderous IS ideology and turned into killers, with no future other than to fight.

Women in IS had few rights. Their dress, movements and communications were strictly controlled and IS banned women from owning many items. In Mosul, the dress codes got gradually stricter until women had to be completely covered, including a film of black cloth across the eyes. Tripping was common as women couldn't see and if they lifted the veil to eat or look they were targeted by the morality police. The ultimate aim of this policing was to keep women within their homes. Punishments for offences were harsh, including significant fines, public whippings, stonings and beheadings. As conditions of life in war-torn Syria gradually deteriorated, recruits discovered leaving IS was far harder than joining it. At least one jihadi bride, the Australian Samra Kesinovic, was killed for trying to escape. As IS's territory was retaken through 2017, the utopia finally collapsed.

Fighting Islamic State online

Western governments were surprised both by the rapid military successes of Islamic State in 2014 and by its open online presence and the support it attracted. On 3 November 2014, GCHQ Chief Robert Hannigan complained US technology companies weren't doing enough to combat terrorism, arguing that Web 2.0 platforms had become 'a command and control network for terrorists' and saying that security services were engaged in a constant battle to remove terrorist content (Quinn, Ball and Rushe, 2014). Criticism also came from victims of terrorism. On 19 December 2016, for example, families of three people killed in the 12 June Orlando nightclub shooting sued Google, Facebook and Twitter for providing 'material support' to Islamic State. Under S. 230(1) of the 1996 US Communications Decency Act, the platforms weren't legally responsible for what they carried, and the lawsuits that were brought were later dismissed, but the case highlighted how anger at the platforms was growing.

As its official social media accounts were quickly suspended, IS relied on a network of supporters to share content and maintain their presence. This included tactics such as piggybacking on major sporting developments or political events, with hashtags for the 2014 World Cup and Scottish Referendum being hijacked to distribute IS material. Abdulrahman al-Hamid (@Abu_Laila), for example, told his 4,000 Twitter followers, 'We need those who can supply us with the most active hashtags in the UK. And also the accounts of the most famous celebrities. I believe that the hashtag of Scotland's separation from Britain should be the first'. Replies suggested using #andymurray, #scotland, #scotlandindependence, #VoteNo and #VoteYes and linking to David Cameron's twitter handle. At the same time @With_baghdadi told Isis supporters to 'invade' the #voteno hashtags 'with the video of the british prisoner' (Malik, Laville, Cresci, and Gani, 2014).

Facebook, Twitter and YouTube had dedicated teams working to remove flagged content and soon security services and specialist police units began working with the technology and social media companies. To speed up the process, YouTube gave some agencies 'trusted flagger status' for priority attention. The UK's Counter-Terrorism Internet Referral Unit (CTIRU) was working with the US companies to take down material, removing 1,100 items a week in 2014, including removing back-up accounts before they could be used (and before they had broken the site's T&Cs). The suspensions clearly irritated IS. Al-Hamid criticized his followers who were worried about suspension, saying, 'I swear to god there are other people willing to sacrifice themselves for their religion … You should be ashamed of yourself that your account might be attacked even 10 times' (Malik, Laville, Cresci, and Gani, 2014). Following a wave of account terminations the previous week, he tweeted on 14 September 2014,

> We talked a lot about the deletion of accounts and the means of staying steadfast and to push people to continue if their accounts were deleted or suspended … We have to admit that this is a disaster and we have to be patient.
>
> Malik, Laville, Cresci, and Gani (2014)

The authorities were now engaged in an ongoing, real-time game of cat-and-mouse. The 2015 Brookings Institute study of IS Twitter use suggested a minimum of 1,000 pro-IS accounts were suspended from September to December 2014, although it noted that even this 'aggressive' level of suspensions couldn't be expected to eliminate their activities. The battle continued through the following year and on 26 September 2015 Twitter suspensions forced IS to shift its propaganda output to Telegram, where it opened the Nashir channel as the primary outlet for their propaganda. Pressure on Telegram built after the 13 November 2015 Paris attacks, when it was widely criticized as being a hub for IS organization and propaganda. On 18 November, the app was used to publish *Dabiq* issue 12, entitled 'Just Terror', which celebrated the downing of the Russian Metrojet Flight 9268 over Sinai on 31 October that killed 224 and the 130 deaths in the Paris attacks. The following day Telegram shut down 78 IS channels, covering 12 languages.

On 4 February 2016 Twitter announced it had shut down 125,000 accounts linked to IS, al-Qaeda and the al-Nusra Front since mid-2015. It evidently had a psychological impact on IS. On 23 February, pro-IS hackers released a video threatening the founders of Facebook and Twitter as revenge for the suspensions, with photos of Mark Zuckerberg and Jack Dorsey shown with superimposed bullet holes. The amateurish video claimed to show hackers taking over social media accounts, changing profile pictures and using the accounts to distribute IS propaganda. By April 2016 a report by the threat intelligence community Recorded Future suggested IS Twitter activity was down 40 percent on the year before as the number of accounts using particular IS-related hashtags declined from an average of 24,271 in August 2015 to around 14,700 in March 2016. Accounts were also being shut down faster, with the median life of a pro-IS account now less than two days, compared to a number of weeks the previous summer. On 18 August 2016, it was reported that Twitter had closed down a further 235,000 accounts in the six months since its February announcement.

This wasn't a battle that could be finally won, though. Research by the terror-monitoring group Kronos showed that 95 percent of people suspended immediately opened new accounts, but the gradual attrition suggests that the tactic was exhausting IS supporters and pushing IS onto less public and less effective platforms. Their Twitter tactics continued, however. Following the 16 July 2016 Nice lorry attack, IS supporters again promoted their message and celebrated the attack by hijacking popular hashtags.

As Twitter acknowledged, 'there is no one "magic algorithm" for identifying terrorist content on the internet' (Woolf, 2016). Companies mostly relied on human moderators and flagging, but they were searching for technological solutions. In June 2016, it was reported that Facebook and YouTube were using automation to remove content, employing technology designed to remove copyrighted material. The technique involved classifying identified videos by assigning 'hashes' – a unique digital fingerprint – to them, which subsequently allowed all content with matching fingerprints to be automatically picked up and removed. The tech companies, however, remained strong believers in freedom of speech, didn't want to be arbiters of what was acceptable and were suspicious both of government intervention and of being seen by their users to be too close to government. Political pressure and bad PR, however, forced their hand.

Top-down government attempts to combat the allure of terrorism generally proved ineffective. Although they realized the necessity of combating the ideas online over social media, their campaigns were often misconceived. In December 2013, the US State Department launched their counter-violent extremism (CVE) project, clumsily and badly named 'Think Again Turn Away', with accounts on Twitter, Facebook, Tumblr and Ask.

fm. The Twitter account posted counter-message news stories but often failed to fact check them, leading to widely debunked stories such as the claim that IS had mandated female circumcision throughout its territory being published. Other messages were naïve. Tweets such as 'Dutch #ISIS suicide bomber targets Iraqi police station – another foreigner terrorizing locals' displayed a stunning lack of awareness from a government that had launched major air and missile campaigns against the country in 1991, 1993, 1996, 1998 and 2003. As Rita Katz, the director of the SITE intelligence group also argues, the Twitter account's decision to directly address prominent jihadist accounts was a huge mistake, as engaging them in debate only gave them a bigger platform. Pro-IS accounts were thrilled to be noticed by the US government, were boosted by the endorsement and enjoyed the chance to debate political issues with the USA such as Abu Ghraib. 'Arguing over who has killed more people while exchanging sarcastic quips', was a 'ridiculous' and 'embarrassing' strategy, she says (Katz, 2014).

Their Ask.fm strategy didn't work any better. Asked 'How much $ has the US spent on ISIS', the account replied, 'you can't put a price on ridding the world of terrorists', then immediately did so, saying by mid-late October 2014 they'd spent '$424m'. Asked what their favourite restaurant in the world was, they stated their support for restaurants that donated to the homeless and accused al-Baghdadi of not sharing their concern. Asked about their favourite ice cream flavor they posted a photo of a stars-and-stripes covered cone. When faced with IS defenders trolling them they piled in to the argument, with an impassioned but rather futile fury: the message 'Our death are in paradise while your death are in hell. Bloody greetings, Jihadi John' was met with the less-than-impressive come-back, 'As appalling as his actions are, Jihadi John likely has better English-language skills than what you have demonstrated' (Knibbs, 2015).

In the UK, the primary CVE project was called 'Prevent'. Introduced in 2003 and expanded upon in 2011, it was one of four linked strategies in the government's Counter-Terrorism Strategy (CONTEST). In 2017, the government claimed that it had removed 220,000 pieces of terrorist material since February 2010, had supported over 1,000 people since 2012 and in 2015 disrupted over 150 attempted journeys to Syria/Iraq, protecting over 50 children from being taken to the conflict zone, and in 2015–16 had delivered 142 anti-radicalization projects to 42,000 participants. Prevent has been criticized, however, as an intrusive scheme that treats all Muslims with suspicion. It includes a legal duty for schools, colleges and universities to report those suspected of holding extremist views.[1] Notable failures of the policy include a 7-year-old Muslim child reported to the police by St. Edwards Catholic School in Birmingham in November 2016 for bringing in a 'bullet' that turned out to be a piece of brass.

If the official response to IS was problematic, a potentially more effective fight was being waged by ordinary internet users. After the murder of James Foley, Twitter users began sharing photos of him in life to honour his life and counteract the images from the video. The hashtag #ISISMediaBlackout spread, being used over 11,000 times after Foley's murder, with users promising not even to mention IS or their activities. In a speech on 20 January 2016, Facebook's COO Sheryl Sandberg expanded upon the power of individual users, and recommended using a 'like' attack on IS, giving the example of protestors who flooded a neo-Nazi page with likes and positive messages so that, 'What was a page filled with hatred and intolerance was then tolerance and messages of hope' (Yadron, 2016).

Others took a more offensive approach. Parody videos mocking IS were popular. 'Allahu Ackbar Washing Machine', for example, showed a spinning washing machine with a brick thrown into it, with its violent self-destruction being accompanied by someone screaming

'ALLAHU ACKBAR!!!!'. 'Allahu Ackbar Cat' similarly shows a cat falling into a bath of water and furiously scrambling to get out, accompanied by the same screaming. 'Jihadi fail' videos were popular on YouTube, with compilations such as 'Terror Fails: Jihadi Dummies and Ackbar Idiots' and 'ISIS Idiots With Guns' showing terrorists shooting each other and blowing themselves up, all culled from their own filmed footage.

The 'ISIS Fails' video shows an inept group of four IS fighters in a truck, misfiring rockets, all set to the Benny Hill TV theme. 'Abu Hajaar. Pass me another rocket', Abu Abdullah asks. 'Which one? The ones for firing at people or armored vehicles?', Abu Hajaar replies. Abu Abdullah asks Abu Hajaar to cover him then shouts 'Abu Hajaar! Watch out! The bullet casings are hitting us'. Abu Hajaar manages to fire a rocket ('Be careful Abu Hajaar'), but the blowback hits the people in the vehicle who tell him, 'Good job, but you roasted us too'. Abu Hajaar is asked to get another rocket and is then asked, 'What is wrong with you Abu Hajaar?' The fighters fail to load and light a bomb before lobbing it a short distance outside the vehicle, whereupon Abu Abdullah exasperatedly says, 'What are you doing Abu Hajaar? That is a rocket for people'. Another fighter tries to fire a rocket but is told, 'Take the safety cap off the rocket' just before their own vehicle is hit by a Peshmerga rocket. They scramble out and retreat. The driver is killed, another is shot whilst fleeing and the cameraman dies soon after (ISISFAILS, 2016).

The clip was edited from four videos captured on a GoPro HD headcam by the IS fighters. Vice News received the retrieved memory card from the Kurdish Peshmerga Higra Agre Fire Force Unit 80 who had fought them near Mosul in March 2016 and published it on 27 April. The video received 2.2 million views in 72 hours and went viral, with Abu Hajaar becoming a meme for IS incompetence. Its depiction of IS's chaos, fear and haplessness proved a powerful informational weapon against IS's self-created myth of unbeatable jihadis. On 8 November 2016, another video made the news, showing IS fighters in Syria. To cries of 'Allahu Ackbar' a truck pulls up with a mounted machine gun and as it fires another fighter climbs on the back and shoots his own machine gun. Suddenly the truck pulls away and the fighter falls off, causing the person shouting 'Allahu Ackbar' to break down in hysterics. In the twenty-first century, participative war comes with its own soldier-filmed 'blooper' reel.

IS has been widely ridiculed online. In March 2015, images emerged of the IS commander of the al-Anbar Lions group, Abu Wahib, standing by a truck in Iraq's Anbar province. His stance and camouflaged battle fatigues led journalists at Raqqa is Being Slaughtered Silently to post images on Twitter comparing him to a kebab, with the hashtag #same. The images went viral and made the mainstream media. The distinctive look of the hirsute commander had already been mocked online, with commentators editing images of him to make it look like he was in a shampoo commercial or on a dating website. Twitter became a major platform for ridiculing the group. On 22 August 2014, UK comedian Lee Hurst began an #AskIslamicState hashtag with the question, 'When do you expect the Caliphate to be opened up to tourism?'. The hashtag began trending. @naeshitsherlock asked them, 'What's your favourite episode of friends; @Maj75red asked, 'Did you cry when Patrick Swayze died in Ghost?'; @TheOncoming asked, 'What is it that meatloaf won't do for love', and @Willmerr pitched in with, 'If I joined you, could I eat what I like or do I have to have what the Caliphate?'.

Anonymous also launched their own war against IS. Building on June 2014's 'Operation NO2ISIS' and the January 2015 #OpCharlieHebdo campaign in response to the attack on the magazine, on 8 February 2015 they posted a video launching #OpISIS, warning, 'ISIS we will hunt you down, take down your sites, accounts, emails, and expose you'. By the third day Anonymous claimed to have taken down over 1,000 websites, 800 Twitter accounts, a dozen Facebook pages and 50 email accounts with pro-IS affinities according to a list they published on Pastebin. On 10 March, they took down the IS social media site Kelafabook ('5elafabook') launched by supporters only two days before. The operation wasn't technologically sophisticated, mostly involving finding and reporting pro-IS pages and accounts, but as Anon Greg Housh explained, this opened it up to anyone:

> One of the things I like about this is anyone can take part. You don't really have to have any hacking skills, and you don't have to break the law to do something here. Just find ISIS talking online and then tell someone about it.
>
> Borchers (2015)

IS, however, had their own cyber-unit, the 'Cyber-Caliphate', run by Junaid Hussein, a former Anon. As Housh pointed out, 'Junaid, kind of taught ISIS everything they know. So both sides have the same toolbox' (Borchers, 2015). In January 2015, the Cyber-Caliphate declared war on the USA, hacking into the Pentagon's Central Command date and taking over its Twitter and YouTube accounts. The online war stepped up again after the 13 November 2015 Paris attack. Anonymous launched #OpParis on the 14, declaring war (again) on IS, leading, over the following days, to thousands more accounts being taken down. By the 21st Anonymous were claiming over 20,000 Twitter accounts had been removed, but this figure was disputed. Both Twitter and the FBI rejected Anonymous' claims, with a Twitter spokesperson saying a review of lists published by Anonymous had found them 'wildly inaccurate and full of academics and journalists' (Cameron, 2015).

A review by Ars Technica found the Anonymous script targeted accounts that had nothing to do with IS. Most of the 4,000 Twitter accounts in one Pastebin list hadn't posted pro-IS messages at all. Some were trolling the group, some were Palestinian and some were simply written in Arabic. They also reported that some joining the #OpISIS IRC channel weren't the most clued up, with one asking, 'Who's ISIS?' (Gallagher, 2015). But Anonymous weren't just using reporting. True to their 4Chan roots they also launched the #RickrollDaesh hashtag and began spamming Twitter with apparently pro-IS links that took people straight to a video of Rick Astley's 'Never Gonna Give You Up'. Rickrolling was an old tactic, but this was the first time anyone had weaponized Rick Astley as part of a cyberwar.

Elsewhere anti-IS sentiment could be expressed through Facebook posts and filters, including the Eiffel Tower peace sign and tricolour flag filters that users could add to their profile. This low-level political signalling has been dismissed as a simplistic and ineffective 'clicktivism' or 'slacktivism' (Morozov, 2011:179–203), but enough people did it on Facebook to rile Islamic State, whose members (such as the account of @SaefAzd14) created their own filter in response showing a French flag with a boot-stamp mark over it. Anonymous weren't alone in attacking IS. A splinter group called Ghost Security (or 'GhostSec') was also active in hacking and taking down pro-IS accounts. On 25 November 2015, it found an IS WordPress TOR Darkweb site set up in response to Anonymous taking down their sites on the open web. GhostSec hacked the page, replacing it with an advert for Prozac and the message, 'Too much ISIS. Enhance your calm. Too many people are into this ISIS-stuff. Please gaze upon this lovely ad so we can upgrade our infrastructure to give you ISIS content you all so desperately crave' (Gibbs, 2015). This was troll warfare.

The spiritual home of Anonymous, the image-board 4Chan, also got involved. Mainlining the trolling spirit its users began superimposing bright yellow rubber ducks' heads onto images of IS fighters. 'How about castrating the image of Isis by replacing the faces on ALL the propaganda photos with bath ducks?', a 4Chan user wrote on Shit4chanSays (/s4s/) and from there the theme spread to Reddit (spawning debates such as 'Would you rather fight 1,000 duck-sized Isis members, or one Isis-sized duck'?) and onto Twitter and Facebook. Some added toilet brushes to the images, or remade the IS flag with a duck-shaped symbol. The image of al-Baghdadi as a duck and the declarations of 'Allahu Quackbar!' were used to undermine the seriousness and fear the group exploited (Gunter, 2015).

On 3 December 2015, a text file was posted to Ghostbin announcing Anonymous was planning a 'Troll ISIS Day'. On 11 December, tens of thousands of participants on Twitter posted images under the hashtags #Daesh and #Daeshbags. Duck-headed fighters were common, but most aspects of internet, pop and anime culture were represented. Trolling IS spread. When, on 26 December 2015, al-Baghdadi released a rare 24-minute audio recording urging believers to join the fight – a call repeated by the @iyad_elbaghdadi Twitter account – the Muslim response wasn't entirely what was expected. @MohsinAr-ain91 replied, saying, 'Sorry mate, I don't want to risk dying before the next Star Wars comes out'; @JayLikesIt said, 'Sorry Amir al-Mushrikeen, I'm busy being a real Muslim, giving to charity etc. Also your dental plan sucks #GoatTeethISIS'; and @Muaijaz responded, 'Nah, I prefer to sleep in and binge watch shows on Netflix'. Trolling IS wasn't just a western phenomenon. On 8 December 2015, the Levant Front, a coalition of Syrian rebel militia, released a battlefield video entitled 'Muslims are not Criminals', showing their black-suited and balaclava-wearing fighters standing behind a line of orange-suited IS prisoners, aiming their guns at their heads. They then take their balaclavas off and put away their guns, as a cleric says, 'This is not our policy. We are not evil'.

IS were mocked again in February 2016, although this time by fellow jihadis. A video released by Hidayah Media, a group associated with AQAP and fighting IS in Yemen, contained claims from an IS-defector that the group had faked battle scenes in their propaganda videos. As well as fake fights and raids, IS fighters had pretended to be dead Houthi rebels, being covered in fake blood made from Vimto. The Arabic hashtag for 'Vimto Caliphate' soon began trending, accompanied by photoshopped images of IS-spokesman Abu Muhammad al-Adnani as a Vimto delivery man. The trolling of IS continued after the IS-inspired shooting at Pulse, a gay nightclub in Orlando, Florida, on 12 June 2016, which killed 49 and wounded 58. By the next day, Anon WauchulaGhost had hacked into many pro-IS Twitter accounts, posting the words 'Jacked by a Ghost' and filled them with gay pride images and messages, LGBT slogans (such as rainbow flag profile pictures with 'I'm gay and I'm proud' written on them) and even gay pornography.

None of this defeated IS but it did appear to have an effect upon it as they regularly responded to these provocations and issued threats to Anonymous. Arguably, trolling was a more effective response to IS than government propaganda schemes, as it directly countered 'jihadi cool' and its appeal to the young in a way that 'Think Again Turn Away' didn't. Trolling engaged IS at its level – on the internet and using internet culture – in a far more relatable way, undercutting its seriousness with enjoyable internet-snark: to IS's grand, apocalyptic threats, troll-face asked, 'Are you mad bro?'

This trolling was also healthy for our own societies. Since 9/11 western governments have responded to the terrorist threat by hardening their own societies, increasing security checks and patrols, limiting freedom of speech through anti-terrorism legislation and developing mass-surveillance programmes to spy on their own populations, all of it justified under the rubric of 'security'. But if 'terror' is a state of mind, then its opposite isn't a militarized, securitized surveillance state with anti-vehicle defences and black-suited, kevlar-clad armed counter-terrorist police on the streets who look identical to IS fighters, but rather humour, satire and a lack of seriousness. Trolling not only undercut the group's cool and youth appeal, it was an effective weapon against their 'terror' and the over-estimation of their threat.

After Islamic State?

The USA began bombing IS targets in Iraq from 8 August 2014, expanding this to Syria on 23 September. The campaign was formalized in October 2014, with the creation of 'Combined Joint Task Force – Operation Inherent Resolve' (CJTF-OIR), a US-led coalition with military forces from over 30 countries aiming to 'degrade and destroy' IS. By November 2017, Airwars estimated 28,256 coalition airstrikes had been carried out (14,029 in Iraq, 14,227 in Syria), killing a minimum 5,637 civilians. This was only one of many campaigns engaging Islamic State. In Iraq, 'Operation We Are Coming Ninevah' to free Mosul was launched on 16 October 2016, by a coalition of 54–60,000 Iraqi Security Forces (ISF), 40,000 Kurdish Peshmerga, 14,000 Popular Mobilization Forces (Iranian and Hezbollah supported Shia militias) and a small number of CJTF special forces personnel supplemented with CJTF drones and air strikes. The Iraqi PM, Haider al-Abadi declared victory on 10 July 2017, although at a huge cost for Mosul and its Sunni civilian population. There were further ISF victories in the Battle of Tal Afar by 2 September and the Hawija Offensive by 8 October 2017, effectively ending IS as a state in the north.

Iraqi operations coincided with operations against IS in Syria. Russian intervention in support of Assad began in September 2015 and proved decisive in shoring up his regime. By September 2017 the Syrian Observatory for Human Rights (SOHR) claimed Russian airstrikes had killed 5,703 civilians, about a quarter of them children. Meanwhile CJTF airstrikes on IS continued, also impacting on their captive population. From late 2016 IS were in retreat. They were defeated by Turkish forces, Syrian rebel groups, the Syrian Democratic Forces (SDF) and the Syrian Army in the Battle of al-Bab in the north by 23 February 2017 and were defeated again by the Syrian Army's Central Syria campaign by 21 October 2017, and their October 2017 Mayadin Offensive in the Deir ez-Zor Governate. On 6 November 2016, the SDF launched their Raqqa campaign, finally taking the city of Raqqa from IS on 17 October 2017.

By early 2018, therefore, the caliphate had been destroyed, IS had lost the ability to hold territory and had been reduced as a military threat. It hadn't, however, been defeated. A secret deal let its fighters leave Raqqa and al-Baghdadi remains alive (for the moment). IS has come back from near-defeat in the past and, although it is now unpopular with its own Sunni constituency, the ascendency of Shia powers in Syria, Iraq and Iran means that radical Islamists will continue to have an appeal as defenders of the regional Sunni populations unless the governments in Iraq and Syria improve their treatment of them.

As IS's hold on physical territory declined, so too did its online hold on the global public imagination. Military defeat proved to be the most effective counterweight to its online recruitment. The romantic appeal of life in the state and of fighting for the caliphate disintegrated as it was pushed back and besieged. The military operations also impacted upon IS's ability to keep up the same volume of propaganda and transformed the content of that propaganda, leading to a less celebratory tone. Once IS lost its real-world invincibility, therefore, its online presence and impact was significantly diminished. Hence, most of the current fears in the west now focus on real-world threats, such as the risks from returning fighters, the existence of a diaspora of IS jihadis around the world, the danger of revenge terrorist attacks and the future emergence of new Salafi-Jihadi movements.

All of these are important but it shouldn't be assumed that the online threat has been reduced. As Burke points out, hardline Islamist ideas that were once rare are now commonplace among many Muslims: in the 1970s to 80s

> the language of extremism and violent action was restricted to a tiny fringe of Muslims, largely in the Middle East and South Asia. Today, if actual violence is still limited to a negligible proportion of the world's Muslims, the language is spoken by a much larger number.
>
> Burke, 2015:208

This isn't simply the result of Islamist propaganda and activity. An endless 'War on Terror', numerous military interventions, innumerable drone strikes and other operations, incalculable Muslim civilian deaths and misery, the destruction of entire nations, the terrorist blowback in the west and the resulting rises in Islamophobia, racism, hate crimes and hate speech all helped prepare the soil for these ideas to flourish, even in the heart of the west. Islamic State may not have succeeded in eradicating 'the grey zone', to drive all Muslims towards its purified Islam, but it succeeded in exploiting and feeding a polarization and internal reaction to western policies and culture that we should take seriously. Islamic State is probably finished, but after 17 years of a 'War on Terror' the popular appeal of Salafi-Jihadism is far greater now than it was in 2001.

Key reading

Post-invasion Iraq attracted only a limited attention from publishers until 2014 when the spectacular success of Islamic State – a group seemingly arising out of nowhere for the western public – forced the country into the news again. Over the next two years there was a minor publishing explosion with many books explaining the origins and develop-ment of this movement and exploring contemporary Salafi-Jihadism. The first part of this chapter describes the rise of Islamic State and this is well-covered in a number of excellent books, including (in alphabetical order) Atwan (2015), Cockburn (2015), Gerges (2016), Hosken (2015), Griffin (2016), Moubayed (2015), Stern and Berger (2015), Warrick (2015) and Weiss and Hassan (2015). These are all excellent on the history and politics of Islamic State and I would recommend reading any or all of them. Wood's (2015) article is also worth reading. Few of these, however, examine IS's use of communications or digital media in much detail. There are chapters in Griffin (2015, ch. 15), Atwan (2015, ch. 1), and Weiss and Hassan (2015, ch. 11), but the best account is in Stern and Berger (2015, chs. 5–7). Far more detail about IS and digital media, however, is available online. Koerner's (2016) article is recommended, but there are obviously a lot of similar commentaries available. I've covered a lot here, but there's far more to be found if you spend time digging. Google my examples for more information about them.

For background information about the broader Salafi-Jihadi movement I would recom-mend starting with Kepel (2006), which traces the history of political Islam, then reading Maher (2016) and his discussion of the features of the Salafi-Jihadist movement. Lacy (2008) is an important collection of key jihadi texts, including sections from Naji's *The Management of Savagery*. Qutb's *Milestones* (2002) remains the most important modern

Islamist text, explaining the key ideas that influenced much of the anti-western Salafi Jihadist movement and should be read. For discussions of the Afghan Jihad and the origin and development of al-Qaeda see the key reading for chapter 4. The Wahhabist background to jihadism is well-covered in Moubayed (2015, ch.1) and Burke (2015, ch. 2), whilst Stern and Berger's appendix (2015:257–80) provides a recommended introductory primer to the history of Islam, issues around the caliphate, Salafism and Wahhabism, and issues around the practice of Salafi-Jihadism. For an important discussion of the contemporary mindset and appeal of Islamic Militancy, see Burke (2015, especially chs. 7–9).

Note

1 The Counter Terrorism and Security Act 2015 made it a statutory duty for UK universities to engage with the Prevent agenda, requiring staff to report or challenge extremist ideas. In the eight years of teaching my module on 'Digital War' I have never come across any Islamist extremism. Every year, however, I set my predominantly white, western class the moral conundrum of whether they would launch a drone strike on a high-value target if there was a certainty of innocent civilian casualties. Every year the majority happily vote to launch the drone's missiles, regardless of the number of civilians I suggest are present. When I ask if they'd launch if those civilians were their own families most hands go down. These are the only students I've met who want to inflict harm on a large number of civilians.

References

Atwan, A. B. (2015) *Islamic State: The Digital Caliphate*, London: Suqi Books.

Berger, J. M. and Morgan, J. (2015) 'The ISIS Twitter Census: Defining and Describing the Population of ISIS Supporters on Twitter', *The Brookings Project on U.S. Relations with the Islamic World*, Analysis Paper, No. 20, March, https://www.brookings.edu/wp-content/uploads/2016/06/isis_twit ter_census_berger_morgan.pdf.

Borchers, C. (2015) 'Operation ISIS: Anonymous Member Discusses how Group is Waging War on Militant Group', *Independent*, 28 November, http://www.independent.co.uk/life-style/gadgets-and-tech/news/operation-isis-anonymous-member-reveals-how-they-are-waging-war-on-the-militant group-a6752831.html.

Brooking, E. T. and Singer, P. W. (2016) 'War Goes Viral', *The Atlantic*, November, https://www.theatlantic.com/magazine/archive/2016/11/war-goes-viral/501125/.

Burke, J. (2015) *The New Threat from Islamic Militancy*, London: Bodley Head.

Cameron, D. (2015) 'Twitter: Anonymous' Lists of Alleged ISIS Accounts are "Wildly Inaccurate"', *The Daily Dot*, 21 November, https://www.dailydot.com/layer8/twitter-isnt-reading-anonymous-list-isis-accounts/.

Cockburn, P. (2015) *The Rise of Islamic State*, London: Verso.

Cottee, S. (2015) 'The Challenge of Jihadi Cool', *The Atlantic*, 24 December, https://www.theatlantic.com/international/archive/2015/12/isis-jihadi-cool/421776/.

Diab, K. (2015) 'The Jihadist Selfie is Changing the Image of Holy War', *Al-Jazeera*, 11 March, http://www.aljazeera.com/indepth/opinion/2015/02/jihadist-selfie-changing-image-holy-war-150224053159076.html.

Diaspora* (2011) 'Diaspora*', https://web.archive.org/web/20111002003516/http://blog.diasporafoun dation.org:80/2011/09/21/diaspora-means-a-brighter-future-for-all-of-us.html.

Fears, D. (2014) 'ISIS Gives Tips on How to be a Good Jihadi Wife', *New York Post*, 1 November, https://nypost.com/2014/11/01/isis-media-wings-tips-on-how-to-be-a-good-jihadi-wife/.

Gallagher, S. (2015) 'Anonymous' #OpParis Campaign Against ISIS Goes Horribly Awry', *Ars Technica*, 22 November, https://arstechnica.com/tech-policy/2015/11/whos-isis-anonymous-opparis-campaign-against-islamic-state-goes-awry/.

Gerges, F. A. (2016) *Isis: A History*, Princeton: Princeton University Press.

Gibbs, S. (2015) 'Anonymous Swaps ISIS Propaganda Site for Prozac ad in Trolling Fight', *Guardian*, 26 November, https://www.theguardian.com/technology/2015/nov/26/anonymous-swaps-isis-propaganda-site-for-prozac-ad-in-trolling-fight.

Griffin, M. (2016) *Islamic State: Rewriting History*, London: Pluto Press.

Gunter, J. (2015) 'Isis Mocked with Rubber Ducks as Internet Fights Terror with Humour', *Guardian*, 28 November, https://www.theguardian.com/world/2015/nov/28/isis-fighters-rubber-ducks-reddit-4chan.

Hosken, A. (2015) *Empire of Fear: Inside Islamic State*, London: Oneworld Publications.

ISISFAILS. (2016) 'ISIS Fails (Bennie Hill Theme)', *YouTube*, 27 April, https://www.youtube.com/watch?v=8mTyDRyn17A.

Katz, R. (2014) 'The State Department's Twitter War With ISIS is Embarrassing', *Time*, 16 September, http://time.com/3387065/isis-twitter-war-state-department/.

Kepel, G. (2006) *Jihad*, London: I.N. Tauris and Co. Ltd.

Klausen, J. (2015) 'Tweeting the Jihad: Social Media Networks of Western Foreign Fighters in Syria and Iraq', *Studies in Conflict and Terrorism*, 38(1), pp. 1–22.

Knibbs, K. (2015) 'The State Department Tried to Fight ISIS on Ask.fm and It Didn't Go Well', *Gizmodo*, 12 April, https://gizmodo.com/the-state-department-tried-to-fight-isis-on-ask-fm-and-1746071495.

Koerner, B. I. (2016) 'Why ISIS is Winning the Social Media War', *Wired*, April, https://www.wired.com/2016/03/isis-winning-social-media-war-heres-beat/.

Lacey, J. (ed.) (2008) *The Canons of Jihad*, Annapolis, MD: The Naval Institute Press.

Maher, S. (2016) *Salafi-Jihadism. The History of an Idea*, London: Penguin.

Malik, S., Laville, S., Cresci, E. and Gani, A. (2014) 'Isis in Duel with Twitter and YouTube to Spread Extremist Propaganda', *Guardian*, 24 September, https://www.theguardian.com/world/2014/sep/24/isis-twitter-youtube-message-social-media-jihadi.

Morozov, E. (2011) *The Net Delusion*, London: Allen Lane.

Moubayed, S. (2015) *Under the Black Flag*, London: I. B. Tauris.

Quinn, B., Ball, J., and Rushe, D. (2014) 'GCHQ Chief Accuses US Tech Giants of Becoming Terrorists' "Networks of Choice"', *Guardian*, 3 November, https://www.theguardian.com/uk-news/2014/nov/03/privacy-gchq-spying-robert-hannigan.

Qutb, S. (2002) *Milestones*, New Delhi: Islamic Book Service Ltd.

Rose, S. (2014) 'The Isis Propaganda War: A High-Tech Jihad', *Guardian*, 7 October, https://www.theguardian.com/world/2014/oct/07/isis-media-machine-propaganda-war.

Seib, P. and Janbek, D. M. (2011) *Global Terrorism and New Media*, Oxon: Routledge.

Siddique, H. (2014) 'Jihadi Recruitment Video for Islamist Terror Group Isis Features Three Britons', *Guardian*, 21 June, https://www.theguardian.com/world/2014/jun/20/jihadi-recruitment-video-islamist-terror-group-isis-features-britons.

Steele, C. (2015) 'The Social Media Sisterhood of ISIS', *PC Mag UK*, 25 March, https://www.pcmag.com/article2/0,2817,2478301,00.asp.

Stern, J. and Berger, J. M. (2015) *Isis: The State of Terror*, London: Harper Collins.

Svirsky, M. (2014) 'Isis Releases "Flames of War" Feature Film to Intimidate West', *Clarion Project*, 21 September, https://clarionproject.org/isis-releases-flames-war-feature-film-intimidate-west/.

Thornhill, T. (2014) 'Isis use Top Video Game Grand Theft Auto 5 to Recruit Children and Radicalise the Vulnerable', *Daily Mail*, 22 September, http://www.dailymail.co.uk/news/article-2765414/Isis-use-video-game-Grand-Theft-Auto-5-recruit-children-radicalise-vulnerable.html.

Warrick, J. (2015) *Black Flags. The Rise of ISIS*, London: Transworld Publishers.

Weiss, M. and Hassan, H. (2015) *Isis: Inside the Army of Terror*, New York: Regan Arts.

Winter, S. (2016) 'ISIS Issue CAT Fatwa: Islamic State Jihadists Target West by Forbidding CATS', *Daily Express*, 7 October, https://www.express.co.uk/news/nature/718833/Islamic-state-jihadists-target-West-forbidding-cats-banned-breeding-ISIS-terrorists-pigeon.

Wood, G. (2015) 'What ISIS Really Wants', *The Atlantic*, March, https://www.theatlantic.com/magazine/archive/2015/03/what-isis-really-wants/384980/.

Woolf, N. (2016) 'Twitter Suspends 235,000 Accounts in Six Months for Promoting Terrorism', *Guardian*, 18 August, https://www.theguardian.com/technology/2016/aug/18/twitter-suspends-accounts-terrorism-links-isis.

Yadron, D. (2016) 'Facebook's Sheryl Sandberg: "Likes" Can Help Stop Isis Recruiters', *Guardian*, 21 January, https://www.theguardian.com/technology/2016/jan/20/facebook-davos-isis-sheryl-sandberg.

12 Augmented war

Wearables, phones, soldier-systems,
AR, simulations, sensors,
exo-skeletons and BCI

The problem of presence

Perhaps the central problem of war is the problem of presence. In order to impose your will militarily you need a physical presence, but that introduces the risk of your forces being killed. The key question then is how do you eliminate that risk to forces and remain able to impose your will?

This was the problem the USA faced in the 1991 Gulf War. At issue was the 'Vietnam syndrome' – the fear that the USA was too weak to risk its own servicemen's lives as mounting casualties would turn the public against any war. To overcome this the USA employed a new system of media management to control the public understanding and overwhelming military and technological superiority to remove their forces from danger. The Gulf War established a model of 'top-down', close control of the media and military theatres that was replicated in Kosovo in 1999, where the air campaign achieved a zero casualty rate. An essential part of this military superiority was informational superiority, with the USA aiming at a total awareness of the battlefield and the positions and activities of all the forces. One solution to the problem of presence, therefore, was *hyperpresence*: an augmented sensory capacity aiming at an omniscient vision of the battlefield. In the years following the Gulf War, the US military retheorized warfare to emphasize issues around communication and information. Much of this was devoted to the idea of controlling all battlefield information, such as the concept of 'full spectrum dominance', although developments in digital technology would, as we've seen, make this dream of full control unrealizable. What did survive, however, was the idea of full awareness.

This was central to the concept of 'network-centric warfare' (NCW), proposed in a 1998 article by Cebrowski and Garstka. NCW argued that a small, highly mobile, informationally networked unit, supported by an information back-plane could gain enough 'situational awareness' as to have a significant advantage over their enemy and hence be able to outmaneuver and defeat them. A superior information position would enable them to use 'speed of command' and bottom-up 'self-synchronization' among the units to foreclose enemy action, shocking and disrupting their strategies and (as in the new wired economy) locking out enemy competition. That is to say, a traditional, larger, top-down-commanded, conventional force could be pulled apart by a much more efficient and effective, smaller, faster force with the informational advantage of an omniscient vision of the battlefield.

The 2003 Iraq War was the first real test of NCW. Military and technological supremacy again ensured the USA's success, with the Iraqi army defeated in 21 days at a cost of only 139 troops. Network-centric warfare proved less effective, however, in keeping order in Afghanistan and Iraq, due to the small size of the forces and the emergence of networked insurgencies. US soldiers, therefore, proved far more vulnerable in the aftermath of wars than during them. Hyperpresence was a central part of counter-insurgency (COIN) operations, through drone ISR and cyberwarfare hacking, but the insurgencies remained deadly for those fighting them. Drones, however, also offered a new solution to the problem of presence. Traditionally, distance had been the best way of removing combatants from danger whilst retaining capability, but most projectiles still required forces on or near the battlefield and hence still at risk. Drones went further by replacing presence with telepresence, allowing forces to operate remotely through sensors and weapons systems. For the first time in conventional war you could fully abstain from deploying physical bodies and be able to implement your will.

The US military continues its attempts to solve the problem of presence. Two main research strands in particular may impact significantly on future wars: continuing research into *hyperpresence* through networking and augmenting down to the level of individual soldiers, and continuing research into *telepresence*, to move beyond current drone technology with closer connections through brain–computer interfaces.

Wearables, phones, and soldier-systems

As the USA upgraded its military through the 1980s there was a realization that the infantryman was being left behind, remaining highly vulnerable to attack. Two main research strands developed, aiming to augment the individual solider and integrate them better into these systems. These aimed at either sensory hyperpresence, with wearable technology and complete soldier-systems expanding informational awareness, or physical hyperpresence, with exo-skeletons expanding the body's capabilities.

Sensory hyperpresence revolved around the dream of a next-generation infantryman combat system. This would be a soldier equipped with wearable and hand-held computing technologies and sensors, all linked to the military data-net to give individuals unparalleled 'digital situational awareness', allowing more decentralized and networked operations. The first major US programme was the Solider Integrated Protective Ensemble (SIPE), an Advanced Technology Demonstration (ATD) begun in 1989, inspired by Steve Mann's pioneering work on 'wearable computing' and based on the emerging concept of the 'soldier as a system'. SIPE's aim was to coordinate and integrate all aspects of a soldier's functioning and equipment such as communications, protective suit, weapons, climactic protection, load-carrying equipment, etc. to improve their effectiveness and survivability. In addition to long-term goals such as 'chameleon camouflage', the 2005-10 SIPE timeframe envisioned a power-assisted combatant, life-support system, exo-skeleton, weapon system integration and power-generation system.

The underlying concept of SIPE fed into the next major project, the Land Warrior Integrated Fighting System (LW), which began in 1994 with a contract run by Hughes Aircraft Company (now Raytheon Systems). The technologies of the day were bulky, heavy and limited by the state of computing; hence they were continually modified through the 1990s as equipment got lighter, more powerful and cheaper. In February 2003 General Dynamic Decision Systems (now General Dynamic C4 Systems) were awarded a contract to integrate LW into the US Army digital communications and make it interoperable with the Stryker Brigade Combat System (a mounted system, it was argued, would enable suits to be more easily recharged in the field). The LW Stryker

Interoperable systems ('Mounted Warrior', or MW) were delivered for testing in 2005. In February 2005, the Army merged Land Warrior with its Future Force Warrior (FFW) Advanced Technology Demonstration, which was itself part of the Future Combat Systems Project (which ran from 2003 until 2009 and was succeeded that year by the Army Brigade Combat Team modernization programme). The FFW similarly wanted to create a light-weight, fully integrated infantryman combat suit, but also envisioned the use of radical technologies such as nanotechnology and exo-skeletons.

A late 2004 side-by-side test of LW-equipped soldiers with ordinary troops showed the new system did improve combat effectiveness, leading to battalion-level testing of Land Warrior/Mounted Warrior by the Army at Fort Lewis from May to September 2006. LW was cancelled in February 2007, although was renewed upon review after a LW-equipped force, deployed before the system could be withdrawn, reported favourably on it. 229 LW and 133 MW ensembles were deployed with the 4th Battalion, 9th Infantry Regiment to Iraq from May 2007 to June 2008. A Stryker Brigade later deployed with the system in Afghanistan and LW remained in use until Spring 2012. A decision was made in 2011 to continue the programme and LW remains listed as an available technology.

Land Warrior is a modular, wearable system in two configurations: for the soldier and the squadron leader. Its aim is to improve the soldier's lethality and survivability and provide full command communications at an individual level. It comprises a number of subsystems. The soldier wears a fighting load vest that includes a ten-point hub, connecting the electronic systems, a GPS employing five satellites that tracks the user to within 10m, a power system of batteries providing 8–24 hours of sensor and computer operation, and a computer containing battlefield software that downloads the soldier's sensor data. The helmet contains the communications equipment, including an antenna that connects via the hub to the soldier's radio. The helmet includes a head-mounted display (HMD) over the eye that provides command and control information and that receives information from the M4 Carbine daylight video scope (DVS) or weapon-mounted infrared thermal scope as well as topographical and satellite maps with friendly positions indicated which updates every 30 seconds. The HMD can also use the DVS to capture battlefield images and send and receive these images and data. The soldier can switch screens and select among the menus either using a button on the rifle stock, or through the computer on his body which uses a joystick to move mouse and cursor buttons. The computer includes six software packages, including mapping, which can generate live satellite maps within minutes, and selection control software, which allows the soldier to control the amount of data they receive so they remain operationally effective. The rifle contains control buttons, including push-to-talk, switch screen and take-a-picture commands, allowing the soldier to operate without lowering their weapon. Targeting uses a laser rangefinder and digital compass. Data from the rifle and about the soldier can be used for identification and targeting when support is called in. A BBN Technologies sniper-detection system uses microphones to detect muzzle blast and a display giving the distance and direction of the sniper. The squadron leader configuration includes a keyboard and a handheld flat-panel display.

The LW project suffered from its weight (16 pounds in 2007), which added considerably to the infantryman's 80-pound combat gear, as well as from issues around battery life, the poor quality of its computing and maps, and its spiralling cost, with the equipment for one soldier costing $85,000. Initially, soldiers using the equipment in 2007 weren't impressed. 'It's just a bunch of stuff we don't use, taking the place of useful stuff like guns', one soldier reported, adding, 'It makes you a slower, heavier target' (Page, 2007).

Those using the system in Iraq, however, became fans, although they modified it in use and found it functioned better if only the patrol leaders had the full kit. By disposing of equipment they didn't need (such as the useless gun-cam), they eventually helped get the system down to seven pounds.

Research also continued under the merged LW/FFW programme, becoming the basis for the next-generation Ground Soldier System (GSS), an integrated, modular, dis-mounted system with the aim again of increasing lethality and survivability and situational awareness. GSS entered development in 2009. GSS Increment 1 was renamed 'Nett Warrior' (NW) in June 2010, after World War II Medal of Honour winner Robert B. Nett, and was demonstrated in Spring 2011 as essentially the LW system with software improvements. In October 2011, however, the Army unveiled a new 'End User Device', replacing LW's HMD with a commercial off-the-shelf (COTS) hand-held display utilizing in-house software and integrated with the Rifleman radio. It consider-ably simplified the LW system, reducing it to three pounds.

Nett Warrior, therefore, differs from LW in using a chest-mounted Android cellphone device in a ruggedized wrapper. It is designed for navigation, command and control communications and friendly-force tracking, showing graphics on a small, digital moving map that displays the positions of forces and enables the identification and targeting of enemy forces. As Lt. Col. Adrian Marsh explains, 'It provides unprecedented situational awareness at the dismounted level through the map display. The icons show where all the other users are on the battlefield, and the device allows for battlefield messaging. Every-one sees the same picture … The battle changes in real-time and information can transmit across the force in real-time' (Osborn, 2016). NW has worked with a number of smart-phones, including the Motorola Atrix and Samsung Galaxy Note 1 and 2. The military continue to develop Nett Warrior: in October 2016, it was reported that the Army was upgrading and widening its NW deployments, and in July 2017, it was reported that NW was now able to stream live video feeds from the Dragon Runner 20 UGV and the hand-launched Raven UAV. Nett Warrior will also be deployed to Afghanistan in mid-2018 with the 1 Security Force Assistance Brigade.

Nett Warrior highlighted the US military's growing interest in phones and tablets. It had begun by trying to develop its own soldier-system computers but commercial technologies quickly outpaced them and the 2007 iPhone and 2010 iPad were game-changing technol-ogies. Their improved processing and video capacities, cheaper cost, lighter weight, ease of use and familiarity now made COTS more attractive to the military. One opportunity the military saw to harness these developments was through smart-phone apps. In March 2010, it launched an 'Apps for the Army' competition. Fifty-three apps were submitted, with 25 cleared for military development. The winner, in August 2010, was a multi-media app digitizing the Army's huge, physical training manual. Another app, for the iPhone, was the 'Sigacts' programme, which allowed individual soldiers to access the Army's 'Command Post' software. Following this, the Army opened the 'US Army Marketplace' in April 2011 for iOS and Android, making software and apps available to buy for servicemen for their computers and phones. Personalized pages allowed comments, reviews and calls for apps to be developed. Private developers with DoD community membership could respond to these calls.

One problem that remained was security, with no commercial phone certified by the government as secure enough to receive military data. That might explain why, in April 2011, the Army announced it was developing its own smart-phone, to be produced by MITRE and using Android OS. The hand-held would run their new Joint Battle Command

Platform (JBC-P), an information system for Blue Force (friendly-force) tracking and critical messaging. The JBC-P was tested in January 2012 and deployed in the field in May 2015. It has had problems (with a damning January 2015 report describing JBC-P Software build 6.0 as 'not operationally effective'), although it continues to be developed and is employed on Nett Warrior. In October 2011, the Army also unveiled its Common Operating System (COE), a set of standards that would allow developers to produce apps for army smart-phones.

By then, however, soldiers had their own smart-phones. In 2009, it was reported that soldiers preferred their own iPhones and iPod Touches to military equipment, using, for example, the translation software available. As *Newsweek* commented, the military liked Apple products:

> Apple gadgets are proving to be surprisingly versatile. Software developers and the U.S. Department of Defense are developing military software for iPods that enables soldiers to display aerial video from drones and have teleconferences with intelligence agents halfway across the globe. Snipers in Iraq and Afghanistan now use a 'ballistics calculator' called BulletFlight, made by the Florida firm Knight's Armament for the iPod Touch and iPhone. Army researchers are developing applications to turn an iPod into a remote-control for a bomb-disposal robot (tilting the iPod steers the robot). In Sudan, American military observers are using iPods to learn the appropriate etiquette for interacting with tribal leaders.
>
> Sutherland (2009)

In 2010 the US Army gave soldiers in Afghanistan and Iraq iPod touches with language-learning apps on to help with Iraqi Arabic, Kurdish, Dari and Pashto, whilst in October 2011 Textron Systems released a Blue Force map app for the iPad. The commercial battle of the operating systems, therefore, was also being fought out within the military, with the Army opting for Android, but military personnel, and many producers and researchers, favouring Apple.

Android began to win as it was an open system that could be used on many platforms and the military feared being locked-in to a closed ecology owned by one company with their fortunes tied so closely to its success. In July 2016, however, the US Army began turning towards Apple again, replacing its fleet of Android phones with iPhone 6s as the Army Special Operations Command found its Android 'Tactical Assault Kit' (TAK) slow, glitchy and prone to freezing. The TAK are modified versions of popular smart-phones such as the Samsung Galaxy Note that are positioned on the chest and connected to a networked Harris radio and designed to run custom-made military apps. The Android systems were found to be inferior to iOS, especially when running live video-feeds such as from the Instant Eye UAV. Android didn't refresh fast enough and had to be regularly restarted.

Phones and tablets, therefore, had a lot of value for the military, aiding especially in terrain navigation, friendly-force tracking, language and translation services, giving access to technical manuals and live drone feeds and ISR information, and enabling soldier-unit communication. There were, however, problems. While early military wearable/hand-held technology was heavy and cumbersome, rapid developments in phones and tablets led to the opposite problem. Smaller devices could be too easily dropped or lost and wouldn't survive the rigours of a combat situation. Devices had to be large enough and screens had to be protected, have simplified interfaces and be usable with gloves. There was also a security issue with commercial technology. Ideally, the phone's production, OS, apps, services and pages accessed would be certified and secured to military standards but no company could achieve that with their current systems. A proliferation of phone or tablet-like devices also introduced new security risks from malware or hacking – imagine, for example, feeding false enemy position data to a unit – whilst any lost devices could seriously compromise operational security and allow retro-engineering by an enemy. The devices also required considerable technical support, not only to maintain them and produce the software for them but also to provide the connectivity they depended upon. Looking to the future, there was also the problem of bandwidth in a battlefield where every soldier might have a range of connected devices.

Some also questioned the value of the devices in combat as the screen would take up time and attention needed for offence/defence. If soldiers followed their civilian counter-parts, the screens would also distract the soldier and disconnect them from the real world. Others warned of the danger of information overload, or of becoming too reliant on data to take any action without it. The military have had to confront the issue of getting the right data at the right moment to the people who need it and teach critical, independent military skills. One solution to this problem is to limit the screen devices to those leading a unit. A further problem for the military with these devices is the speed of technological development. Wearable and hand-held military computing will quickly become obsoles-cent and a significant commitment to particular systems or interfaces could backfire on a force. Non-state actors, of course, don't have this problem, being able to utilize any COTS system they can find.

Experimental Future Soldier Systems like Land Warrior and Nett Warrior are now common around the world. They include the Italian Soldato Futuro Future Soldier System, Spain's COMFUT, Australia's LAND 125, Sweden's MARKUS, Switzerland's IMESS, Poland's Projekt TYTAN, India's FINSAS, Singapore's ACMS, Russia's Ratnick, Iran's SARV and Norway's NORMANS future soldier system. The UK has the Future Infantry-man Soldier Technology (FIST) system, which includes five areas of capability: a HMD with night vision, a communications system, a weapons system, lightweight protection and GPS navigation. It is still at an early stage of development and is expected to evolve. FIST V2 (version 2) began trials in October 2005, with more following in November 2006. The British Army, RAF Regiment and Royal Marines are expected to procure 35,000 FIST sets which will enter service by 2020. Not every soldier will be equipped with FIST, with unit commanders tailoring the system's use to operational and mission requirements.

Germany has its Rheinmetall Group-produced Gladius infantryman of the future (Infanterist der Zukunft – IdZ) system, which includes an assault rifle, weapon-mounted laser system, a command control communications computers and information (C4i) system integrated into the load-carrying vest, eye and ear protection systems, nuclear-biological-chemical (NBC) protection, a ballistic and stab protection subsystem and night vision subsystem. The prototype was tested in Kosovo in 2002 and IdZ1 was deployed in 2006–07 in Afghanistan, Kosovo and the Congo. In September 2017 Rheinmetall publically unveiled their next-generation Gladius 2.0 system, which includes a goggle-projection system and a chest-mounted 5-inch PDA and 7-inch tablet for the section and platoon commanders, respectively.

France, similarly, has the Sagem-produced FELIN (Fantassin à Équipements et Liaisons Intégrés) Future Infantry Soldier System. Development began in 1996, with testing from 1999 to 2001 and the first production units being delivered in 2009. All French infantry units were equipped with FELIN V1 in 2010, with production of V2 beginning in 2015. FELIN includes a portable computer, a voice and data radio, a GPS system, combat clothing with body armour, a helmet with two LED displays and an optronic system with a camera, microphone and vibrating speaker linking the soldiers and commander through the Rif (infantry information network). The weapons system includes image-intensifying and thermal-imaging sights, all linked to the communications system so the targets can be transmitted digitally in real-time through the FELIN communications network. The assault rifle has push controls for other systems and a video sight linked to the helmet displays that also allows the soldier to extend the weapon around corners to fire without exposing themselves. This proliferation of systems led to NATO's Army Armament Group setting up the Topical Group 1 (TG/1) to coordinate soldier system interoperability, prevent duplication and standardize protocols, for example for tactical display symbols.

Other programmes continue alongside these systems. From 2009 to 2015 the Army Research, Development and Engineering Command (RDECOM) ran the speculative 'Future Soldier 2030' project, designed 'to ignite the imagination and spark discussion across the Army as to how current and emerging research will culminate on the soldier' (HQAMC Public Affairs, 2009). The project discussed a range of possible technologies including linking real, augmented and virtual environments; powered exo-skeletons; real-time sensors embedded in uniforms for status monitoring; AR and VR interfaces; cognitive enhancing smart drugs; neural prosthetics; permanent prosthetics; robotic battlefield telepresence; a personal intelligent agent AI 'digital buddy'; heads-up displays (HUD) with virtual and physical interfaces with robots; wrist-mounted displays; biometric facial recognition systems; electronic textiles; smart fibers able to respond to identification requests and biometrically matched to individual soldiers; improved blast protection through carbon nanotube based composites; wound and bleeding management; a micro-climate conditioning system; smart weaponry; and small robotic systems to aid the warfighter.

On 16 September 2015, the UK MOD released its own 'Future Soldier Vision' (FSV) project. Based upon emerging technologies, it was an image of what the infantryman could be wearing in the 2020s. This included a head subsystem, with sensors for information sharing and an integrated power supply, feeding information to the soldier and others in the unit; a torso subsystem with integrated connectors and power supply and sensor-laden body armour for improved survivability and situational awareness; smart watch-style wearable communications including sensors to monitor the soldier's health data and data-display capacities; smart glasses with bone-conducting headphones, a HUD and integrated camera, displaying images to soldiers and able to send video to commanders; a touch-screen, personal role computer with radio and camera capabilities, allowing information sharing and communications; a weapon system allowing the sharing of targeting information and digital sights sending data to the visual displays and radios. As the MOD state, 'The Future Soldier Vision is designed to work as an integrated system with survivability, enhanced situational awareness and network capability all central to the concept' (UK Government, 2015).

These speculative visions include emerging technologies with military applications. Armour, for example, might be improved with nanotechnology, which allows material to be manipulated on an atomic, molecular or supramolecular level. This would lead, it is claimed, to more resistant and stronger materials as they could be designed without imperfections, employing a perfect lattice structure of carbon fibre infused ceramics with perfect integration. Other uniform-based technology includes experiments into 'liquid armour', which remains flexible in use but which hardens on impact. The Polish body armour company Moratex, for example, is working on a 'magic liquid' called Shear Thickening Fluid that instantly hardens and disperses energy over a larger area to minimize bullet damage. In 2015, the US Army also expressed an interest in uniforms that would make its soldiers invisible, putting out a call for proposals from companies manufacturing 'metamaterials' – adaptive structures first demonstrated in 2006 that can bend light around the wearer, making them 'invisible' from certain angles. Although complete invisibility is impossible, partial camouflage is achievable and some companies may be on their way to fulfilling the military's requirements. The Canadian form HyperStealth Biotechnology, for example, clams it demonstrated metamaterial camouflage to US military scientists in 2014.

Another development is BAE Systems' 'Broadsword' range, which is built around a 'Spine' that uses e-textiles to wirelessly charge military equipment, with the energy use being monitored by a smart-phone app. Spine gets its energy from an inductive charging seat in a vehicle, thus solving one of the problems of soldier systems – that of power generation and batteries – as soldiers can be kept powered throughout a mission. The seat powers up the vest, which can then, in turn, power up and transfer data to and from radios, cameras, smart helmets, torches and smart weapons. A total of eight devices can be plugged in and charged at a time and the vest's electrically conductive yarns can also charge other gadgets wirelessly.

Smart weapons are also central to the smart-soldier system. In December 2014 DARPA released test footage of a new 'smart bullet' that can change direction in mid-air called EXACTO (Extreme Accuracy Tasked Ordnance). The sniper fires through a laser sight and an optical sensor on the bullet detects where the laser is. As the target moves, so the sniper can move the laser to follow it. A small actuator motor on the bullet receives the data from its sensors about the laser's new location and fins on the bullet change its trajectory 30 times a second to compensate, allowing it to hit the new target.

Augmented reality

Although many militaries are using chest-mounted/hand-held screen devices to display information, HMDs offer another possibility that is just beginning to be exploited. Instead of simply displaying information for the soldier they can also be used as Augmented Reality (AR) systems, showing digital information on top of the real-world view. BAE Systems, for example, produces the Q-Warrior HMD, a full-colour, lightweight display that clips onto existing in-service helmets. As BAE explain, 'With Q-Warrior HMD, waypoints, points of interest, and targets are all displayed, overlaid on the real world, reducing workload'. This includes Blue Force tracking overlaid onto the HMD vision, with AR bringing 'Complete awareness of all assets, friend or foe'. The 'Entire 3D battle space is permanently displayed in a simple manner', enhancing situational awareness (BAE Systems, 2018b).

Over recent years CERDEC, ARL and DARPA have been working on the technologies to make AR work on the battlefield, including developing a system, unveiled in October 2015, called Heads-Up Navigation, Tracking and Reporting (HUNTR). Able to serve as a heads-up extension to Nett Warrior, this, again, overlays digital information onto the field of view. Continued research led to another US Army system being unveiled in May 2017 called Tactical Augmented Reality (TAR). Designed to offer enhanced navigation, Blue Force tracking, information sharing and weapons targeting, TAR represents another attempt to increase situational awareness. It uses GPS, helmet camera data and sensors to 'geo-register' a user's field of view, so that symbols and information can be projected onto it. Augmented reality, however, remains an experimental technology and it will be years before such systems are significantly developed and deployed. They have the potential, however, to free the soldier from hand-held screens that distract attention and require physical manipulation, thus increasing situational awareness and responsivity.

Simulations

Whilst augmented reality is now being developed for combat use, military uses of virtual reality (VR) are primarily limited to training simulation technologies. By the late 1990s commentators such as Bruce Sterling were talking about the 'Military-Entertainment Complex', whilst James Der Derian's 2001 book *Virtuous War* described what he called 'MIME-net', or the Military-Industrial-Media-Entertainment Network' (2009:83), noting the growing links between the military, Hollywood and electronic entertainment companies and the development of electronic, immersive training simulators. 9/11 and the 'War on Terror' spurred this on, with defence spending on MS&T (Modelling, Simulation and Training) rising from $3bn to $6bn from 2000 to 2006. There was a move especially to 'mixed-reality' trainers designed to hone cognitive and muscular reactions such as the 'Engagement Skills Trainer' (EST) from 2000, which trained soldiers in simulated combat. In January 2012, the US Army announced a plan for a new immersive system that would develop an accurate avatar for each soldier based on their own physical attributes and developing skills that would be used throughout training, and in December 2013 Northrop Grumman unveiled its VIPE Holodeck – the Virtual Immersive Portable Environment system – a room you enter, to interact with 360 degrees of video walls.

One recent development has been to move away from immersive systems to individual technologies. By May 2010 Combatredi by Cubic had developed a helmet-mounted 360 virtual headset system, winning a $4.8m contract to supply 27 of these systems to the Florida Army National Guard. Combatredi comprises a high-definition OLED video display, an integrated headset for stereo sound and a wireless, connected rifle that performs like a real one, including the need for fresh magazines and various firing modes. The user's movements are tracked by a computer on their back, while a suite of sensors can tell where the user is looking and whether they're standing, kneeling or lying down.

By 2014 the US Army was considering the possible uses of the Oculus Rift HMD virtual reality system, exploring how lifelike battlefield simulations could be used to train medics to treat wounded soldiers under fire. In May 2014 DARPA also revealed it was experimenting with the Oculus Rift to give cyberwarriors a new way to visualize three-dimensional network systems, to improve their attacks. Allowing users to look around data, it would be intended as a more intuitive way for US Cyber Command and US military hackers to visualize, plan and carry out their defensive and offensive operations. 'Say we want to turn out the lights in some place where we have boots on the ground, but it's on a subnet connected to a hospital', Plan X Program manager Frank Pound explains, 'With the Oculus you have that immersive environment. It's like you're swimming in the internet' (Greenberg, 2014).

By 2014 the US Army was also displaying its new, mass-produced, full-body, immersive VR system, the Dismounted Soldier Training System (DSTS), which allowed up to nine soldiers to train together in VR. The system uses a HMD, headphones, microphones, a backpack with a computer and batteries and a power distribution system for the system's elements, using a joystick to simulate movement within a virtual environment whose game engine can run up to 5,000 non-player characters (NPCs). Virtual equipment can be grabbed and used from around the body (represented by passive RFID tags placed where the equipment would be) and the system simulates M4 rifles, M320 grenade launchers and M249 light machine guns, with the correct weight and distribution. Other militaries are also using VR. Polish soldiers train with a VR system with integrated feedback, delivering small electric shocks when they're 'shot'.

Another line of development is whole battlefield simulation. Real-world war-game exercises are often now combined with war-game software to simulate broader battles, but, building on earlier attempts such as SIMNET in the 1990s, there is a move towards fully digital battlefield simulations. In January 2013 the UK staged 'Exercise Urban Warrior 5', its largest virtual battlefield simulation at the Warminster Land Warfare Centre. A mixed real-life and computer simulation it used 230 soldiers in training exercises as well as the new 'Urban Warrior 5' virtual system, built using VB52 software. BAE has also developed a Virtual Battlefield System, using its Dedicated Engineering Network (DEN) simulators, which link together simulators such as the E-3D Sentry simulator at New Malden, Warton's Typhoon simulator and Broad Oak's Type 45 Destroyer simulator. The simulators can be accessed through numerous bases, combining networked simulators together to play out different military scenarios.

In December 2013, Northrop Grumman carried out a unique virtual aerial refueling demonstration using a C-17 Globemaster simulator in Texas, a KC-135 Tanker simulator in Florida and a boom operator simulator in Oklahoma, connected by the Air Force's Distributed Missions Operations Centre (DMOC) at Kirklands AFB, with all platforms functioning together as if on a real mission. As the C-17 pilot said, 'It is hard to believe the boom operator is not just 50 feet away from you, but that he is two states away. There were no delays; the mission was real-time and very realistic' (Turnbull, 2014). The USA's DMOC also runs a quarterly exercise called 'Virtual Flag' which simulates real-world aerial war games. Employing the latest simulation technology, Virtual Flag incorporates personnel from other countries including the UK, Canada and Australia. In September 2013, for example, a UK RAF E-3D AWACS crew participated in an exercise, providing airborne warning assets to coalition forces, without leaving their RAF base. Such simulations allow aircrew to experience major combat operations and large force deployments that are impossible to create in peace time.

One of the key motivations for VR simulations is budgetary. Real-world training exercises are expensive, and so are real-world flying hours. Ongoing budget cuts have reduced the time a pilot spends in the air, so the USAF is looking to simulators as a cheap way of maintaining combat readiness and preventing 'skill-fade'. But simulations aren't simply cheap training options: their aim is to change warfare itself. The core idea of a simulation is to model the real so as to be able to predict and control it. As Baudrillard's philosophy demonstrates, simulation isn't creating something false, but creating instead something that functions as and replaces the real as an *efficacious* reality. Essentially, whether they involve full battlefield simulations or small-scale unit operations, simulations attempt to impose order through game-play. They aim to gain the power to manipulate and create reality itself.

These simulations are limited, however, by the lack of fear and danger, by the stilted reactions they produce (being watched, few soldiers would commit war-crimes such as the Afghanistan 'Kill-Team' or Abu Ghraib), and by their programming. Soldiers can only respond to what has been programmed. This reveals that simulations are essentially reactive, reducing the complexity of life, over-simplifying situations and failing to see emergent threats. But their aim isn't to understand the future, but rather to control the present, hence they fit easily into the hyperpresent soldier-systems we've been discussing, in augmenting soldiers to enable them to better control the real. Simulators, of course, aren't just used by the military, with cheap COTS systems democratizing their possibilities. The 9/11 hijackers, for example, trained on Microsoft's *Flight Simulator*. They didn't need to learn how to take off or land, they only needed to know how to steer an already flying plane.

Sensors and 'Big Data'

Today's armies are getting smaller, smarter and more powerful. As a result, training is also getting more expensive, leading to a realization that militaries need to protect their assets more. Hence the development of technologies such as soldier-systems that augment and protect the individual soldier. Uniform sensors are central to this augmentation, in monitoring the soldier's position, functioning and health. In January 2017, for example, it was reported that, as part of their Future Naval Capability programme, the US Navy Office of Naval Research had developed BLAST (Blast Load Assessment Sense and test), a system detecting brain injuries in soldiers exposed to blast pressure and shock waves. Coin-sized sensors in the helmet and suit detect blast-waves and an algorithm converts the data into a 'go or no go' injury assessment to determine if the soldier needs medical treatment. The three-part system is able to measure blast pressure, establish injury thresholds for the brain and analyse potential traumatic brain injury (TBI). It was expected to be trialed with marines by mid-2018. Sensors, therefore, are becoming more important to the military, enabling them to increase their knowledge of their soldiers and improve their survivability and performance.

In the USA, the Army, Navy, Air Force, Marine Corps and special operations forces are all funding research to collect biophysical data from their forces. As Tucker explains, 'The goal is to improve troops' performance by understanding what's happening inside their bodies, down to how their experiences affect them on a genetic level' (Tucker, 2017). Programmes such as the Army Research Lab's (ARL) The Human Variability Project (THVP) aim to outfit soldiers with wearable, interactive sensors to produce a range of data that could be machine-read and analysed, quantifying bodily processes and reactions in order to understand their operation. THVP is interested in how different aspects of the body, such as size, weight, height, health, alertness, affect how different people interact with their environment as well as with their vehicles and equipment. As they explain, 'by understanding and predicting human variability across multiple time scales, we will enable adaptive system designs that are dynamic and capable of eliciting the full potential of the humans with which they interact' (The Human Variability Project, 2016).

From 2015 to 2017 the US military spent $2m on Fitbits and other COTS systems, but as these didn't fulfill its biotracking needs it began creating its own wearables, including fabrics with electronic components embedded into them and electroencephalographic (EEG) helmets recording continuous brain-wave activity. One application is for fighter pilots, whose bodies are exposed to dangerous physical forces. As well as helmet sensors, Air Force researchers are working on a comprehensive cognitive monitoring system to gather as much data about their pilots as possible. Tucker points out that cameras can capture many elements, such as cerebral oxygenation, which can be measured by shining an infrared light on the pilot's forehead. 'Another research project configured simple laptop camera lenses to detect whether a person's hemoglobin is oxygenated, which makes blood shows up slightly redder, or de-oxygenated, which is slightly bluer. Essentially, this lets you read a person's heart rate from a distance' (Tucker, 2017). Stress is one of the most important signs to detect and can be measured through mask sensors monitoring breath or through eye-tracking cameras. Sensors can now detect metabolic changes that indicate weariness and stress before the person realizes, with military research showing it's possibly to detect stress responses from at least 100m away.

In February and March 2017, the Air Force successfully tested a new helmet with 'physiological monitoring capabilities'. Tucker says, 'The goal is to give every pilot a slightly different experience based on their unique physical and mental strengths and weaknesses, as well as their physical condition at the moment' (Tucker, 2017). Information from this research will guide future fighter jet design, but biophysical data will also be used for more individualized designs, with the aim of creating technologies and weapons that are based on the individual and their strengths and will adapt to their users. As Air Force research psychologist Glenn Gunzelmann commented at a National Defense Industrial Association in March 2017: 'The basic goal here is: we want to get greater precision and accuracy in predicting which people will succeed in particular job areas or missions' (Tucker, 2017).

Some projects hope to go further. Col. Kirk Phillips, associate chief for bioenvironmental engineering at the US Air Force, with his colleague Dr. Richard Hartman are leading a programme called Total Exposure Health: 'The goal is simple: collect and analyze as much data as possible about what happens to soldiers beyond the battlefield, right down to the kinds of molecules to which they are exposed' (Tucker, 2017). The aim is to use that data to produce new insights into how individuals interact with their environment, in real-time and in unprecedented detail. This includes giving a fuller picture of how specific sets of experiences affect specific sets of micro-RNA (ribonucleic acid) inside a specific soldier. Ultimately, a total awareness of an individual's biology and environmental exposure may enable a precise awareness of their capabilities and specific healthcare needs. As Tucker says, 'For the military, this opens up new choices that are pulled directly from dystopian science fiction: anticipating what soldier is best suited for what assignment or mission' (Tucker, 2017). The USA isn't yet considering genetic engineering military personnel, he notes, but some have pointed to that as a future possibility.

In fact, a future of sensor-equipped soldiers opens up more possibilities than these. As well as allowing more precise, individually targeted missions, equipment and health treatment, it also opens up the possibility of employing Big Data analytics. A stream of real-time, continuous data from every member of a mission would soon build up a significant data-set that might have significant analytical value for the military, informing future deployment and tactics. Big Data analysis techniques, therefore, may eventually lead to algorithmic commands (or commanders).

The proliferation of sensors, however, will open up new security issues. Anyone hacking battlefield sensor technology would acquire considerable information about the numbers and locations of troops. This isn't hypothetical. In November 2017 Strava released a data visualization 'heat map' of the users of its fitness app showing every single activity ever uploaded to the app – more than 3 trillion individual GPS data points. In January 2018 military authorities noticed the map gave away important information about military personnel using it. With US personnel abroad in places such as Afghanistan, Djibouti and Syria often being the only app users, US bases were clearly identifiable and mappable. Zooming in even showed the internal layout of the bases.

Sensors, along with a range of wearable technologies, may have another future too. Instead of being worn or carried, some technology may eventually go *inside* the body, as part of a coming 'implant revolution'. In 1998 Professor Kevin Warwick had a silicon chip transponder surgically implanted in his arm that was able to interact with his nervous system and brain and also interact with his surroundings, including opening doors, switching on lights and welcoming him as he approached. A future military force with GPS and health sensors implanted inside the soldiers would provide much more data and improve the tracking and location of forces. Prof. Warwick continued his experiments in 2002 using Brain-Computer Interface (BCI) implants that enabled him to move an electronic arm, thousands of miles away, through the internet. This technology is explored below.

Exo-skeletons

In addition to augmenting the human body to create a sensory hyperpresence, another strand of research has attempted to create a physical hyperpresence, with 'exo-skeletons' giving the ability to perform tasks beyond normal human capabilities. This man–machine coupling has often been described as a 'cyborg' (cybernetic-organism), a term coined by Manfred Clynes and Nathan S. Kline for their 1960 paper 'Cyborgs and Space', which discussed the possible technological adaptation of the human body for extraterrestrial environments. The cyborg, therefore, was an enhanced, adapted human, functioning as a unitary, homeostatic system with their technology. The idea of man–machine similarity or symbiosis has a much longer philosophical history, but it has also had a significant impact on popular culture, with science fiction especially exploring its military implications.

H.G. Wells' 1898 novel *War of the Worlds* had aliens using three-legged, walking, 'fighting machines' whilst Robert A. Heinlein's 1959 novel *Starship Troopers* included soldiers in 'mobile infantry power suits', or powered, armoured exo-skeletons, an idea most famously developed in Marvel's 'Iron Man', a hero whose superpowers were wholly located in his armour. The idea of giant, walking suits appeared in James Cameron's *Aliens* (1986) with Ripley donning the 'cargo-loader' to fight the alien queen, whilst weaponized suits appeared in the Wachowski's *The Matrix: Revolutions* (2003), which had Armoured Personnel Units (APUs) battling the machinic enemy, and Cameron's *Avatar* (2009), which featured the armed Amplified Mobility Platform (AMP) walker. Perhaps the most realistic military exo-skeleton appeared in Doug Liman's 2014 film *Edge of Tomorrow* (also named *Live. Die. Repeat*), whose 'battle suits' were directly inspired by ongoing DARPA research and featured potential capabilities. The 2014 video game *Call of Duty: Advanced Warfare* attempted to extrapolate military trends and included exo-suits as a key technology, helping boost human abilities such as movement and speed.

Real exo-skeletons were slower to develop and suffered significant limitations. The pioneer for these was General Electric engineer Ralph Mosher who (building on his experimental 1956 GE Yes-Man teleoperator arms) began work on a 'cybernetic amorphous machine' (CAM), a manipulator with force-feedback 'touch' designed to safely service an experimental nuclear-powered aircraft. Developed from 1958–59, Handyman was a pair of mechanical 'remote anthropomorphic arms' controlled by a linked, human-worn and manipulated exo-skeleton harness (Hoggett, 2010b). Its successor was the

Hardiman, a DoD-funded General Electric suit developed by Mosher from 1965 and named after the project, the 'Human Augmentation Research and Development Investigation'. A full suit was developed but it was considered too dangerous. It was attached to the user's back, with the user inserting themselves into the mechanical arms and feet, with the legs attached at their sides. It had a slow walking speed, weighed three-quarters of a ton and its movement was uncontrollable. Researchers focused on the arm alone, but the prototype was expensive and too heavy to be used. Military interest in Mosher's work led to a request for a walking device that could transport loads through difficult conditions such as the Vietnamese jungle. Mosher developed the test 'Pedipulator' in 1964, a two-legged tank that was replaced by the 'CAM Walking Truck' in 1968. The quadruped truck used a hydraulic feedback system to transfer limb force to the legs, but proved exhausting to control. It worked, but weighed 3,000 pounds, had a maximum speed of 5mph and was so limited that only a prototype was made.

Other exo-skeletons have been designed for medical purposes. Influenced by Heinlein's power-suits, in 1986 US Army Ranger Monty Reed came up with the idea of the 'Lifesuit': an air-powered exo-skeleton enabling the disabled to walk. LS1 was built in 2001 and the current model is LS14. Other medical 'exos' include the Hal5 (Hybrid Assistive Limb) created by Cyberdene Inc. in 2012 (with its earliest prototype in 1997); Esko Bionic's eLegs (or ESKO bionic suit), unveiled in 2010, and ReWalk Robotics' ReWalk suit, created in 2011.

The US military continues its interest in exo-skeletons. An exo-skeleton is an external, mobile structure, suit or armour worn by the individual, employing powered hydraulics or motors to boost limb movement and increase strength, endurance and even speed. Prototypes have been used to carry additional loads but the eventual military aim is a combat-enhanced infantryman. They remain at an experimental stage and haven't been deployed. Their success depends upon solving three issues: (1) they need to be light enough not to burden the soldier whilst being strong enough to enhance performance; (2) they need to allow natural movement without causing strain on the muscles; and (3) they need a power source that is compact, long-lasting and not an additional physical burden. Given the increasing weight of soldier kits, exo-skeletons may become a necessary military technology.

In 2000 DARPA began funding research into exo-skeletons. One design, originating at UC Berkeley, evolved by 2009 into Esko Bionics and Lockheed Martin's 'Human Universal Load Carrier' (HULC). HULC is a pair of lightweight, powered, titanium, computer-controlled exo-skeleton legs with a built-in power supply, strong enough to increase load-carrying capacity and intended to improve metabolic efficiency, allowing the soldier to operate better, for longer (also including a chest swing-mount to take the weapon's load too). A micro-processor linked to the skeleton's sensors detects the wearer's movements and calculates movements in the exo-skeleton to match them. Hydraulics in the limb-joints work like pistons, providing the motion, without impeding agility. An electric motor at the rear drives the legs, powered by replaceable, rechargable lithium-ion batteries that provide 4–5 hours of functioning. With it, a soldier's carrying load could increase from 119 pounds to around 201 pounds. The initial hype was substantial: a Lockheed press release announced HULC 'will enable soldiers to do things they cannot do today, while helping to protect them from musculoskeletal injuries' (Cornwall, 2015).

Another DARPA-funded system was Raytheon Sarcos's XOS, a load-carrying exo-skeleton, giving significantly enhanced strength. The latest model, 2010's XOS2, was trialled by the Army for deployment in 2020. In 2015, however, problems with HULC became apparent. Soldiers using it on a treadmill burned more energy than they did walking unaided, with heart-rates jumping 26 percent and oxygen consumption rising 39 percent, as a lack of coordination between man and machine forced people to walk in an unfamiliar way. Work on HULC and XOS2 was subsequently paused. Efforts then moved onto the Tactical Assault Light Operator Suit (TALOS, or 'The Iron Man suit'), begun in 2013 by Admiral William McCraven for Special Operations Command (SOCOM) with a brief to be bulletproof, weaponized, able to monitor vital signs and give enhanced strength and perception. The suit's development is a collaboration between 56 corporations, 16 government agencies, 13 universities and 10 National Laboratories, with a prototype to be delivered for testing in 2018.

TALOS is intended to be a battery-powered robotic exo-skeleton, with strength and power-increasing systems, but it also merges with soldier-system ideas and technologies. It will include MIT's research into next-generation Shear-Thickening Fluid (STF), 'liquid body armour' that transforms from liquid to solid in seconds when a magnetic field or electrical current is applied or when hit, as well as a physiological subsystem that lies against the skin embedded with sensors to monitor core body temperature, skin temperature, heart-rate, body-position and hydration levels. GA will provide a tiny combustion engine running 10,000 revolutions a minute, using Liquid Piston's 'X' engine, which would recharge the batteries that power the suit.

TALOS remains a 'rigid' exo-skeleton design. These take the weight off the soldier but they are heavy and lock users into particular joint movements. Current designs suffer from a slow response and require a lot of power. They can be exhausting to wear and have often impacted upon performance. As MIT-engineer Hugh Herr commented, 'it's very difficult from a design perspective to augment human walking and running, because we're so good at it' (Cornwall, 2015). For critics, rigid designs fail to address fundamental physiological issues and require a greater knowledge of biomechanics than we currently have. The alternatives are 'soft' suit designs. These are lightweight, efficient and easy to wear, potentially improving performance, but they don't significantly enhance their strength or load-carrying capacity.

DARPA began its Warrior Web programme to fund multiple research projects with the aim of producing a soft, lightweight under-suit that would enhance performance and reduce injuries. In 2014, for example, it awarded $2.9m to Harvard's Wyss Institute to develop exo-skeleton trousers – a smart fabric using a low-power microprocessor and straps with flexible sensors that mimics how muscles and tendons work to offer support as the wearer walks. A Warrior Web prototype 'soft exo-suit' combining different research elements was tested in July 2017. Weighing just 9kg and using only 140 watts of power, it used computer-controlled textiles and wires to offer orthopedic support as well as powered robotics systems in the legs to reduce strain on muscles and tendons. Those using the suit, carrying loads equal to 30 percent of their bodyweight, were an average of 7 percent more efficient than those without it.

Other research focuses on more specialized technologies. In August 2014 MIT's d'Arbeloff Laboratory showed off its new shoulder-mounted robotics system. Worn at the hips it gives the wearer two additional robotic limbs to increase their capabilities (much like Marvel's Doctor Octopus). In September 2015 Arizona State University announced they had created prototypes of 4MM: a lightweight, powered jetpack that would allow soldiers of the future to sprint in and out of battle, giving 'faster movement and agile motions' (Vincent, 2014), even whilst carrying heavy weapons and armour using its thrust. The project is named after the goal of getting soldiers running a 4-minute mile.

In November 2017, it was reported that the US Army is testing a military variant of Lockheed's FORTIS, a human-powered exo-skeleton designed to aid industrial tasks. Army evaluators are testing a FORTIS knee-stress-release-device (K-SRD) performance-enhancing exo-skeleton. Using independent actuators, motors and lightweight conformal structures, the lithium-ion powered FORTIS allows soldiers to carry 180 pounds up five flights of stairs, whilst expending less energy. The K-SRD uses B-Temia's Dermoskeleton technology to counteract overstress on the lower back and legs. The Dermoskeleton's conformal upper structure works on a belt attached to the waist that connects to flexible hip sensors that tell a computer where the soldier is in space together with the speed and velocity of their movements. The computer uses advanced AI to sense the user's mobility intentions and generate synchronized movements at the user's knees, replicating an individual's walk-patterns. Lockheed claim FORTIS provides additional torque, power and mobility and enables heavier load-carrying whilst reducing the energy required for tasks by 9 percent. Intended as Lockheed's next-generation exo-skeleton, FORTIS hopes to overcome the inefficiencies of HULC. A 2017 University of Michigan Human Neuromechanics Laboratory study confirmed its benefits, finding that on a treadmill task all participants showed a statistically significant reduction in exertion as compared to performing that same task without the K-SRD unit.

Exo-skeletons, of course, are not just being developed by the USA. Like other technology, research is international and successful devices will proliferate. China unveiled its own exo-skeleton, the L70, designed by EEAE, at Zuhai 2014, which is flexible enough to let a user stand on one leg, kneel or side-kick. An improved version was shown at Zuhai 2016. In June 2017, Russia unveiled its own concept, model exo-skeleton and soldier-system at the National Museum of Science and Technology in Moscow. Developed by the State-owned Central Research Institute for Precision Machine Building, the prototype combat-suit features a powered, titanium computer-controlled exo-skeleton, ballistic protection from bullets and shrapnel, a digital helmet with visor display and torch and night-vision camera and battery packs near the boots. The individual technologies are currently being developed by a range of companies.

Brain–computer interfaces

As we have seen, drones represent another solution to the problem of presence, allowing telepresent action at a distance, without the deployment of physical bodies. Marshall McLuhan saw electronic technology as extending the human central nervous system and consciousness itself 'in a global embrace' (McLuhan, 1994:3) and although drones represent this well, current modes of telepresence remain limited to cameras and basic sensors. Some suggest, therefore, that digital technology may one day offer a more realistic, embodied and experiential 'presence', fulfilling McLuhan's claims of an extended CNS and consciousness.

Robert Vendetti's 2006 comic, *The Surrogates*, offered a vision of this. Set in 2054 it showed a world where everyone remains at home, plugged into chairs that project their consciousness into 'surrogate' telepresent robotic bodies that enable them to live how they like, without fear of harm (Vendetti, 2006). The 2009 film adaptation added an extra scene, showing a troop carrier spilling soldiers out onto a battlefield. When one is hit, we cut again to their operator sitting in a drone station who takes off his headset and curses the loss of his surrogate. Similar ideas of telepresent surrogate bodies were used in Cameron's 2009 film, *Avatar*. These ideas are obviously science fiction, but the technology of brain–computer interfaces (BCI) is real.

Current BCI technologies have their roots in the 1920s in the discovery that neurons convey information through electrical 'spikes' that can be recorded with a thin metal wire, or electrode. One consequence is that neurons can be accessed through electrodes to produce behaviours, as seen in the work of the pioneering neuroscientist Jose Delgado, who experimented with electrode implants in bulls, primates and humans from the 1950s. He demonstrated radio-control of a charging bull in 1963 and even foresaw a military application of the technology, arguing, 'Someday armies and generals will be controlled by electric stimulation of the brain' (Veronese, 2011). A second consequence is that these connected electrodes can be connected to other devices to control them. The pioneer here was Eberhard Fetz, who published a 1969 paper describing how a monkey learnt to control a dial connected to a single neuron in his brain by thought alone. The term BCI was coined by Jacques J. Vidal in 1971. Today, there are two forms of BCI: invasive techniques, using electrode implants, and non-invasive techniques, usually based on electroencephalograph (EEG) recording from the scalp using a helmet or cap.

BCI has seen significant successes. In 1999 Duke University's Miguel Nicolelis trained a rat to swing a cantilever with its mind; in 2002 Brown University's John Donoghue succeeded in training a monkey to move a cursor across a screen using thought; in 2002 Prof. Kevin Warwick used his BrainGate technology to move a robotic arm; in 2004 the BrainGate team replicated the cursor experiment with a quadriplegic human subject, Matt Nagle; and in 2006 the University of Washington demonstrated a non-invasive, brain-controlled robot avatar. In 2008 Andrew Schwartz at the University of Pittsburgh had a monkey feed itself marshmallows with a mind-controlled robotic arm; in 2011 Nicolelis demonstrated monkeys could move an avatar arm and 'feel' objects in a virtual world. In 2012 Schwartz's University of Pittsburgh team enabled a paralysed woman to feed herself chocolate using a robotic arm.

There are many problems with BCI. Critics point out the technology is crude and records only a tiny selection of the brain's activity, that the electrode's recording deteriorates over time, and that implants damage the brain and often move, leading to scar tissue and risk infections. They argue we don't know enough about how the brain works, lack sensitivity in the equipment and the results remain laboratory based, requiring specialist equipment and expert operators. BCI, therefore, is a long way from being a viable, safe, useful, affordable technology. Nevertheless it has found some mainstream uses, in gaming (with the Emotiv EPOC EEG video-game controller) and in health (with cochlea implants).

Advances, however, continue to be reported. In February 2013 Nicolelis published a paper describing the connection of two rats in a brain-to-brain interface allowing them to collaborate in real-time on tasks. In August 2013, human brain-to-brain communication was reported by the University of Washington, with brain signals being sent via the internet to make another man's arm move. In February 2015, the world's first mind control of a prosthetic hand was reported. In March 2015, the University of Pittsburgh enabled a paralysed woman to fly an F-35 jet simulator by mind control. In July 2015, it was reported that a 'brain-net' had been created of four linked monkey brains, enabling them to synchronize neuronal activity to collaborate on tasks. In September 2015, John Hopkins University reported they had demonstrated a mind-controlled robotic, prosthetic hand that used electrical signals to mimic feeling and sensation. In September 2015, a paraplegic man was able to walk again when brain signals were sent to electrodes that triggered the muscles to move his legs. In March 2016 Duke University published a paper describing how monkeys could move a wheelchair by mind control, and in November the Swiss

Federal Institute of Technology enabled a paralysed monkey to walk through a wireless brain–spine interface that stimulates its nerves. In 2017:

> Researchers at the University of Pittsburgh Medical Center connected touch sensors from a robot's fingertips to a paralyzed man's sensory cortex so he could feel what it was touching. At Case Western, scientists linked a paralyzed man's motor cortex to a computer that electrically stimulated muscles in his arm, enabling him to bring a forkful of food from a dish to his mouth. At Brown, Borton's team implanted electrodes and a wireless transmitter in a monkey's motor cortex and connected it to a receiver wired to the animal's leg, restoring its walking motion.
>
> Wise (2017)

Much of the BCI work has been carried out with medical aims in mind, but others are thinking of broader applications, including military ones.

Many of these projects are recipients of DARPA funding. DARPA has a long-standing interest in neuroscience research and has been funding BCI projects since the 1974 Close-Coupled Man/Machine programme. As Miranda et al. argue, DARPA's primary interest lies in two areas of research: BCI technologies that might help restore neural or behavioural functioning (such as research into mobility, activity, prosthetics and perception etc.) and BCI technologies that might improve training and performance. In 2002 DARPA took a closer interest in BCI, launching the Brain-Machine Interface (BMI) programme, followed soon after by the Human Assisted Neural Devices (HAND) programme. Recent programmes aimed at restorative BCI include Revolutionizing Prosthetics (from 2006), Reorganization and Plasticity to Accelerate Injury Recovery (REPAIR, from 2010), Restorative Encoding Memory Integration Neural Device (REMIND, 2009–14), and Reliable Neural Interface Technology (RE-NET, 2010–11). Programmes aiming at improving performance include, the Accelerated Learning programme (2007–12), Narrative Networks (N2, from 2011), Neurotechnology for Intelligence Analysts (NIA, 2005–14), and Cognitive Technology Threat Warning System (CT2WS, from 2007).

In April 2013 President Obama launched the Brain Research Through Advancing Innovative Neurotechnologies (BRAIN) initiative, and most continuing BCI programmes now take place under this rubric. Alongside ongoing projects such as RE-NET, which seeks 'to develop the technologies needed to reliably extract information from the nervous system, and to do so at a scale and rate necessary to control complex machines, such as high-performance prosthetic limbs', and Revolutionizing Prosthetics, which seeks to improve the functionality of prosthetic arms, there are numerous other programmes running (DARPA, 2018). These include Electrical Prescriptions (ElectRx), which aims 'to help the human body heal itself through neuromodulation of organ functions using ultraminiaturized devices'; Hand Proprioception and Touch Interfaces (HAPTIX), which aims 'to create fully implantable, modular and reconfigurable neural-interface microsystems' to deliver naturalistic sensations to prosthetics; Neuro Function, Activity, Structure and Technology (NeuroFAST), which seeks 'to enable unprecedented visualization and decoding of brain activity to better characterize and mitigate threats to the human brain'; Restoring Active Memory (RAM), which aims to develop computational methods to aid memory recall; Systems-Based Neurotechnology for Emerging Therapies (SUBNETS), which seeks 'to create implanted, closed-loop diagnostic and therapeutic systems for treating neuropsychological illnesses; and Targeted Neuroplasticity Training (TNT), which seeks 'to advance the pace and effectiveness of cognitive skills training through the precise activation of peripheral nerves that can in turn promote and strengthen neuronal connections in the brain' (DARPA, 2018).

Another BRAIN programme is Neural Engineering System Design (NESD), unveiled in January 2016, which aims to develop a high-bandwidth, implantable neural interface connecting humans directly to computers, effectively turning them into cyborg soldiers. Phillip Alvelda, NESD manager for DARPA, says: 'Today's best brain–computer interface systems are like two supercomputers trying to talk to each other using an old 300-baud modem. Imagine what will become possible when we upgrade our tools to really open the channel between the human brain and modern electronics' (Gibbs, 2016). DARPA hope to be able to connect individual neurons to give finer control reduce noise and, in theory, speed up communications between man and machine, although this will require break-throughs in a number of related fields.

Efforts to improve performance have achieved faster results. In November 2016, researchers at Air Force Research Laboratory at Wright-Patterson AFB and Wright State University published a paper describing how electrical brain stimulators could enhance the mental skills of staff. The brain stimulation kit uses transcranial direct current stimulation (tDCS), employing five electrodes to send weak electrical currents into particular parts of the cortex, potentially boosting the performance of aircrews, drone pilots and others in demanding roles. Similarly, in October 2017, it was reported that DARPA's Restoring Active Memory (RAM) programme had successfully tested a non-invasive EEG device using tDCS on monkeys that saw a 40 percent increase in their learning speed. The USA has also had success with the BrainGate technology, fitting soldiers who have lost limbs with powered prosthetics that respond to their thoughts and even provide sensation. The US military reports 167 amputees have returned to full, active duty, some deploying to the battlefield.

Beyond restorative and performance-enhancing BCI, however, DARPA has longer-term hopes for the technologies involving the possibility of direct computer coupling and modes of robotic/UAV telepresence that fulfil the dreams of science fiction. In February 2012, it was revealed DARPA had awarded $7m for research into 'Project Avatar'. Named after the film, DARPA hoped to develop similar BCI-based 'mind-control' technologies for remote robotic bodies. According to its 2013 budget, 'The Avatar programme will develop interfaces and algorithms to enable a soldier to effectively partner with a semi-autonomous, bi-pedal machine and allow it to act as the soldier's surrogate' (Drummond, 2012). The surrogate robots should be able to perform all normal human tasks. Although DARPA doesn't spell out how, it does refer to 'key advancements in telepresence and remote operation of a ground system'. DARPA is still lavishly funding projects whose aim is the integration of cybernetic devices with soldiers, thought-controlled robots and vehicles and even network-enabled telepathy.

There is, however, another solution to the problem of presence that doesn't involve either augmented sensory hyperpresence, or brain-computer linked telepresence, but instead revolves around the *absence* of the human and their will. This is the possibility of humans being replaced by artificial intelligence (AI) systems and Lethal Autonomous Weapons Systems (LAWS), or robots.

Key reading

For the background to the US military's interest in and re-theorization of information and communication in warfare during the 1990s, see chapter 4 and the key reading indicated there. In particular, Cebrowski and Garstka's (1998) article on 'Network-Centric Warfare' is essential reading for this chapter as it sets out the theory of informational hyperpresence and its military advantages. Most of the information about wearables and future soldier systems can be found online. Discussions of SIPE, Land Warrior, Nett Warrior, FIST, FELIN and others can be found in Gourley (2013) and Army Technology (2018a, 2018b, 2018c). BAE Systems (2018a, 2018b) provide company information about Broadsword and Q-Warrior, whilst Gallagher (2017) discusses HUNTR and TAR. Information about the military use of VR can be found through the references and by Googling specific systems. Most of my discussion of sensors comes from Tucker (2017). The concepts of the 'cyborg', cybernetics and early exo-skeletons and walkers are covered in Rid (2016 ch.4) and Hoggett (2010a, 2010b, 2010c, 2010d, 2010e). Those interested in the longer history of thought about the relationship between man and machines should read Channell (1991). Information about HULC, XOS2, TALOS and FORTIS is, again, easiest found online, distributed across a lot of web pages by the military and developing companies as well as articles. YouTube has numerous videos showing these exo-skeletons at various phases of development or as speculative technologies. Cornwall (2015) has a useful discussion of the issues around these exo-skeletons. Information about BCI can be found in Wise (2017), Rao and Wu (2017) and Economist (2018) and by Googling the examples given. YouTube has many videos about developments in BCI and the images, for example of monkeys using robotic arms are astonishing to watch. Vendetti's graphic novel (2006) and Mostow's film of *The Surrogates* (2010) are worth following up on. Information about DARPA's funded BCI research can be found in Miranda et al (2015) and DARPA (2018). It is also worth returning to chapter 2's discussion of the work of Chris Hables-Gray and his work on the 'cyborg soldier'. Although academic interest in the 'cyborg' waned soon after, the ideas and processes have become central to the military.

References

Army Technology. (2018a) 'Land Warrior Integrated Soldier System', *Army Technology*, http://www. army-technology.com/projects/land_warrior/.

Army Technology. (2018b) 'FIST: Future Infantry Soldier Technology System', *Army Technology*, http://www.army-technology.com/projects/fist/.

Army Technology. (2018c) 'FELIN (Fantassin à Équipements et Liaisons Intégrés): Future Infantry Soldier System', *Army Technology*, http://www.army-technology.com/projects/felin/.

BAE Systems. (2018a) 'Broadsword Spine', *BAE Systems*, https://www.baesystems.com/en-uk/pro duct/broadsword-spine.

BAE Systems. (2018b) 'Q-Warrior Helmet-Mounted Display', *BAE Systems*, https://www.baesystems. com/en/product/qwarrior-helmet-mounted-display.

Cebrowski, A. K. and Garstka, J. J. (1998) 'Network Centric Warfare: Its Origin and Future', US Naval Institute Proceedings, January, available at: http://www.kinection.com/ncoic/ncw_origin_future.pdf.

Channell, D. F. (1991) *The Vital Machine*, Oxford: Oxford University Press.

Cornwall, W. (2015) 'Feature: Can We Build an "Iron Man" Suit that Gives Soldiers a Robotic Boost?', *Science*, 15 October, http://www.sciencemag.org/news/2015/10/feature-can-we-build-iron-man-suit-gives-soldiers-robotic-boost.

DARPA. (2018) 'DARPA and the Brain Initiative', *DARPA*, https://www.darpa.mil/program/our-research/darpa-and-the-brain-initiative.

Der Derian, J. (2009) *Virtuous War*, London: Routledge.

Drummond, K. (2012) 'Pentagon's Project "Avatar": Same as the Movie but with Robots Instead of Aliens', *Wired*, 16 February, https://www.wired.com/2012/02/darpa-sci-fi/.

Economist, The. (2018) 'Thought Experiments', *Technology Quarterly*, 6 January, https://www. economist.com/technology-quarterly/2018-01-06/thought-experiments.

Gallagher, S. (2017) 'Heads-Up: Augmented Reality Prepares for the Battlefield', *Ars Technica*, 25 May, https://arstechnica.com/information-technology/2017/05/heads-up-augmented-reality-pre pares-for-the-battlefield/.

Gibbs, S. (2016) 'US Military Aims to Create Cyborgs by Connecting Humans to Computers', *The Guardian*, 20 January, https://www.theguardian.com/technology/2016/jan/20/us-military-cyborg-connecting-humans-computers.

Gourley, S. R. (2013) 'The Rise of the Soldier System', *DefenseMediaNetwork*, 25 June, https://www. defensemedianetwork.com/stories/the-rise-of-the-soldier-system/.

Greenberg, A. (2014) 'DARPA Turns Oculus into a Weapon for Cyberwar', *Wired*, 23 May, https:// www.wired.com/2014/05/darpa-is-using-oculus-rift-to-prep-for-cyberwar/.

Hoggett, R. (2010a) '1956 – GE Yes-Man Teleoperator – Ralph Mosher (American)', *Cyberneticzoo. com*, 11 April, http://cyberneticzoo.com/man-amplifiers/1956-ge-yes-man-teleoperator-american/.

Hoggett, R. (2010b) '1958-9 – GE-Handyman – Ralph Mosher (American)', *Cyberneticzoo.com*, 25 May, http://cyberneticzoo.com/man-amplifiers/1958-9-ge-handyman-ralph-mosher-american/.

Hoggett, R. (2010c) '1965-71 – G. E. Hardiman/Exoskeleton – Ralph Mosher (American),' *Cyberneticzoo.com*, 4 April, http://cyberneticzoo.com/man-amplifiers/1966-69-g-e-hardiman-i-ralph-mosher-american/.

Hoggett, R. (2010d) '1962-64 – GE Pedipulator – Ralph Mosher (American)' *Cyberneticzoo.com*, 16 February, http://cyberneticzoo.com/walking-machines/1962-64-ge-pedipulator-ralph-mosher-ameri can/.

Hoggett, R. (2010e) '1969 – GE Walking Truck – Ralph Mosher (American)', *Cyberneticzoo.com*, 30 January, http://cyberneticzoo.com/walking-machines/1969-ge-walking-truck-ralph-mosher-american/.

HQAMC Public Affairs. (2009) 'AMC to Showcase Future Soldier 2030, Other High- Tech Equipment at the AUSA Winter Meeting', *U.S. Army*, https://www.army.mil/article/17226/amc_to_showcase_future_soldier_2030_other_high_tech_equipment_at_the_ausa_winter_meeting.

Human Variability Project, The. (2016) 'The Human Variability Project', *Centre for Adaptive Soldier Technologies*, https://www.arl.army.mil/cast/?q=THVP.

McLuhan, M. (1994) *Understanding Media*, London: MIT Press.

Miranda, R.A. et al. (2015) 'DARPA-Funded Efforts in the Development of Novel Brain-Computer Interface Technologies', *Journal of Neuroscience Methods*, 244, 15 April, pp. 52–67, https://www. sciencedirect.com/science/article/pii/S0165027014002702.

Mostow, J. (2010) *Surrogates*, Walt Disney Home Entertainment [DVD].

Page, L. (2007) 'US "Land Warrior" Wearable-Computing Headed for Iraq', *The Register*, 19 April, http://www.theregister.co.uk/Print/2007/04/19/us_fist_to_iraq_even_though_cancelled/.

Osborn, K. (2016) 'Army to Upgrade and Widen Deployments for Nett Warrior', *Defense Systems*, 20 October, https://defensesystems.com/articles/2016/10/20/nett.aspx.

Rao, R. P. N. and Wu, J. (2017) 'Melding Mind and Machine: How Close are We?', *Scientific American*, 11 April, https://www.scientificamerican.com/article/melding-mind-and-machine-how-close-are-we/#.

Rid, T. (2016) *Rise of the Machines. The Lost History of Cybernetics*, London: Scribe Publications.

Sutherland, B. (2009) 'U.S. Soldiers' New Weapon: An iPod', *Newsweek*, 17 April, http://www. newsweek.com/us-soldiers-new-weapon-ipod-77563.

Tucker, P. (2017) 'Tomorrow Soldier: How the Military is Altering the Limits of Human Performance', *Defense One*, 12 July, http://www.defenseone.com/technology/2017/07/tomorrow-soldier-how-military-altering-limits-human-performance/139374/.

Turnbull, G. (2014) 'Digital Battlefield: Connected Simulators Let Pilots Fly a Full- Scale War', *Air Force Technology*, 18 March, http://www.airforce-technology.com/features/featuredigital-battle field-connected-simulators-let-pilots-fly-a-full-scale-war-4198082/.

UK Government (2015) 'MOD Unveils Futuristic Uniform Design', *Gov.UK*, 16 September, https:// www.gov.uk/government/news/mod-unveils-futuristic-uniform-design.

Vendetti, R. (2006) *The Surrogates*, Atlanta: Top Shelf Productions.

Veronese, K. (2011) 'The Scientist Who Controlled People Through Brain Implants', *Gizmodo*, 28 December, https://io9.gizmodo.com/5871598/the-scientist-who-controlled-peoples-minds-with-fm-radio-frequencies.

Vincent, J. (2014) 'Soldiers of the Future Could Sprint in and Out of Battle Powered by Lightweight Jetpacks', *The Independent*, 15 September, http://www.independent.co.uk/life-style/gadgets-and-tech/soldiers-of-the-future-could-sprint-in-and-out-of-battle-powered-by-lightweight-jetpacks-9733663.html.

Wise, J. (2017) 'Brain-Computer Interfaces are Already Here', *Bloomberg*, 7 September, https://www.bloomberg.com/news/features/2017-09-07/brain-computer-interfaces-are-already-here.

13 Algorithmic war
The AI and robotic RMA

I fought the LAWS

The last chapter considered two responses to the problem of presence – the inevitable risk to forces that comes from putting them into the field. One was a mode of hyperpresence, super-equipping individual soldiers with sensory and protective technologies to improve their situational awareness, lethality and survivability; another was the use of telepresence, removing the bodies of humans whilst enabling them to still impose their will. A third possibility is the absence of human bodies and will. This is the possibility – some argue certainty – that future battlefields will be dominated by artificial intelligence (AI) systems and independently acting robots, or Lethal Autonomous Weapons Systems (LAWS). Although battlefield robotics is currently dominated by remote-controlled technologies, we are already seeing the emergence of battlefield AI systems and improvements in AI autonomy, with significant consequences for the future of warfare.

The rise of the robots

The opening scenes of the 2014 film *Robocop* show a media unit accompanying a US Army patrol through a neighbourhood of occupied and pacified Iran as part of 'Operation Freedom Tehran'. The patrol force is largely robotic. Humanoid bipedal robots are the main force, backed by huge, heavily armoured and armed, two-footed walking robots and aerial drones in the sky, scanning the area for threats. Soldiers wearing future soldier-system exo-skeletons accompany them, identified, along with the journalists, by their security tags. 'May peace be upon you', a robot voice ironically declares, 'Please exit your homes with your arms raised'. The robots scan the iris and bodies of each civilian to establish they aren't a threat (exposing the women's bodies under the chador), and the hostility to the occupying force and fear of the robots is obvious. Back in the studio, the show's presenter is jubilant: 'Incredible. Incredible. Not long ago that would have been American men and women risking their lives to pacify these people ...'. A General agrees, saying that now the USA can carry out its objectives without any loss of American lives. The film cuts to a resistance group who attempt an attack on the patrol but are easily killed. One of their children runs out with a knife in response and is annihilated by a robot as a 'threat'.

This was a surprisingly realistic opening for a science-fiction film, but then many commentators see the scenario less as science-fiction than as emerging science. Peter Singer, in his 2009 book *Wired for War*, warns, for example, of the coming rise of the 'robots'. Robotics, he says, will constitute a Revolution in Military Affairs (RMA), equivalent to the arrival of gunpowder or the atomic bomb. Robots will be 'one of the most fundamental changes ever in war', he argues, because 'robotics alters not merely the lethality of war, but the very identity of who fights it' (Singer, 2009:7, 10). Robotics represents, therefore, 'The end of humans' monopoly on war' (Singer, 2009:10). It is 'a revolution in warfare and technology that will literally transform human history' (Singer, 2009:11).

Singer points to the rise of military robots. In 2003 the USA had a handful of unmanned aerial vehicles (UAVs) and no ground robots. By 2008 it had 7,000 UAVs in the battlefield and 12,000 unmanned ground vehicles (UGVs), being used for intelligence, exploration, bomb disposal and combat missions. Just as World War II spurred the development of modern computing, so, Singer says, the War on Terror has spurred R&D into unmanned and robotic technologies. Singer's claims, of course, are slightly premature as most of what he discusses as 'robotics' and most 'robots' deployed to date remain telepresent technologies remotely controlled by human operators. However, the years since his book was published have seen a movement towards greater autonomy and genuine robotic weaponry. Singer is right, therefore, that these early forms presage full robotic autonomy and military capability.

Unmanned remote-controlled technologies come in a range of forms. Although drones (UAVs) are the most famous, there are also unmanned ground vehicles (small robots or self-driving vehicles) (UGVs), unmanned sea vehicles (USVs) and unmanned undersea vehicles (UUVs). One of the most famous early UGVs was iRobot's PackBot, created in 2001. PackBot was first used in Afghanistan in 2002 to clear caves and bunkers, search buildings and cross live, anti-personnel minefields, being used again in Iraq from 2003 in urban warfare scenarios and vehicle searches. By 2008, PackBot had disarmed over 10,000 improvised explosive devices (IEDs). In February 2010, iRobot delivered its 3,000th PackBot to the US military.

The Packbot was a small, remote-controlled, tracked UGV. It had a speed of 14km/h and used 'flippers' for 360-degree rotation and negotiation of rough terrain. It could climb gradients of up to 60 percent and could survive submersion in up to 2m of water, a drop of 2m on a concrete surface, falling downstairs or being thrown through a window. Weighing 18kg, it could be carried in a backpack and deployed in minutes. It had an integrated GPS, electronic compass, orientation and temperature sensors. The control station used laptop or eyepiece displays and a hand-held video-game controller with PackBot's cameras also operating in wide-angle and close-up. The basic model could carry a range of payloads, including bomb disposal kit, 'explorer' kit, hazmat detection kit and sniper detection kit. iRobot also produced a larger, faster variant called the Warrior, a 129kg UGV capable of carrying up to 500 pounds and of throwing PackBots into action with its manipulator arm. iRobot teamed up with Metal Storm to produce an electronic firing system to weaponize the Warrior and also produced a smaller version of PackBot, the 13kg XM1216 small unmanned ground vehicle (SUGV) for the Army's Future Combat Systems project, designed to integrate with a future soldier-system. The current Packbot model is the PackBot 510, which comes in numerous variant forms.

iRobot's main competitor was Foster-Miller whose Talon, created in 2000, performed similar functions, including IED disposal. Again, it could operate in all weathers, day and night, with amphibious capabilities, and carried a variety of payloads and sensors, including 'multiple cameras (color, black-and-white, infrared, thermal, zero-light), a two-stage arm, gripper manipulators, pan/tilt, two way communications, NBC (nuclear, biological, chemical) sensors, radiation sensors, UXO/Countermine detection sensors, grenade and smoke-placing modules, breaching tools, communications equipment, distractors and disruptors' (Global Security, 2011). Talon was deployed at 'ground zero' after 9/11, operating within the remains for 45 days without failure and was subsequently deployed to Bosnia in 2002, Afghanistan in 2002 and Iraq in 2003.

In 2005 Foster-Miller produced an upgraded version called Special Weapons Observation Reconnaissance Detection System (SWORDS), an armed UGV that could be fired by the operator. Three were deployed in Iraq in 2007, protecting a site, although the weapons weren't used as the Army appeared concerned with armed robotics systems and stopped funding SWORDS after the deployment. A subsequent comment about SWORDS from Army Program Manager, Kevin Fahey, that 'the gun started moving when it was not intended to move' attracted considerable attention, although Qinetiq, who own Foster-Miller, denied any such occurrence (Popular Mechanics, 2009). Foster-Miller began work on a replacement, the Modular Advanced Armed Robotic System (MAARS), being tested by the military from 2013. Other small tracked UGVs are available, including Tialinx's Cougar 20-H, a small, remote-controlled surveillance robot announced in February 2011, whose distinguishing feature is an RF sensor array that can scan through concrete walls and detect human breathing. It can detect still or moving humans, pinpoint their location in real time and send an image back to the operator.

PackBot/Talon-style small, tracked UGVs are now common. BAE Systems produce similar unmanned vehicles. In September 2017, it released images of its new driverless mini-tanks, codenamed 'Ironclad'. These are small, tracked vehicles modifiable with different attachments and able to carry out reconnaissance and tasks such as bomb disposal. Although operated by wireless control, BAE expects the systems to be capable of full autonomy in the future. In April 2017, the Navy began tests on a small, tracked or wheeled, load-carrying, armed UGV produced by General Dynamics called the Multi-Utility Tactical Transport (MUTT), with a view to marines using them as a forward force for beach invasions. The USA is also testing larger UGVs, including, since May 2015, the 'Ripsaw' remote-controlled and armed light tank.

Another key US robotic company is Boston Dynamics. With DARPA-funding it created the 'BigDog', a quadruped, packhorse robot capable of carrying loads over difficult terrain. Boston Dynamics released several videos showing its impressive performance, including its self-righting capability. In March 2012 video was released of an improved 'AlphaDog', and in February 2013 video was released of a BigDog with a robotic arm, hurling objects. Despite its capabilities, the Legged Squad Support System (LS3) was dropped by the military in late 2015 due to its noise giving away positions, the challenge of repairing them and the issue of integrating them within marine patrols. A smaller version, 'Spot' couldn't carry the loads and lacked sufficient autonomy, although research continues on 'SpotMini' to perfect it.

Boston Dynamics also produced the quadrupedal Cheetah robot, which, in August 2012 set a robot speed record of 28.3mph, which is faster than the human speed record of 27.8mph set by Usain Bolt on 16 August 2009. Cheetah is a tethered prototype but an untethered WildCat robot was unveiled in October 2013. On 29 December 2017, season 4 of *Black Mirror* was released, which included the episode 'Metalhead' that simply showed a group of people hunted down and killed by a remorseless running robot called a 'dog'. Creator Charlie Brooker later admitted the entire episode was inspired by Boston Dynamic's videos of the clunking, running Cheetah robot and BigDog's resilience (although the design is clearly based on their LittleDog). Boston Dynamics also produce the DARPA-funded Atlas, a 6ft bipedal robot able to traverse obstacles and even jump and do back-flips, unveiled to the public in July 2013, based on their PETMAN humanoid robot. Its capabilities continue to be updated and demonstrated in company videos.

DARPA is also funding the MIT/Harvard/Seoul Meshworm project, unveiled in August 2012. This is a small, tube-shaped, soft-bodied, flexible robot, moving through artificial muscles, with possible reconnaissance uses. Others are experimenting with 'swarm robotics', where large numbers of tiny, insect-sized robots operate individually or in unison. They could be used for reconnaissance, or to cause fear or, (with a 'sting' or explosives) as a weapon as themselves. The US Army is researching swarms under its Micro Autonomous Systems and Technology (MAST) programme, running from 2008. One example has been the 'autonomous quadrotors' produced by the University of Pennsylvania's GRASP with Kemel Robotics. These are small, flying robots, able to perform a range of stunts, including flipping, changing direction, a figure-of-eight formation, and coordinated flight through windows and hoops. Initial versions were 'dumb', requiring a nearby sensor array, but by 2016 they were able to use their own sensors and on-board computation for localization, state estimation and path planning.

In March 2013, BAE won a $43m contract to lead the extended MAST programme, which aims to provide a soldier with a swarm of insect-sized robots that can operate without supervision and provide intelligence about what is around corners or in buildings. The programme envisaged a range of robots, from spider-like robots providing initial reconnaissance to dragonfly-type, camera-equipped robots for more information. MAST concluded in August 2017 with 17 live demonstrations by researchers at the Army's Aberdeen Proving Ground. As Allison Mathis, the programme's deputy manager, commented,

> We have created advances in everything. There are new platforms, new algorithms, new sensors. Not all of this will be ready next year, or even the next five years, but we have absolutely advanced technology. We are making an impact right now.
>
> McNally (2017)

When MAST began in 2008 the idea of hand-held micro-drones was science-fiction. By the time it ended, the technology had advanced so far that they had become a commercial reality as a child and hobbyist's toy flown by millions. MAST had numerous successes, including developing insect-like cyclocopters under 30 grams that could outperform polycopters and were quieter; another was Salto, a 98-gram hopping monopod that could move at 2m a second, can operate in spaces where flying robots couldn't and travel over terrain a tracked robot couldn't. It was developed by University of California, Berkeley's Biomimetic Millisystems Laboratory, which is also working on cockroach-like robots able to crawl into small spaces and right themselves if they are tipped over. DARPA are continuing MAST's work on navigation with the Fast Lightweight Autonomy (FLA) programme, which aims to develop small drones that can 'ingress and egress into buildings and navigate within those buildings at high-speeds' (Economist, 2017).

MAST was replaced in 2017 by the Distributed and Collaborative Intelligent Systems and Technology (DCIST) programme (running till 2022) which is interested in distributed intelligence, heterogeneous group control and adaptive and resilient behaviours. In the final MAST demonstration, GRASP was able to show three autonomous quadrotors coordinating their activities using their own sensors and computers, and the next phase is coordinating heterogeneous elements, including robots of different sizes and other military systems including soldiers. By December 2016 the US Office of Naval Research held a month-long swarm robotics demonstration in the lower Chesapeake Bay, Virginia. Four drone boats with control and navigation software, patrolling an area of 4x4 nautical miles, showed they could cooperate collectively to defend a harbour, collectively deciding which boat would track and trail an intruder vessel. It is believed swarms of drone boats could help prevent attacks such as the suicide bombing of the USS *Cole* on 12 October 2000.

Basic swarm robotics may be an existing battlefield technology. On 5 January 2018, two Russian bases in Syria were attacked by what Russia described as a 'swarm' of drones operated by Syrian rebels. Russia claimed they were 'aircraft-type' drones, with ten in one attack and three in another, each armed with a 10-pound bomb. Syrian rebels and IS have been known to use commercial drones weaponized with bombs but these were custom-built and flew pre-planned routes to precise GPS coordinates. Russia's Pantsir-S air-defences and electronic-warfare specialist hacking brought most of the drones down, but a future attack containing huge numbers of micro or nano-drones operating with advanced swarm software could cause significant problems for any defence system, overwhelming it with a 'cloud' of robots.

Another line of development in the USA is cyborg robots, applying robotics to insects, to exploit their existing biological movement, energy system and intelligence. As part of DARPA's Hybrid Insect Micro Electromechanical Systems programme, in November 2011 University of Michigan researchers reported a project to add computer power, cameras and sensors to the back of insects. These 'bugged' insects could hypothetically be sent into places where humans couldn't get or which were too dangerous, with obvious military surveillance applications. In January 2017 Draper Labs with the Howard Hughes Medical Institute unveiled a genetically modified cyborg dragonfly, the DragonflEye, containing electronics in a backpack that activate a messenger neuron within the insect, which carry steering command to its wings to direct its flight. Although not designed for military use, it demonstrates the possibility of cyborg insect control. Other experiments are more speculative. In September 2017, researchers at MIT's Computer Science and Artificial Intelligence Lab (CSAIL) created a new, adaptable, experimental cube robot called 'Primer', with the ability wrap itself with new exo-skeleton outfits to equip itself with particular powers such as walking, swimming or flying. It isn't yet a military technology but it is a proof of concept, with future potential.

US robots haven't just been deployed abroad. They are also employed domestically by many authorities for a range of tasks, including bomb disposal. On 8 July 2016 Micah Johnson became the first person killed by robot by law enforcement officers in the USA. Johnson, who had shot dead five police, and wounded nine others and two civilians, was killed by Dallas police who used a remote-controlled Northrop Grumman ANDROS tracked-UGV to approach him, with explosives held in its robotic arm.

The AI arms race

The most important development in contemporary robotics is the ongoing improvement in artificial intelligence (AI) systems, which is creating increasingly autonomous technologies. Many argue that these improvements in autonomy will one day lead to a range of fully independent, artificially intelligent robotic technologies. The topic of 'artificial intelligence', however, remains controversial.

The term was coined in 1956 by John McCarthy at the Dartmouth conference that established AI as a field. The concept attracted government interest and funding, leading to new university research departments such as the Stanford Artificial Intelligence Laboratory (SAIL), founded in 1963 by McCarthy. Progress in robotics, however, proved painfully slow compared to other fields such as computing. Since Alan Turing's 1950 paper, 'Computing Machinery and Intelligence', debates around AI have also been hampered by the philosophical question of whether machines could ever be 'intelligent' or 'think', or ever achieve 'human-level' intelligence or 'consciousness'. For many, 'Moore's Law' – the doubling of transistors on an integrated circuit every two years, bringing ever more-powerful, smaller, cheaper computing – seemed to offer a solution. Critics, however, argue that more powerful processing won't necessarily lead to intelligence, plus there are physical limits to the application of this 'law'.

For critics, no amount of computing power can lead to 'intelligence' as what this leaves out is the entire realm of neurobiology. Many argue, therefore, that consciousness and intelligence can only exist biologically. As John Searle's 'Chinese room' argument suggests, the best we can hope for with technology is just a more efficient processor of information – that is, a non-cognitive, task-oriented 'weak AI'. Defenders of the idea of real intelligence, or 'strong AI', point out, however, that human consciousness is a material phenomenon and the product of basic biological elements following rules, and hence there is nothing spiritual or transcendent about it: if it is a material phenomenon then it remains, theoretically, understandable, and potentially reproducible. For some AI researchers, biology offers a template for success. Instead of trying to write an AI line-by-line, they suggest either modelling organic brains neuron by neuron to create a virtual brain, or creating an AI kernel that would assimilate information and 'grow' into a full AI.

None of this, however, matters for the military. Questions of 'intelligence' and 'consciousness' are irrelevant: all that matters is whether the 'AI' *works*: whether it can carry out programmed instructions and perform desired tasks such as identification, analysis, targeting and accurate firing. Hence, AI systems and LAWS that fall far short of human-level 'intelligence' are militarily useful, are already being deployed on the battlefield and will proliferate in the future. What this demonstrates is that AI doesn't need human-level intelligence as it has its own machinic intelligence that can already massively outperform us in the tasks it can do. Anthropocentric, philosophical concepts, therefore, have little military value and even if we wanted to retain a distinction between human and machinic intelligence it is likely that, given enough time, incremental advances in AI may one day make such distinctions impossible to prove.

It is worth, however, looking again at the military question of AI. Early texts such as Singer's *Wired for War* framed the issue of future autonomous technologies around robotics and since then the question of 'killer robots' and military hardware has dominated the discussion. For Singer, for example, this is a *robotic* Revolution in Military Affairs. But developments in robotics are linked to and are part of a broader AI revolution and it is better to think in terms of *an AI Revolution in Military Affairs*. The US military is already talking of an 'AI arms-race' between the US and other nations (Department of Defense, 2017), and AI has become central to their future military development, as seen in their discussion of the 'Third Offset Strategy'.

In November 2014, the Pentagon publicly revealed its 'Third Offset Strategy', an idea developed through their Strategic Capabilities Office (SCO), set up in 2012. An 'offset strategy' is a technological means to compensate for a military disadvantage, aiming at ensuring victory, or at least deterring attack. In the 1950s, the USA used nuclear deterrence to compensate for greater Soviet military forces and from 1975 to 1989 they promoted the development of intelligence, surveillance, reconnaissance technologies, precision-guided munitions, stealth and space technologies to offset Soviet conventional superiority. The power of these systems was demonstrated in the 1991 Gulf War. The third offset strategy is a new focus on next-generation technologies such as autonomous AI, Big Data and robotics, as well as miniaturization and manufacturing, all of which could offset foreign (primarily Russian and Chinese) advantages in 'anti-access and area-denial' (A2/AD) systems (a family of military capabilities used to prevent or constrain the deployment and freedom of movement of enemy forces in a theatre of operations).

The US military recognized that this AI arms race was already happening domestically in the competition between the major technology companies such as Google, Apple, Amazon, Netflix and Facebook, all of which were heavily investing in AI systems capable of aiding search and purchasing, presenting individually targeted recommendations, information, news and ads, organizing stock and warehousing operations and enabling voice and home assistants. What was driving all these advances was the application of 'machine learning' techniques.

If AI is the broader field of computer science which tries to get machines to perform tasks that we would consider 'smart', then machine learning is the current, leading application of AI, based on the idea that we should give machines access to data and let them learn for themselves. Hence software can now respond to data and adapt itself without being explicitly programmed. The key to machine learning has been the development of neural networks. These are computer systems designed to classify information the way a human brain does, working by probabilistic analysis, 'learning' through being informed of its errors, and modifying its future behaviours based on this. The cutting-edge field of machine learning is 'deep learning'. Deep learning is based in feeding a computer system a lot of data so that it can make decisions about other data. This enables the development of deep neural networks, able to deal with the complexity of the largest data-sets. The result, as Marr says, is that, 'Deep Learning can be applied to any form of data – machine signals, audio, video, speech, written words – to produce conclusions that seem as if they have been arrived at by humans – very, very fast ones' (Marr, 2016b). One hope of machine learning and deep learning is to be able to move beyond 'narrow' or 'applied' AI, where computer systems can only carry out one highly specific task, to a 'general' AI able, in theory, to perform any required task, 'thinking' about it as a human would.

Machine learning and deep learning, therefore, form the basis for the development of next-generation autonomous military systems able to receive data, respond to it by selecting from the options and adapt their behaviour based upon the results of this process. Developments in machine learning AI guidance have led to new pilotless drone technologies such as the X-47B Autonomous Unmanned Combat Air System (UCAS), and the forthcoming X-47-C. Another AI-based pilotless system is BAE System's Taranis Unmanned Combat Autonomous Vehicle (UCAV) which began tests in 2014. It is designed to have an intercontinental range, with a capability for air-to-air refuelling and the ability to respond autonomously to threats against it with a 'quick-draw' response that requires no human input. BAE stress that it still has a human operator, by which they mean humans remain in control of the weapons systems and decision to fire, although that would itself be on the basis of machine-provided information. Taranis isn't expected to be operational until 2030, by which time, as Sayle says, 'the fiction of humans having a meaningful place in the loop may have passed' (Sayle, 2014).

Full robotic autonomy does, however, already exist. The Samsung SGR-A1, developed between 2003 and 2006, is a South Korean DMZ border sentry gun that includes surveillance, tracking, firing and voice-recognition capabilities. At the moment, a human has to authorize it to fire, but it is capable of autonomous target-detection and firing should it be desired. It is one of the first deployed examples of a LAWS. There have also been significant developments in autonomous UGVs. On 13 March 2004, DARPA ran the first 'Grand Challenge' competition for driverless vehicles on a 150-mile off-road course. No one completed that challenge, with the most successful vehicle only managing 7.32 miles. The following year, on 8 October 2005, five vehicles completed the course and all but one of the 23 entrants travelled further than the previous year's most successful entrant.

One off-shoot is the Unmanned Ground Combat Vehicle and Perceptor Integration System ('Crusher'), an all-terrain prototype capable of carrying 8,000 pounds and autonomous driving developed for DARPA by Carnegie Mellon University's National Robotics Engineering Centre. The IDF have their own UGV, the Guardium – a tele-operated border patrol vehicle capable of functioning autonomously. Autonomous vehicles, however, remain at a prototype stage with the complexity of large vehicle perception and control needing more work. However, their rapid development, combined with advances in processing power and AI, suggest a significant future role once costs reduce. Another DARPA-funded autonomous vehicle is the ASW Continuous Trail Unmanned Vessel (ACTUV), or 'Sea Hunter'. Begun in 2010, with successful trials in 2016, this is a 132ft self-driving, unmanned sea-vehicle (USV) designed as a surveillance drone for finding and tracking hostile submarines. It is able to launch from and return to port, operate for 30–90 days without human control, and avoid other ships through radar and cameras. In February 2018, DARPA delivered a prototype to the Office of Naval Research (ONR) with naval deployment expected within the year.

Although the future of the 'third offset strategy' is in doubt under Trump (as an Obama-era project), the commitment to advanced AI research continues under the new president. In April 2017, the DoD set up the Algorithmic Warfare Cross-Functional team, which began work on Project Maven. One of the major problems the military faces today is the overwhelming amount of data it receives, especially still and moving drone camera imagery, which is dealt with by human intelligence operatives who can only analyse a fraction of what is being produced. Project Maven's aim was to automate this process. After building up a computational infrastructure and a labelled data-set using motion video data from tactical and medium-altitude drones, it was able to use machine learning to teach an AI to distinguish human and inanimate objects and even between different types of objects. The programme was a success and in December 2017 the system was being employed to analyse drone feeds from the battlefield in Syria and Iraq against Islamic State.

The future military applications of machine learning and deep learning are potentially revolutionary. These techniques can be applied to any form of data being produced by the military. That includes all intelligence, surveillance and reconnaissance data produced by drones, satellites, radar and any other sensor-based detection system, as well as the USA's massive electronic surveillance and computer network exploitation activities. When the military sensor revolution discussed in the last chapter gathers steam it will also include all the information about combat operations, movements and activities from units and even individuals, with personal health data especially being collected from wearable or implanted sensors and equipment. As I suggested last chapter, therefore, military 'Big Data' analytics will form an increasingly important part of military understanding, planning and prosecution.

AI, therefore, will permeate every aspect of the future military, forming the basis for future technological superiority. US military officials stress that this will mean a closer coupling of humans and machines (or 'human-machine teaming'), with battlefield AI complementing the human operators and remaining under their control, but these claims are unrealistic. Machine learning AI will almost certainly lead to fewer humans being needed as it automates many roles and it will also hasten the development of LAWS as it has the potential to improve target identification and operation and make LAWS less susceptible to error. As a 'thinking' technology able to 'learn' from its errors, machine learning weapons systems are likely to be trusted one day to make autonomous decisions.

The pros and cons of robotics

Robotics have numerous military benefits. As with drones, the primary benefit is the removal of the soldier's body from combat. LAWS would mean troops don't have to be committed and their specialist capabilities, such as surveillance, mean they can make situations safe for troops. Importantly, they don't suffer from the limitations of the human body and mind. They don't get hungry, and don't feel fear or get nervous. They don't disobey orders or run, have no 'morality' or question orders they are given, and they don't have emotional issues regarding their personal relationships or family. They don't need to check their phone or update Facebook, and can concentrate for long periods of time, doing dull, dirty or dangerous tasks. They can operate in hazardous environments, are capable of movement that a human couldn't undertake and have the potential to be more militarily effective and lethal than humans, due to the speed of their reactions, accuracy of firing and connected, sharable intelligence. As Gordon Johnson of the Pentagon's Joint Forces Command points out: 'They don't get hungry. They're not afraid. They don't forget their orders. They don't care if the guy next to them has just been shot. Will they do a better job than humans? Yes' (Sayle, 2014).

Speed is a major factor in robotic superiority. As Sayle notes, 'Defensive robot systems can recognize threats and respond to them in the time a human operator would take to utter the first syllable of whatever fear-induced expletive comes to mind' (Sayle, 2014). Another factor is the sheer capability of these systems. The Sea Viper air defence system used on the Royal Navy's Type 45 destroyers can track over 2,000 targets out to 400km and operate multiple defensive systems ranging from Aster missiles to Phalanx machine guns to engage supersonic targets from high altitude to the sea's surface, from 120km to point-blank range, simultaneously and autonomously. The system is only currently limited by the need for human authorization.

Just as drones are cheaper than aircraft, so robots are potentially cheaper than tanks and fighters. Since the Joint Strike Fighter programme began in 1996, the unit cost has almost doubled, figures of $229m per aircraft have been reported and the total lifetime costs for the fleet have been estimated at $1.5 trillion. 'By contrast, the unmanned X47-B programme, started in 2006, has cost $813m to develop and is expected to deliver battle-ready drones by 2019' (Sayle, 2014). Indeed, the more autonomous systems become, the cheaper they are as they don't need to factor in the space and survivability of the human body.

Robots are also seen as a possible solution to the problems of asymmetrical warfare. They remove troops from harm, can enter, reconnoitre and patrol inaccessible or danger-ous environments, and help fight an enemy when the problem is as much about finding and identifying them. UAVs and UGVs have both proven their value to date in urban combat and against IEDs. There are also potential psychological benefits from robots. It is argued that ideological motivations may be reduced when fighting robots as risking one's life or suicide becomes less attractive if it will have no effect. Robots will also induce an obvious fear as they are remorseless in their operation. As Kyle Reese says of the titular robot in the influential 1984 film *Terminator*: 'Listen, and understand. That Terminator is out there. It can't be bargained with. It can't be reasoned with. It doesn't feel pity, or remorse, or fear. And it absolutely will not stop, ever, until you are dead'.

For all these reasons, robots appear to be perfect military technologies. If the aim of military training has always been to produce a robot-like obedience and automatic operation in recruits, then robots themselves represent the fulfilment of that ideal. There are, however, concerns about their development and use. Firstly, even military advocates of robots are unhappy with the idea of full autonomy. There is a fear of humans being 'out-of-the-loop' and losing control over key military decisions. These fears may be justified as the decision to use armed force can be much highly nuanced, requiring an evaluation of a range of broader factors as well as potential consequences. An algorithm is unlikely to be able to weigh humanitarian, political and diplomatic issues as well as they can target and fire. Others argue, however, that the speed, confusion and information overload of the modern battlefield means that humans are less and less able to respond in time or make informed decisions. Asking humans for permission slows robots down and eventually militaries will realize they will gain an advantage from allowing robots to be what they are and do what they can do.

The US military is currently determined to keep humans 'in the loop' (with humans monitoring operations and in control of major decisions) or 'on the loop' (supervising machines and ready to intervene and assume control). In 2012, the Pentagon issued a policy directive saying: 'These systems shall be designed to allow commanders and operators to exercise appropriate levels of human judgment over the use of force. Persons who authorize the use of, direct the use of, or operate, these systems must do so with appropriate care and in accordance with the law of war, applicable treaties, weapons-systems safety rules and applicable rules of engagement' (Economist, 2018). As Sayle points out, however, assuming humans are 'in the loop' now is wrong:

> Consider modern air-to-air combat; BVR (Beyond Visual Range) missile engagements are conducted at a minimum of 37km. The MBDA Meteor missile about to enter service with the RAF can engage targets at 100km or more. The pilot is physically incapable of seeing the target. All his assessments are based on information fed to him by his aircraft's systems – in effect, the machines tell the organics what to do.
>
> Sayle (2014)

As the fighter is designed for instability to enhance its agility in dogfights, he adds, then the entire machine is so unstable that without computer-assisted flying it would plummet to the ground. The human component is already less important than we might assume.

These fears of autonomy are linked to fears of human redundancy. AI and robots are already destroying human jobs, for example in warehouses, and delivery drones and driverless vehicles threaten to decimate transportation employment. The rise of UAVs has already led to deskilling and unemployment for military fighter pilots and it is certain that increases in robotic and unmanned forces will shrink the human military component. Another problem with computer-based and robotics systems is deference, as we trust computers are right. On 3 July 1988, the USS *Vincennes'* Aegis system on semi-automatic identified an 'assumed enemy', misidentifying the plane as an Iranian F-14 Tomcat. No one in the 18-strong command overrode the computer's assessment, authorizing it instead to fire, shooting down Iran Air Flight 655 in Iranian space, over Iranian territorial waters, killing all 290 people on board.

There also remains the problem of computer errors and breakdowns. On 25 January 1979, Robert Williams apparently became the first person killed by a robot in an accident at Michigan's Ford plant, when he was hit by a robot arm that lacked safety features (although, predating this by a century, a Mr Maybrook was killed in 1876 when he was hit by a bell-chiming hammer wielded by a wooden, carved automaton in a clock he was inspecting in the church in Glogau, Silesia, Germany). On 12 October 2007, a South African National Defence Force automated anti-aircraft system malfunctioned during a training exercise, killing nine and injuring 14. Other times computers have made the wrong decision based on data, such as on 6 October 1960, when a NATO early warning system detected a Soviet nuclear launch, with 99 percent certainty. NATO went into full retaliation procedures before it was realized the computer had detected the rising moon. Human errors can also feed into computer systems. On 9 November 1979, a test programme containing war games simulating an attack was mistakenly loaded into a missile warning system that interpreted them as real. US Command had scrambled its alert bombers before the mistake was realized.

A future robotic battlefield would be incredibly complex. The command and coordination problems would be significant, considerable technical support would be required and systems would need to be in place for repairing or extracting equipment. A new 'doctrine' of robotic war needs to emerge to realize their military potential, especially as a future battlefield could comprise dozens of different systems, ranging from the human, to the telepresent, to cyborgs, to fully autonomous robots of different sizes and capabilities, either acting alone, or together, or in swarms, all of which require control, communication, coordination and bandwidth. This complexity introduces new security risks. Any connected system is vulnerable to hacking and autonomous robotic technologies have the potential to be damaged, made to exhibit different behaviours and even be turned against their own forces. Human soldiers cannot be hacked in this way.

Critics also suggest that, like drones, robots will be seen as the weapon of cowards: of nations too afraid to risk their own lives. This may not matter if the only mission is combat but many military operations today involve ongoing relationships with a country and its people and a state of insurgent conflict that falls short of open war. Thus, the hostility an 'occupying' force experiences today may be nothing compared to that produced by the sight of robot patrols in villages and neighbourhoods. Rather than being a perfect COIN technology, robots may be dangerously inflammatory: a robot walking around Sangin, in the Helmand province of Afghanistan, is going to appeal even less than US marines.

There are also fears of the moral distance created by autonomous technologies. Whereas claims of a moral distance for UAV operators have been rebutted, due to the telepresent nature of the work, autonomous robots would certainly remove human experience, sympathy, understanding, morality and decision making from the combat zone. This has important implications for questions of war-crimes. Who would be responsible for these – the commander, the robot, the maintenance technician or the programmer? And how would charges be prosecuted against the ever-ready defence of technical failure or accident?

This point leads directly to the most important criticism of autonomous systems: that their potential lethality is so great that they represent either too dangerous a force for warfare, or they will lead to a massive increase in civilian casualties through a disregard for human life. The threat may be greater than we think. On 12 November 2017, the Californian Future of Life Institute uploaded a video to YouTube called 'Slaughterbots' (Stop Autonomous Weapons, 2017). It showed a company presentation of AI micro-drones with cameras, tactical sensors, facial recognition systems and 3g of shaped explosives inside. The presenter throws one in the crowd and it immediately identifies a dummy on stage and penetrates its skull, exploding inside. The presenter then shows 'real' footage of the drones being used to kill 'bad guys' with surgical precision, with the drones working as a team to penetrate any building or space. 'A $25m order now buys this', he says, to footage of a plane releasing a drone swarm, 'Enough to kill half a city … The bad half'. 'Nuclear is obsolete', he concludes. 'Take out your entire enemy, virtually risk-free. Just characterize him, release the swarm and rest easy'. The video then shows the weapons leaking and being used in terrorist attacks on congressmen and student activists targeted through their politics, social media posts and facial recognition. As the company says, today you can find your enemy through data. The authorities are powerless and recommend staying indoors as thousands are killed by the 'slaughterbots'. Created by University of California AI professor, Stuart Russell, to highlight the dangers of LAWS, the video's ideas closely resemble the 2016 *Black Mirror* episode, 'Hated in the Nation', which also showed weaponized drones (bees) using facial recognition and social media posts for targeting. In recent years these warnings about LAWS have grown more common.

Anti-robot warnings

In November 2012, Human Rights Watch published their report, 'Losing Humanity: The Case Against Killer Robots', which warned of the dangers of autonomous systems. It argued that 'fully autonomous weapons would not only be unable to meet legal standards but would also undermine essential non-legal safeguards for civilians'. The report concluded that 'fully autonomous weapons should be banned and that governments should urgently pursue that end'. In particular it recommended that states should, 'Prohibit the development, production, and use of fully autonomous weapons through an international legally binding instrument', 'adopt national laws and policies to prohibit the development, production, and use of fully autonomous weapons', and 'commence reviews of technologies and components that could lead to fully autonomous weapons', suggesting these reviews 'should take place at the very beginning of the development process and continue throughout the development and testing phases.' It also called upon roboticists to 'establish a professional code of conduct governing the research and development of autonomous robotic weapons', to ensure 'that legal and ethical concerns about their use in armed conflict are adequately considered at all stages of technological development' (Human Rights Watch, 2012).

In April 2013, the 'Campaign to Stop Killer Robots' was launched. Overseen by Human Rights Watch, this was a global coalition of 64 international, regional and national non-governmental organizations (NGOs) from 28 countries, calling for a pre-emptive ban on LAWS. It warned that 'killer robots' would be here within a decade and that their development was unregulated, with little thought given to ethics or international law, and called for a global treaty to control their development and use. The campaign webpage warns of a 'robotics arms-race', saying 'agreement is needed now to establish controls on these weapons before investments, technological momentum, and new military doctrine make it difficult to change course'. In particular they argue that allowing life-or-death decisions to be made by machines 'crosses a fundamental moral line'. Autonomous robots 'would lack human judgment and the ability to understand context' and hence 'would not meet the requirements of the laws of war'. Replacing humans with robots 'could make the decision to go to war easier', they say, commenting that there is no clarity yet as to 'who would be legally responsible for a robot's action' (Campaign to Stop Killer Robots, 2017).

The military themselves are not all in favour of these systems. The March–April 2013 edition of *Military Review* included an essay by US Army military intelligence officer, Lt. Col. Douglas A. Pryer. Most of the article is concerned with existing UAV technology, but he also begins to consider the problem of emerging autonomous systems. In any confrontation between *Terminator*-style robots and humans, he says, we will always morally be on the side of the humans. One problem with US military technologies, therefore, is how 'they ignore this moral reality and promote moral dehumanization'. The enemy today is at war with and fires at 'machines', and 'America – home to a proud, vibrant people – has effectively become inhuman' (Pryer, 2013:21). This is only going to get worse, Pryer argues: 'It seems heart-breakingly obvious that future generations will someday look back upon the last decade as the start of the rise of the machines' (Pryer, 2013:23). The future will see armed robots 'so advanced that they make today's Predators and Reapers look positively impotent and antique'. They will all share one thing in common, however, he warns: 'with remorseless purpose, they will stalk and kill any human deemed a "legitimate target" by their controllers and programmers' (Pryer, 2013:23).

The UN has begun to take an interest in LAWS. On 15 November 2013, at the UN, states party to the 1983 Convention on Certain Conventional Weapons (CCW) voted unanimously to begin discussions in March 2014 on LAWS and their development. In August 2014, Angela Kane, the UN's high representative for disarmament also warned autonomous systems were a 'small step' from the battlefield and should be outlawed, arguing

> I personally believe that there cannot be a weapon that can be fired without human intervention. I do not believe that there should be a weapon, ever, that is not guided and where there is not the accountability clearly established by whoever takes that step to guide it or to launch it. I do not believe that we could have weapons that could be controlled by robots.
>
> Farmer (2014)

On 9 April 2015, Human Rights Watch and Harvard Law School published the report, 'Mind the Gap: The Lack of Accountability for Killer Robots', ahead of UN talks on LAWS from 13 April. It argued that under existing laws, computer programmers, manufacturers and military commanders would all escape liability for deaths caused by such machines, and that there is unlikely to be a clear legal framework in the future that would establish responsibility. The report said:

> The hurdles to accountability for the production and use of fully autonomous weapons under current law are monumental. The weapons themselves could not be held accountable for their conduct because they could not act with criminal intent, would fall outside the jurisdiction of international tribunals, and could not be punished. Criminal liability would likely apply only in situations where humans specifically intended to use the robots to violate the law. In the United States at

least, civil liability would be virtually impossible due to the immunity granted by law to the military and its contractors and the evidentiary obstacles to products liability suits.

Human Rights Watch (2015)

Hence, it concludes, 'The limitations on assigning responsibility thus add to the moral, legal, and technological case against fully autonomous weapons and bolster the call for a ban on their development production, and use'.

On 28 July 2015, an open letter at the International Joint Conference on Artificial Intelligence (IJCAI) called for a ban on LAWS, arguing:

If any major military power pushes ahead with AI weapon development, a global arms race is virtually inevitable, and the endpoint of this technological trajectory is obvious: autonomous weapons will become the Kalashnikovs of tomorrow. Unlike nuclear weapons, they require no costly or hard-to-obtain raw materials, so they will become ubiquitous and cheap for all significant military powers to mass-produce. It will only be a matter of time until they appear on the black market and in the hands of terrorists, dictators wishing to better control their populace, warlords wishing to perpetrate ethnic cleansing, etc. Autonomous weapons are ideal for tasks such as assassinations, destabilizing nations, subduing populations and selectively killing a particular ethnic group. We therefore believe that a military AI arms race would not be beneficial for humanity.

In summary, the letter said, AI has the potential to benefit humanity in many ways and that should be the goal of the field. To prevent an AI arms race, there should be 'a ban on offensive autonomous weapons beyond meaningful human control' (Future of Life Institute, 2015).

Another letter calling on the UN to ban LAWS was put forward on 21 August 2017, signed by the founders of 116 AI and robotics companies from 26 countries, including Elon Musk and Google DeepMind's Mustafa Suleyman. The letter argued that weapons able to kill without human intervention were 'morally wrong' and their use should be controlled under the CCW. It claimed this technology would usher in a 'third revolution in warfare', equalling the invention of gunpowder and nuclear weapons. Once developed, autonomous systems, 'will permit armed conflict to be fought at a scale greater than ever, and at timescales faster than humans can comprehend', they wrote. 'These can be weapons of terror, weapons that despots and terrorists use against innocent populations, and weapons hacked to behave in undesirable ways' (Gibbs, 2017; Future of Life Institute, 2017). It was published ahead of the IJCAI, which was intended to coincide with UN talks exploring such a ban.

On 1 September 2017, President Putin, speaking at a meeting with students, said the development of AI created 'colossal opportunities and threats that are difficult to predict now'. He said it would be undesirable for any nation to have 'a monopolist position' and promised Russia would share its knowledge of AI with other nations (CNBC, 2017). On 2 November 2017, hundreds of AI experts from Canada and Australia added their voices to the growing anti-LAWS movement, submitting open letters to their governments, urging them to support a UN ban on autonomous weapons. On 6 November, the late physicist Stephen Hawking, speaking at the Lisbon Web Summit technology conference, joined in the criticism of unrestrained AI, saying,

Unless we learn how to prepare for, and avoid, the potential risks, AI could be the worst event in the history of our civilization. It brings dangers, like powerful autonomous weapons, or new ways for the few to oppress the many.

Kharpal (2017)

From 13–17 November 2017, the first meeting of the CCW Group of Governmental Experts (GGE) on LAWS took place at the UN in Geneva, but the participating nations achieved little more than an agreement to continue formal deliberations the following year. Twenty-two countries now call for an outright ban, but most of the leading robotics nations don't support this. Critics also wonder whether a ban could be adequately defined and imposed. Importantly, however, governments and militaries don't necessarily want fully autonomous weapons. Responding in 2014 to Angela Kane's comments, Huw Williams, Unmanned Systems Editor at Jane's International Defence Review, said he knew of no programmes to make killer robots and that even the most advanced machines had little ability to act on their own. Taking human control away would have a big military disadvantage, he concluded, in representing 'a real loss of control over the battlefield for commanders' (Farmer, 2014)

The 2015 IJCAI letter led to an article by Evan Ackerman entitled 'We Should Not Ban "Killer Robots", and Here's Why'. Ackerman argues that 'no letter, UN declaration, or even a formal ban ratified by multiple nations is going to prevent people from being able to build autonomous, weaponized robots. The barriers keeping people from developing this kind of system are just too low'. Consider, for example, how smart-phone controlled quadcopters have become cheap toys, so 'just imagine what you'll be able to buy tomorrow'. Hence, he says, 'What we really need, then, is a way of making autonomous robots ethical, because we're not going to be able to prevent them from existing.' In particular, he argues, robots can be better than humans at war, since there are rules that can be followed. They could be more cautious than soldiers: 'you could program them to not engage a hostile target with deadly force unless they confirm with whatever level of certainty that you want that the target is actively engaging them already'. Unlike humans, a robot could 'just sit there, under fire, until all necessary criteria for engagement are met'. Hence, Ackerman says, 'The real question that we should be asking is this: Could autonomous armed robots perform better than armed humans in combat, resulting in fewer casualties (combatant or non-combatant) on both sides? I believe so' (Ackerman, 2015).

In 2009 Ronald C. Arkin, Patrick Ulam and Brittany Duncan published a paper entitled 'An Ethical Governor for Constraining Lethal Action in an Autonomous System', which was concerned with how to program a robot to operate within the Laws of War and Rules of Engagement (ROE). Arkin believes there is no scientific limitation to achieving the goal of machines being able to discriminate better than humans can, arguing, 'If that standard is achieved, it can succeed in reducing non-combatant casualties and thus is a goal worth pursuing in my estimation' (Ackerman, 2015). Hence, Ackerman asks, 'if autonomous robots really do have at least the potential to reduce casualties, aren't we then ethically obligated to develop them?' (Ackerman, 2015). Robots may make mistakes but could be programmed to learn from these. Blaming the technology isn't going to solve the problem if the problem is the willingness to use technology for evil. To stop that, he says, 'we'd need a much bigger petition'.

Ackerman's point is important: robots could be better at war than humans. Not only could they follow the ROE better, it is also unlikely that they would engage in the systematic rape of women and children, photograph themselves with a leash around a prisoner on the floor or giving a thumbs up behind a pyramid of naked prisoners, or take a selfie holding a decapitated head. The dangers of robots, therefore, may have been over-estimated compared to the historical dangers of humans. The problem is that any attempt to program robots to follow the ROE or act with ethical limitations will impact on their military effectiveness. However much one military commits itself to that goal there is no guarantee others will and they may gain an advantage from not doing so. Such additions will also add cost and complexity to systems, making them a luxury that could be easily dispensed with. The development of LAWS, therefore, will inevitably tend towards the development of the most-lethal systems.

Robotic and AI proliferation

We may already be en route to the world of slaughterbots. Islamic State and Syrian rebels are already using COTS micro-drones packed with explosives. Given the speed of development of robotics and drone technology and their commercial spread, no international weapons agreement will be able to prevent non-state actors gaining access to and using these systems. Even if the most complex high-end systems are controlled, drone-robotics is out of the bag, will continue to improve and will become a significant weapon. High-end robotics, of course, is not just the preserve of the US military. By 2013 it was reported that 76 countries had military robotics programmes. Today, basic, armed military robots are no longer scarce. In November 2016, the Iraqi government employed its own remote-controlled UGV, in Mosul against Islamic State. Designed by two brothers in Iraq, the armoured, four-wheeled Alrobot is armed with a heavy machine gun and rockets and carries multiple cameras. A month later an Estonian firm showed off its latest invention, a 3ft tall, remote-controlled, armed, shape-changing mini-tank, the THeMIS ADDER.

China, especially, is devoting considerable resources to developing robotics systems. In November 2015, they unveiled their latest robots, including small, tracked 'armed attack' UGV robots armed with guns and grenades. In December 2016, the Chinese Army released footage of a range of robotic systems including UGVs and a version of Boston Dynamics' 'Big Dog' quadruped as part of its 'Overcoming Obstacles 2016' competition. In October 2017, China's Rocket Force Research Centre developed an automated missile system, with robots moving and loading the missiles, helping it fire warheads three times faster and halving the number of soldiers needed. As with drone production, China is expected one day to surpass the USA as a major robotics nation.

Chinese authorities are also prioritizing research into swarm robotics, as a key future military technology. On 11 June 2017, the state-owned China Electronics Technology Group demonstrated the largest coordinated micro-drone swarm to date, with 119 drones flying in formation. China sees swarm robotics as a cheaper option, and, in sufficient numbers, as potentially able to defeat more expensive military systems such as those deployed by the USA. Swarm drones would also allow China to project force without risking a major confrontation, exploiting the 'grey zone' of limited military operations.

More importantly, China recognizes that the underlying revolution is in AI. In March 2016 in Seoul, the world's best Go player, Lee Sedol, was beaten four games to one by Google DeepMind's AlphaGo AI, the first time the best human had been defeated and an event not expected to happen for another decade. The victory was barely noted in the west, but China's best computer scientists were present and for them it was a 'Sputnik moment' – equal to the impact the USSR's launch of the first artificial satellite on 4 October 1957 had on the USA

(Allen, 2017). Just as the USA responded to that wake-up call with ARPA and renewed military research and funding, so China seized the AI initiative. On 20 July 2017, China published its 'Next Generation Artificial Intelligence Development Plan', which designated AI as the transformative technology that will secure its future economic and military power. China sees AI as a disruptive technology in war that will enable it to gain an advantage over their rivals, hence their stated aim to have overtaken the USA in AI by 2025, whilst 'by 2030, China's AI theories, technologies, and applications should achieve world-leading levels, making China the world's primary AI innovation center' (State Council of China, 2017). Simply put, this is a plan for global AI dominance. China is expected to devote huge sums to this AI project, comparable at least to the $150bn it allocated in 2014 to its 10-year national semiconductor plan.

It hope to achieve its success through 'military-civil fusion' (Allen, 2017). China has already established joint military-commercial research laboratories and testing facilities in areas such as deep learning and autonomous vehicle development and expects all AI developments to have a 'dual use' both for commercial purposes and for the state and military. The PLA and military research centres are investing in a range of AI projects – in January 2018, for example, Chinese media reported the recruitment of 120 top AI specialists to work in the Chinese Academy of Military Sciences to push the development of military AI, – whilst Chinese companies are leading the push to improved AI. The search engine company Baidu is already one of the global leaders in AI research, Tencent is focusing on AI research and opened its own AI lab in 2016, whilst Chinese programmers now routinely win coding and machine learning competitions. Asked about US-Chinese AI competition in his 1 November 2017 Centre for a New American Security 'Artificial Intelligence and Global Security Summit' keynote address, former Google CEO Eric Schmidt admitted their capabilities: 'China will catch up extremely quickly', he said (Scharre, Cho, Allen and Schmidt, 2017).

Russia has also invested heavily in unmanned systems, sees AI as central to future military dominance and has expressed the desire to replace soldiers with robots. Lt. Gen. Andrey Grigoriev, head of the Advanced Research Foundation, told RIA Novosti in July 2016:

> I see a greater robotization. In fact, future warfare will involve operators and machines not soldiers shooting at each other on the battlefield. It would be powerful robot units fighting on land, in the air, at sea as well as underwater and in outer space. They would be integrated into large comprehensive reconnaissance-strike systems. The soldier would gradually turn into an operator and be removed from the battlefield.
>
> RT (2016)

In November 2016, Russia displayed their latest, next-generation military robotics, including the Flight surveillance robot, the Alpha Device, a remote-controlled grenade launcher with a firing range of 400m, and their Platforma-M remote-control tank, armed with a 7.62mm gun and four rocket launchers, all designed for deployment on the Russian border. In March 2017, Russia held a 'Robotization of the Russian Armed Forces' exhibition, bringing together over 30 Russian defence companies and 500 participants, and displaying numerous robotic vehicles and armaments, including a reconnaissance drone fired from a hand-held grenade launcher, anti-mine robots, rescue robots and unmanned boats.

Russia now has a number of UGVs, including the Nerekhta, tracked and armed robot, the larger, armed Uran-9, and the Vikhr remote-controlled tank. In April 2017, Russia released footage of the Vikhr on a military testing ground. Weighing 14.7 tonnes, 6.7m long, 3.3m tall and with a maximum speed of 60 km/h, the Vikhr ('Whirlpool') is armed with a 30mm automatic cannon, as well as machine gun and six anti-tank guided missiles. It is also capable of autonomous driving and can automatically detect targets and prepare to attack them. The same month, Russia also released video of its humanoid robot FEDOR (Final Experimental Demonstration Object Research), standing up, holding two guns and shooting at a target. Russia insisted it wasn't building a real-life 'Terminator' and that this was coordination practice for its space-bound robot, but the military potential was obvious. Like China, Russia recognizes the importance of AI to its plans. As Putin explained in his talk to students on 1 September 2017, 'the one who becomes the leader in this sphere will be the ruler of the world' (CNBC, 2017).

The battlefield singularity

Few now doubt that the related combination of robotic technologies and machine learning AI will completely transform human life, society, work, economics and warfare in the near-to-far future. Indeed, the best warning about this threat came nearly 150 years ago in Samuel Butler's 1872 book *Erewhon*, in which he noted 'life' had arisen from nothing once before on the planet and could do so again. Influenced by Darwin, Butler suggested the machines he saw evolving around him contained the spark of that possible life:

> Reflect upon the extraordinary advance which machines have made during the last few hundred years, and note how slowly the animal and vegetable kingdoms are advancing. The more highly-organized machines are creatures not so much of yesterday, as of the last five minutes, so to speak, in comparison with past time. Assume for the sake of argument that conscious beings have existed for some twenty million years: see what strides machines have made in the last thousand! May not the world last twenty million years longer? If so, what will they not in the end become?
>
> Butler (1872)

Many futurists and transhumanists predict the same future of evolved AI, with thinkers such as Moravec, Vinge and Kurzweil suggesting humans would need to respond to the 'singularity' (Moravec, 1990; Vinge, 1993; Kurzweil, 2008) of artificial intelligence – that moment when AI definitively surpasses humanity – by similarly evolving, downloading our consciousness into an electronic form.

We won't, however, have to wait twenty million years to see battlefield AI. The 'hybrid' battlefield of the near future will contain 'combat-teamed' humans and robots, encompassing enhanced and augmented cyborg soldiers, VR and BCI telepresent drone UAVs, UGVs and surrogate robot soldiers, as well as fully autonomous robots. These will be of different scales and levels of intelligence, from nano-scale bots, to insect-size swarms, to larger animal-style robots, to humanoid scale, to giant-sized walkers, tanks and aircraft, each capable of different tasks and roles, and each with a different array of sensors, equipment and weaponry – ranging from simple spikes and shaped explosives, through to major munitions – all capable of working alone or cooperatively, with the whole battlefield networked, with real-time control or oversight. The *Terminator* image of identical, skeletal bipedal terminators strolling over a battlefield, therefore, may end up being far too simplistic and optimistic: robotic warfare will be much more complex and horrific.

But even this is an optimistic scenario, because, as I've argued, the real revolution is in artificial intelligence. Developments in machine learning point to a future beyond Big Data analysis where AI becomes more than a tool or servant, but an active and major military *agent*, involved in the entire range of identification, targeting and attacks, whether directly military and kinetic through robotic systems, or through cyberwarfare (through automated computer network exploitation and computer network offence) or through information war (with automatically generated propaganda). The days when the Internet Research Agency has to hire unemployed Russians to post online are numbered: future infowars will be post-human, created by AI trolls.

Robots, therefore, are *only one* possible physical instantiation of military AI. If we really want to understand the threat of autonomous AI we need to understand the full range of its applications. This includes the development of the battlefield singularity: that moment when military AI supersedes human capabilities, making human control of its systems too slow, too inefficient and too prone to error. At that point AI's most important contribution won't be as military soldiers or weapons, but as military *commanders*. Hence, the anti-killer-robots campaign is too naïve: it is the AIs ordering the killing we will need to oppose. It won't be long before the best AI algorithms – those best able to command swarms and robots and coordinate disparate systems for battlefield dominance – will be more important than the best weapons systems. There is a strong possibility after that that AI will also become the best commanders.

We needn't feel left out, however. There may still be a role for humans in future war. Current discussions of military robots are all predicated on the idea that humans not only could, but *should* be replaced by robotic systems. They imply human life has greater value and should be preserved. There is no guarantee that this view will prevail in the future. Dystopic predictions of the massive impact of automation upon employment, increasing competition for scarce natural resources, global demographic growth, increasing migration and widening inequalities both within and between nations and continents mean we cannot rule out a future in which some human life will have less value than complex, expensive autonomous weapons systems. In this scenario, instead of a future battlefield of surrogate robots controlled by BCI-linked human operators, we may have instead surrogate humans controlled by AI commanders. The French Foreign Legion was established in 1831, accepting recruits from any nation with the reward of the opportunity of French citizen-ship. Imagine, therefore, a global immigrant offered the chance of western citizenship if they submit to AI operation. In such a future, our current conceptions of artificial intelligence, of 'in' and 'on' the loop humans, and of 'combat-teaming' itself may prove to be incredibly naïve. Either way, we should prepare ourselves for a post-human war that will be fundamentally different from the human war that has defined our history to date.

Key reading

The best book on military robotics remains Singer (2009). Its definition of robotics includes remote-controlled UAV/UGV technologies and most of the book focuses on telepresent technologies so that has to be borne in mind. Singer was right, however, to see these as early forms of the more autonomous systems we're already seeing, and his discussion of the implications of robot warfare and its ethics remains essential. The best information about specific robots is found online. Googling Packbot, Warrior, Talon and

Swords will bring up a lot of additional information. Boston Dynamics produce a lot of videos and it's worth watching their demonstrations of their robots (2009, 2010, 2012, 2013, 2015, 2016, 2017a, 2017b, 2017c). Developments in MAST, swarm robotics and other robotic research discussed here can also be found online by Googling the examples. There are numerous books on robotics I would recommend, such as Brooks (2003), Winfield (2012) and Jordan (2016), although rapid developments in the field mean they date very quickly. Good, introductory books on artificial intelligence include Warwick (2012), Kaplan (2016), Alpaydin (2016) and Boden (2018). Marr (2016a, 2016b) also provides a simple overview of AI, machine learning and deep learning. The anti-robot warnings of Human Rights Watch (2012, 2015), Future of Life Institute (2015, 2017) and Campaign to Stop Killer Robots (2017) should be read, together with Ackerman's defence (2015). China's AI ambitions are laid out in the State Council of China (2017) document. The transhumanist ideas of Vinge (1993), Moravec (1990) and Kurzweil (2008) can be explored as additional, if highly speculative, reading.

References

Ackerman, E. (2015) 'We Should Not Ban "Killer Robots", and Here's Why', *IEEE Spectrum*, 29 July, https://spectrum.ieee.org/automaton/robotics/artificial-intelligence/we-should-not-ban-killer-robots.

Allen, G. C. (2017) 'China's Artificial Intelligence Strategy Poses a Credible Threat to U.S. Tech Leadership', *Council on Foreign Relations*, 4 December, https://www.cfr.org/blog/chinas-artificial-intelligence-strategy-poses-credible-threat-us-tech-leadership.

Alpaydin, E. (2016) *Machine Learning*, London: MIT Press.

Boden, M. (2018) *Artificial Intelligence. A Very Short Introduction*, Oxford: Oxford University Press.

Boston Dynamics. (2009) 'LittleDog', *YouTube*, 16 February, https://www.youtube.com/watch?v=UIipbi0cAVE.

Boston Dynamics. (2010) 'BigDog Overview (Updated March 2010)', *YouTube*, 22 April, https://www.youtube.com/watch?v=cNZPRsrwumQ.

Boston Dynamics. (2012) 'Cheetah Robot Runs 28.3 mph; A Bit Faster than Usain Bolt', *YouTube*, 5 September, https://www.youtube.com/watch?v=chPanW0QwhA.

Boston Dynamics. (2013) 'Introducing WildCat', *YouTube*, 3 October, https://www.youtube.com/watch?v=wE3fmFTtP9g.

Boston Dynamics. (2015) 'Introducing Spot', *YouTube*, 9 February, https://www.youtube.com/watch?v=M8YjvHYbZ9w.

Boston Dynamics. (2016) 'Atlas, the Next Generation', *YouTube*, 23 February, https://www.youtube.com/watch?v=rVlhMGQgDkY.

Boston Dynamics. (2017a) 'The New SpotMini', 13 November, *YouTube*, https://www.youtube.com/watch?v=tf7IEVTDjng.

Boston Dynamics. (2017b) 'Introducing Handle', *YouTube*, 27 February, https://www.youtube.com/watch?v=-7xvqQeoA8c.

Boston Dynamics. (2017c) 'What's new, Atlas?', *YouTube*, 16 November, https://www.youtube.com/watch?v=fRj34o4hN4I.

Brooks, R. (2003) *Flesh and Machine: How Robots Will Change Us*, New York: Vintage Books.

Butler, S. (1872) *Erewhon*, https://www.marxists.org/reference/archive/butler-samuel/1872/erewhon/ch23.htm.

Campaign to Stop Killer Robots (2017) 'The Problem', *Campaign to Stop Killer Robots*, https://www.stopkillerrobots.org/the-problem/.

CNBC. (2017) 'Putin: Leader in Artificial Intelligence Will Rule World', *CNBC*, 4 September, https://www.cnbc.com/2017/09/04/putin-leader-in-artificial-intelligence-will-rule-world.html.

Department of Defense. (2017) 'Project Maven to Deploy Computer Algorithms to War Zone by Year's End', *U.S. Department of Defense*, 21 July, https://www.defense.gov/News/Article/Article/1254719/project-maven-to-deploy-computer-algorithms-to-war-zone-by-years-end/.

Economist, The. (2017) 'Military Robots are Getting Smaller and More Capable', *The Economist*, 14 December, https://www.economist.com/news/science-and-technology/21732507-soon-they-will-travel-swarms-military-robots-are-getting-smaller-and-more.

Economist, The. (2018) 'War at Hyperspeed', *The Economist*, 25 January, https://www.economist.com/news/special-report/21735478-autonomous-robots-and-swarms-will-change-nature-warfare-getting-grips.

Farmer, B. (2014) 'Killer Robots a Small Step Away and Must be Outlawed, Says Top UN Official', *The Telegraph*, 27 August, http://www.telegraph.co.uk/news/uknews/defence/11059391/Killer-robots-a-small-step-away-and-must-be-outlawed-says-top-UN-official.html.

Future of Life Institute. (2015) 'Autonomous Weapons: An Open Letter from AI and Robotics Researchers,' *Future of Life Institute*, 28 July, https://futureoflife.org/open-letter-autonomous-weapons/.

Future of Life Institute. (2017) 'An Open Letter to the United Nations Convention on Certain Conventional Weapons', *Future of Life Institute*, 21 August, https://futureoflife.org/autonomous-weapons-open-letter-2017/.

Gibbs, S. (2017) 'Elon Musk Leads 116 Experts Calling for Outright Ban on Killer Robots', *The Guardian*, 20 August, https://www.theguardian.com/technology/2017/aug/20/elon-musk-killer-robots-experts-outright-ban-lethal-autonomous-weapons-war.

Global Security. (2011) 'TALON Small Mobile Robot', *Global Security*, 7 July, https://www.globalsecurity.org/military/systems/ground/talon.htm.

Human Rights Watch. (2012) 'Losing Humanity: The Case Against Killer Robots', *Human Rights Watch*, 19 November, https://www.hrw.org/report/2012/11/19/losing-humanity/case-against-killer-robots.

Human Rights Watch. (2015) 'Mind the Gap: The Lack of Accountability for Killer Robots', *Human Rights Watch*, 9 April, https://www.hrw.org/report/2015/04/09/mind-gap/lack-accountability-killer-robots.

Jordan, J. J. (2016) *Robots*, London: MIT Press.

Kaplan, J. (2016) *Artificial Intelligence*, New York: Oxford University Press.

Kharpal, A. (2017) 'Stephen Hawing Says AI could be "Worst Event in the History of our Civilization"', *CNBC*, 6 November, https://www.cnbc.com/2017/11/06/stephen-hawking-ai-could-be-worst-event-in-civilization.html.

Kurzweil, R. (2008) *The Singularity is Near*, London: Duckworth Overlook.

Marr, B. (2016a) 'What Is The Difference Between Artificial Intelligence and Machine-Learning', *Forbes*, 6 December, https://www.forbes.com/sites/bernardmarr/2016/12/06/what-is-the-difference-between-artificial-intelligence-and-machine-learning/#1d50dac42742.

Marr, B. (2016b) 'What Is The Difference Between Deep-Learning, Machine Learning and AI?', *Forbes*, 8 December, https://www.forbes.com/sites/bernardmarr/2016/12/08/what-is-the-difference-between-deep-learning-machine-learning-and-ai/#2c62363d26cf.

McNally, D. (2017) 'Army Completes Autonomous Micro-Robotics Research Program', *U.S. Army*, 28 August, https://www.army.mil/article/192947/army_completes_autonomous_micro_robotics_research_program.

Moravec, H. (1990) *Mind Children*, Cambridge, MA: Harvard University Press.

Popular Mechanics. (2009) 'The Inside Story of the SWORDS Armed Robot "Pullout" in Iraq: Update', *Popular Mechanics*, 1 October, http://www.popularmechanics.com/technology/gadgets/a2804/4258963/.

Pryer, D. A. (2013) 'The Rise of the Machines: Why Increasingly "Perfect" Weapons help Perpetuate Our Wars and Endanger Our Nation', *Military Review*, March–April, pp. 14–24, http://usacac.army.mil/CAC2/MilitaryReview/Archives/English/MilitaryReview_20130430_art005.pdf.

RT. (2016) 'War Machine: Robots to Replace Soldiers in the Future, Says Russian Military's Tech Chief', *RT*, 6 July, https://www.rt.com/news/349699-russia-future-combat-robots/.

Sayle, J. (2014) 'Terminator to Taranis: Robots on the Battlefield', *UK Defence Journal*, 1 August, https://ukdefencejournal.org.uk/terminator-taranis/.

Scharre, P., Cho, A., Allen, G.C., and Schmidt, E. (2017) 'Eric Schmidt Keynote Address at the Centre For a New American Security Artificial Intelligence and Global Security Summit', *CNAS*, 13 November, https://www.cnas.org/publications/transcript/eric-schmidt-keynote-address-at-the-center-for-a-new-american-security-artificial-intelligence-and-global-security-summit.

Singer, P. W. (2009) *Wired For War: The Robotics Revolution and Conflict in the Twenty First Century*, London: Penguin Books.

State Council of China. (2017) 'A Next Generation Artificial Intelligence Development Plan', *China Copyright and Media*, 20 July, https://chinacopyrightandmedia.wordpress.com/2017/07/20/a-next-generation-artificial-intelligence-development-plan/.

Stop Autonomous Weapons. (2017) 'Slaughterbots', Future of Life Institute, *YouTube*, 12 November, https://www.youtube.com/watch?v=9CO6M2HsoIA&t=.

Vinge, V. (1993) 'The Coming Technological Singularity', 30–31 March, http://mindstalk.net/vinge/vinge-sing.html.

Warwick, K. (2012) *Artificial Intelligence*, Oxon: Routledge.

Winfield, A. (2012) *Robotics. A Very Short Introduction*, Oxford: Oxford University Press.

Conclusion

The clouds of war

War has changed. What 'war' is, how it is fought, how we define and understand it, how we know about it, how we experience it and how we participate in it have all been transformed in recent decades.

On one level, what we have lived through in this time is the maturation of that US model of military operations demonstrated in the 1991 Gulf War, where post-Vietnam developments in precision-guided munitions, stealth technologies, UAVs, and global, electronic intelligence, surveillance and reconnaissance combined together to devastating effect. Following that victory, the US military retheorized warfare to place information and informational operations as at the centre of conflict, with ongoing developments in the key military technologies feeding into their refined model of 'network-centric warfare', which was tested in Afghanistan in 2001 and rolled out in Iraq in 2003. But this kind of overwhelming, technologically superior, set-piece interstate warfare has now become an exception. It has been largely replaced by a different kind of war that has emerged in part as a result of the USA's failure.

In each case, in Afghanistan and Iraq, the immediate success of these closely controlled, media-friendly, globally disseminated *showcase wars* was undermined by the inability to adequately secure the country, effectively defeat the opposing forces, pacify and provide for the public, and reconstruct the economic, civic and political infrastructure. Hence the aftermath of each war collapsed into chaos and violence, with corrupt, unrepresentative governments, and networked, multi-actor insurgencies undoing the dream of limited, achievable goals and quick withdrawals. The neoconservative hope of parachuting in democracy to grateful, pro-western politicians and populations proved a fantasy as, instead, the conflicts exposed deep political and sectarian fractures within each nation and reignited older regional rivalries.

The wars the USA and the west subsequently found themselves engaged in were unofficial, undeclared *non-wars* – 'peace-time' securitization operations that were far more deadly for them than the 'wars' they had confidently declared over. These were new *permawars*, with no sign of an end and no real, possible victory, only compromised, confusing internal politics and divisions, asymmetric resistance, continuing insecurity and atrocities and unending COIN-war. For the west, the result were cyclical *Groundhog Day wars* with declarations of success and public withdrawals followed by ongoing special forces operations and airstrikes, deteriorating security situations and, eventually, renewed deployments, with troops returning again and again to the same theatres. These were also *globalized wars*, with previously separate and unconnected insurgent movements connecting together, communicating, learning from each other and even moving theatres to cooperate or compete. They were also, simultaneously, *multilayered wars*, defined by the multiple internal and external participants, each with their own vision, ideals and agendas, and *fractal wars*, with these agendas breaking down upon closer investigation into ever-more complex local, historical, ethnic, sectarian, cultural, geographical, political, economic and social causes and reasons.

They were also *viral wars*, with the US military actions replicating and splitting to open new fronts in other continents, countries and theatres. They were *clandestine wars*, prosecuted now through special forces, UAVs, computer-network exploitation, electronic ISR and airstrikes, with operations and their effects remaining secretive, deniable and uncertain. They were also *low media-intensity wars*, barely leaving a broadcast footprint, either because of their hidden nature, because the endless parade of insurgent bombings and unsolvable foreign politics had fallen down the news hierarchy, or because a domestic population had grown bored of war and dying civilians. They were also, therefore, seen by many as *hypocritical wars*, built on the concept of the value of innocent civilian lives, designed to avenge and defend American civilians, hoping to spread liberal-democratic political values and processes based upon the idea of inalienable individual rights, whilst remaining indifferent both to the rights of those they targeted and to the massively accumulating civilian suffering and death-toll.

From this perspective, what we are seeing is not simply the temporary eclipse of interstate wars by intrastate and civil wars, but the proliferation and splitting of war itself into many different forms, often occurring simultaneously and often disengaged from each other, with different actors pursuing their own types of conflict. But the situation is even more complex than this as digital technologies have not only transformed how wars are fought but, in doing so, have also impacted upon the very definition and meaning of war. US drones, for example, have turned military incursions into minor events, have transformed warfare into targeted assassination, have withdrawn the bodies of combatants from combat zones, turning soldiers into commuting cubicle-warriors, have been employed by a civilian agency to kill people, and have also engaged a sizable civilian force to work inside the drone analysis 'weapons system' with real, on-the-ground consequences. Ongoing developments in artificial intelligence and autonomous systems complicate even more the what, where and who of warfare.

Similarly, developments in cyberwar have made it difficult to ascertain what might be an attack, when it might have started, what the effect or significance might be, how seriously it should be taken, what the appropriate response would be, whether it constitutes damage and at what point a military response might be justified. Even the identity of the attackers is difficult to prove and it is increasingly likely that they may not be state backed. The Web 2.0 platform revolution, rise of smart-phones and public Wi-Fi connections have also impacted on the concept of 'war'. The military 'information warfare' theorized in the 1990s has now been democratized to anyone with a phone, as governments, militaries and security services, non-state actors such as militias, rebels and terrorist groups, and everyone both within and outside of a specific combat zone is empowered to produce and share propaganda and join in the ideological conflict, reaching deep into enemy nations or battlefields. 'War' today, therefore, involves everyone, everywhere, all the time.

This complexity will only increase with the end of the USA's post-1991 dominance of the international order. The USA, sapped by its post-9/11 global military adventures, is unwilling to extend its forces again in an major interstate conflict (refusing, for example, to significantly intervene in Syria), and this has left room for other nations to assert their interests. Russia has demonstrated a nationalistic resurgence, has reinvested in its military, is pursuing the same technological breakthroughs as the US, is an active cyberpower, engages in aggressive, global information warfare and has asserted itself within its claimed sphere of influence, for example in Georgia, Crimea and the Ukraine, as well as in Syria. With three decades of economic growth behind it, China is also gaining confidence as an emerging superpower, and possibly the future ascendant power. It is already beginning to challenge America's dominance of the international order and is highly active online, with advanced computer network exploitation capabilities. It, too, is building its military capabilities, investing in the most advanced military technologies and asserting itself within its regional sphere, such as in the south and east China seas. The return of 'great-power' rivalry and spheres of influence, the exploitation of the 'grey zone' of activities that fall short of provoking open conflict, and the rise of anti-access/area-denial strategies by Russia and China will, therefore, make the future less certain, less stable and more complex. As more nations invest in and develop drones, robots, artificial intelligence, autonomous systems, and cyberwarfare and information warfare capabilities we can expect threats to the international order, to sovereignty and territorial integrity, to military and domestic stability and security, and to civilian life to significantly increase.

In 2007, Frank G. Hoffmann coined the term 'hybrid war' to describe the new forms of conflict arising in the twenty-first century. As he explains:

> Hybrid threats incorporate a full range of different modes of warfare, including conventional capabilities, irregular tactics and formations, terrorist acts including indiscriminate violence and coercion, and criminal disorder. Hybrid wars can be conducted by both states and a variety of non-state actors. These multi-modal activities can be conducted by separate units, or even by the same unit, but are generally operationally and tactically directed and coordinated within the main battlespace to achieve synergistic effects in the physical and psychological dimensions of conflict. The effects can be gained at all levels of war.
>
> Hoffmann (2007:8)

'Hybrid war' soon became a key military buzzword, with the term being employed to describe the blurring of military, economic, diplomatic, intelligence, and criminal means to achieve political goals. Hoffmann's discussion of the deliberate hybrid blending of tactics and forms of warfare is accurate, as is his prescient claim that hybrid wars will involve the blurring of lines between conventional and irregular combat, physical and virtual combat and even between combatants and non-combatants. But Hoffmann seems to have missed the implications of the Web 2.0 revolution, has little to say about the digital technologies that allow anyone to involve themselves in warfare and doesn't grasp the extent of global participation that this has given rise to. The complexity of war, of its character, forms and participants as well as of its definition, has proven to be even greater than he imagined.

If how wars are fought and what war means has changed, then so too has our knowledge and experience of war today. In 1991 the primary means of informing the domestic or international public about wars was mass-media broadcasting. To solve the problem of how to fight a war in the glare of the world's media and retain the public's support, the US government developed a model of media management in the Gulf War that aimed to control the narrative and experience of war, to make it appear palatable, and even moral. Their success at this meant the military returned to and specifically emphasized this element in its re-theorization of war through the 1990s: its concept of 'full spectrum' battlefield dominance included both the enemy's informational systems and all broadcast media that might offer a counter-narrative to its own. The US dream of total informational control would fail, however, as the ongoing digital revolution, Web 2.0 platforms, improved connectivity, and digital cameras and smartphones made it impossible even to control the media productions of its own soldiers.

This digital explosion of informational access and productivity needs to be understood in relation to the failures and problems of mainstream media journalism. The 2003 Iraq War was professional journalism's nadir. Journalists failed to adequately question the grounds for war, ignored the significant anti-war sentiment and even became active cheerleaders for deposing Saddam Hussein. Once the war began, the willingness of news organizations to don combat fatigues and hop onto tanks or report the war without any critical distance or interest in and understanding of the feelings of the local population was remarkable. Even in the aftermath, few news organizations asked where the WMDs were or devoted enough resources to the increasingly violent, chaotic deterioration of the country and the western responsibility for this. Whilst a live war brought in viewers and readers, these countries quickly fell down the news hierarchy in the months and years after. Afghanistan and Iraq appeared only as a regular drip-feed of bombings, death tolls and military casualties, with little context or critique.

Within a few years of the Iraq War's end, the Web 2.0 revolution led to an explosion of user-generated content and sharing on platforms such as MySpace, YouTube and Face-book. The rise of the iPhone, smartphones and tablets, and domestic and public Wi-Fi also transformed the media ecology. All of this had an impact on newspapers and news organizations, especially upon their business models. Print revenue fell, as did advertising revenue, which shifted online, with a major slice being taken by technology companies such as Google and Facebook who crawled and cannibalized journalistic content, offering it up for free to readers. As a result, newsrooms changed: there were fewer expert foreign correspondents, and less detailed coverage and analysis of international news. With too many sources of news and commentary to choose from, the cultural centrality of news-papers also began to decline, increasing their problems.

At the same time, conflicts were becoming increasingly unreportable. Violent insurgen-cies and ethnic and sectarian conflicts made reporting in Afghanistan and Iraq impossible. The insurgents had no interest in the media or public relations, being interested instead in simply killing or kidnapping westerners, whilst anti-western sentiment was common throughout the warzones. The chaos, murders and threats meant journalists couldn't travel without protection, hence most of the later reporting of the Afghan conflict in the UK was by journalists travelling with the military. Most journalists uncritically accepted the military's own explanations of missions and both politicians and the mass-media con-tinued to frame Afghanistan within the simplistic model of the War on Terror. In all of this, the voices and experiences of ordinary Afghanis were rarely heard. This isn't simply a problem with western wars. The Gaza War and Libyan and Syrian civil wars also proved too dangerous for professional journalism. Today's intrastate and civil wars and all the modes of insurgent, asymmetric war and terrorist warfare cannot be adequately and safely reported.

Technological developments have also made war increasingly difficult to report. The drone campaigns were secretive, occurred far from western media in areas that were among the most remote and inhospitable on earth, and hence were difficult to report on. Official claims regarding the programs were difficult to contradict and proof of who was or wasn't a militant or terrorist was beyond most news organizations. The gradual shift to global, clandestine, deniable, special forces COIN operations employing drones and airstrikes has also made military activities harder to discover, report and verify. The rise of cyberwar is another challenge for journalism. Again, this is a mode of conflict that is highly secretive and journalists have little access to the 'attacks', rely on official sources for information about them and often lack the technical understanding required to discuss attacks and their significance. Future developments in telepresence, interfaces, artificial intelligence and robotics will only increase the problem of journalistic access and knowledge. In conclusion, many of the most important new forms of global conflict take place today beyond the ability of journalists to properly discover, investigate and report them.

One might see the failures and problems of mainstream journalism as being compensated for by the contemporary, bottom-up, digital explosion of information. In theory the rise of global, participative war brings us all closer to the battlefield, exposing the reality of war in ways that were never possible before. In this argument, we pass from the inevitable secrecy and confusion of military action to a new world of transparent war, where video, information and data about conflicts is globally hosted and accessible. The 'fog of war', therefore, gives way to *the clouds of war*.

The concept of 'the fog of war' is usually traced back to Carl Von Clausewitz in his 1832 book *On War*, in which he wrote 'War is the realm of uncertainty; three quarters of the factors on which action in war is based are wrapped in a fog of greater or lesser uncertainty' (Clausewitz, 2008:101). Whilst Clausewitz, was describing the problems of military judgment, the term has been generalized today to mean a general lack of certainty due to the confusion of warfare and lack of knowledge of events on the ground. As we have seen in Gaza and Syria, today's Web 2.0 world has led to much more information about conflicts being produced, with videos, posts, comments, memes, photographs and data of all kinds pouring from and onto the battlefield, all stored in cloud platforms and accounts. As a result, we might hope that the fog of war would be being banished by the clouds of war, with conflicts becoming more open, accessible and understandable.

In fact, the opposite is happening as we face the fractal disintegration of our knowledge. The common social media problem of *too-much-information* ('TMI') takes on here a military significance, as the scale and rapidity of new information overwhelms us. We are left merely with infinite, competing and accumulating snapshots of conflict that cannot be composed into a whole, making it difficult to know what is really happening, calling into question the possibility of establishing any truthful political explanation or of reconstituting events. Writing in 1978, Jean Baudrillard argued that the over-production of information does not add to our understanding, but instead leads to 'the implosion of meaning' itself (Baudrillard, 1983:95–110). This is what we're seeing online as the sheer amount of content exceeds comprehension.

Importantly, it also exceeds collection and research. The dispersal of this information across so many devices and platforms, the surface-level ephemerality of posts as they are buried beneath those coming after, the problem of private accounts and privacy settings, the question of how to find conflict-related material, the problem of how to verify and understand it and of how to evaluate its significance and effects, and the fact that all of this information is held on the servers of private companies and is not a public resource, all creates a fundamental problem for how we – and future historians – make sense of conflict today. Digital war, therefore, poses significant problems for archivists, as to *not* have or be able to access this information will lead to only a partial record of conflicts. It would also risk falsifying the historical record: if you only had access to the mainstream news reports about the 2014 Gaza War, for example, you would fail to understand the extent to which the Israeli narrative failed and you would miss entirely its online defeat. What then should be archived? Can you archive every post, every comment, every tweet, retweet and like? These are issues we have barely begun to confront.

The world of *TMI-warfare* is a world where truth and reality become harder to establish than ever before. Combat zone imagery doesn't simply stand as authentic, unvarnished 'truth', it enters into a complex ecology of content, platforms, actors and ideologies, and hence into the endless informational war around every conflict and political issue. In an era in which 'fake news' is systematically produced and any reportage is subject to claims of being 'fake news', this informational warfare is becoming our everyday political reality. Indeed, actual fakery, too, may be rampant in the future. In early 2018 there was an outcry over 'deepfakes' – AI-assisted, face-swapping software used to produce pornography. The potential of this software isn't limited, however, to pornography and we should expect to see its political, propaganda use very soon.

In January 1961 President Eisenhower warned of the power of a linked 'military-industrial complex'. The concept has been reapplied numerous times since, in claims, for example, of a 'military-entertainment' complex (encompassing the use of cinema, multi-media, video-games, simulations and virtual reality technologies), and a 'military-industrial-media' complex (adding modern, powerful broadcast industries to the traditional linkage). Today, similarly, we might warn of the dominance of the military-social-media complex. Where this differs from earlier complexes, however, is that it isn't a formal linkage, but rather a mode of prosthesis; it isn't confined to one nation, but is globally shared; and it doesn't exist as a demarcated sphere, but includes us all.

The military-social-media complex is a prosthetic relationship. No formal collaboration is needed with the technology companies, rather the devices, platforms, systems and services they create exist outside of governments and the military and are simply appropriated by them and weaponized for their communicational, informational and propaganda purposes. This is a globally-shared appropriation, however, as *any* government, military, movement and militia around the world can seize these prosthetic tools, as Islamic State demonstrated in its continual search for new, more, and better platforms and systems to use and the inability of US technology companies to ever finally push them off the major social media services. The range of social, digital technologies available allows any organization today to bypass traditional broadcast media. Whereas in 1991 they were central to military communications and needed by the authorities, today they are just one element of a broader media and experiential ecology, being available as and when required. Now, governments and military can produce, control and disseminate their messages directly to their audiences.

Unlike the military-industrial complex, the military-social-media complex isn't an exclusionary system. Indeed, it is as globally inclusive as possible, in allowing anyone and everyone to participate. In the world of 'me-dia' (Merrin, 2014) – in which our entire informational experiences are organized *by* ourselves and *around* ourselves, with our own interests, relationships, opinions and desires taking precedence – we take sides all the time, post material, fight for our beliefs and prejudices, and continually self-mobilize as info-warriors. This is war dissolved through everyday life. It is the end of episodic war, in favour of an all-pervasive hegemonic war, permeating our lives and all national boundaries. This is war posted online whilst we pick the kids up from school and shop, a war of 'likes', comments, memes, gifs and thumbs-up stickers. This is now the reality of conflict. There will be few wars from now on that aren't digital, and there will be few wars without digital witnesses and warriors. For good or for ill, this global, citizen-militarism is our future.

We are only a few decades into the life of the web, not even two decades into the world of Web 2.0, and just over a decade into the life of the smart-phone. The books produced in the first fifty years after Gutenberg's invention of printing are called 'incunabula' – Latin for 'swaddling clothes' or 'cradle' – representing, by implication, the infancy of print culture. So too are we still in the infancy of networked, digital technology. And we are also in the infancy of digital war.

References

Baudrillard, J. (1983) *In the Shadow of the Silent Majorities*, New York: Semiotext(e).

Clausewitz, C. V. (2008) *On War*, Princeton, New Jersey: Princeton University Press.

Hoffmann, F.G. (2007) *Conflict in the 21st Century: The Rise of Hybrid Wars*, Arlington, Virginia: Potomac Institute for Policy Studies, http://www.potomacinstitute.org/images/stories/publications/potomac_hybridwar_0108.pdf.

Merrin, W. (2014) *Media Studies 2.0*, Oxon: Routledge.

Index